GB401 APP

Applied Geomorphology

INTERNATIONAL ASSOCIATION OF GEOMORPHOLOGISTS

Applied Geomorphology

Theory and Practice

Edited by

R.J. Allison
University of Durham

JOHN WILEY & SONS, LTD

Copyright © 2002 by John Wiley & Sons, Ltd
    Baffins Lane, Chichester,
    West Sussex PO19 1UD, England

    National   01243 779777
    International (+44) 1243 779777
    e-mail (for orders and customer service enquiries): cs-books@wiley.co.uk
    Visit our Home Page on http://www.wiley.co.uk
                  or http://www.wiley.com

Other Wiley Editorial Offices

John Wiley & Sons, Inc., 605 Third Avenue,
New York, NY 10158-0012, USA

Wiley-VCH Verlag GmbH, Pappelallee 3,
D-69469 Weinheim, Germany

John Wiley & Sons Australia Ltd, 33 Park Road, Milton,
Queensland 4064, Australia

John Wiley & Sons (Asia) Pte Ltd, 2 Clementi Loop #02-01,
Jin Xing Distripark, Singapore 129809

John Wiley & Sons (Canada) Ltd, 22 Worcester Road,
Rexdale, Ontario M9W 1L1, Canada

Library of Congress Cataloging-in-Publication Data

Applied geomorphology : theory and practice / edited by R.J. Allison.
        p. cm.
    ISBN 0-471-89555-5 (alk. paper)
    1. Geomorphology.   I. Allison, R. J. (Robert J.)

  GB401.5 .A67 2002
  551.41—dc21

                                        2001046745

British Library Cataloguing in Publication Data

A catalogue record for this book is available from the British Library

ISBN 0-471-89555-5

Typeset in 9/11pt Times by Mayhew Typesetting, Rhayader, Powys
Printed and bound in Great Britain by Antony Rowe Ltd, Chippenham, Wiltshire
This book is printed on acid-free paper responsibly manufactured from sustainable forestry, in which at least two trees are planted for each one used for paper production.

Contents

List of Contributors

Yasushi Agata, Department of Geography, Graduate School of Science, University of Tokyo, Tokyo 113-0033, Japan.

Robert J. Allison, Department of Geography, University of Durham, Science Laboratories, Durham, DH1 3LE, UK.

Marisa Amadei, Presidenza del Consiglio dei Ministri, Dipartimento per i Servizi Tecnici Nazionali, Gruppo di Lavoro 'Carta della Natura', Via Curtatone 3, 00185 Roma, Italy.

Vittorio Amadio, Presidenza del Consiglio dei Ministri, Dipartimento per i Servizi Tecnici Nazionali, Gruppo di Lavoro 'Carta della Natura', Via Curtatone 3, 00185 Roma, Italy.

Alexandra Angeliaume, Centre de Biogéographie-Ecologie, UMR 180 CNRS-ENS, Le Parc 92 211 Saint-Cloud, Paris, France.

Roberto Bagnaia, Presidenza del Consiglio dei Ministri, Dipartimento per i Servizi Tecnici Nazionali, Gruppo di Lavoro 'Carta della Natura', Via Curtatone 3, 00185 Roma, Italy.

Claude Bernard, Institut de recherche et de développement en agroenvironnement, 2700 rue Einstein, Sainte-Foy, Québec, G1P 3W8, Canada.

M Berti, Department of Earth Science, University of Bologna, Italy.

Marco Bondesan, Dipartimento di Scienze Geologiche and Paleontologiche, Università degli Studi di Ferrara, Via Ercole d'Este 32, 44100 Ferrara, Italy.

G. Bornette, ESA SO23–CNRS, Laboratoire d'ecologie des eaux douces et des grands fleuves, Université C. Bernard, 69100 Villeurbanne, France.

Mauro Casadei, Department of Earth and Planetary Science, University of California, McCone Hall, Room 398, Berkeley, CA 94720, USA.

Filippo Catani, Dipartimento di Scienze della Terra, Università di Firenze, Via La Pira 4, 1-50121 Firenze, Italy.

David Chester, Department of Geography, The University of Liverpool, Roxby Building, Liverpool, L69 3BX, UK.

Chang-Jo F. Chung, Geological Survey of Canada, Spatial Data Analysis Laboratory, 601 Booth Street, Ottawa, Ontario, K1A 0E8, Canada.

Vanda Claudino-Sales, DEPAM, Université Paris Sorbonne, 191 rue Saint-Jacques, 75005 Paris, France and Departamento Geografia, Universidade Federal do Ceará, Brazil.

Arthur J. Conacher, Department of Geography, University of Western Australia, 35 Stirling Highway, Crawley, WA 6009, Australia.

Nicholas J. Cox, Department of Geography, University of Durham, Science Laboratories, Durham, DH1 3LE, UK.

L. D'Alessandro, Dipartimento di Scienze della Terra, Università degli Studi di Roma 'La Sapienza', Piazzale Aldo Moro 5, 00185 Roma, Italy.

L. Davoli, Dipartimento di Scienze della Terra, Università degli Studi di Roma 'La Sapienza', Piazzale Aldo Moro 5, 00185 Roma, Italy.

M. Del Monte, Dipartimento di Scienze della Terra, Università degli Studi di Roma 'La Sapienza', Piazzale Aldo Moro 5, 00185 Roma, Italy.

Daniela Di Bucci, Presidenza del Consiglio dei Ministri, Dipartimento per i Servizi Tecnici Nazionali, Gruppo di Lavoro 'Carta della Natura', Via Curtatone 3, 00185 Roma, Italy.

Silvio Di Nocera, Dipartimento di Scienze della Terra, Università Federico II, Largo San Marcellino 10, Napoli 80138, Italy.

D. Dragovich, Division of Geography FO9, University of Sydney, Sydney 2006, Australia.

Andrea G. Fabbri, International Institute for Aerospace Surveys and Earth Sciences, ITC, Hengelosestraat 99, PO Box 6, 7500 AA Enschede, The Netherlands.

Riccardo Fanti, Dipartimento di Scienze della Terra, Università di Firenze, Via La Pira 4, 1-50121 Firenze, Italy.

Enzo Farabegoli, Dipartimento di Scienze della Terra e Geo-ambientali, Università di Bologna, Via Zamboni 67, 40127 Bologna, Italy.

P. Fredi, Dipartimento di Scienze della Terra, Università degli Studi di Roma 'La Sapienza', Piazzale Aldo Moro 5, 00185 Roma, Italy.

Marco Gatti, Dipartimento di Ingegneria, Università degli Studi di Ferrara, Via Saragat 1, 44100 Ferrara, Italy.

R. Genevois, Department of Geology, Palaeontology and Geophysics, University of Padua, Italy.

G. Gentili, Università di Parma, CNR, Compagnia General Ripreseaesee, Piazzale Aldo Moro 7, 00185 Roma, Italy.

E. Giusti, Università di Parma, CNR, Compagnia General Ripreseaesee, Piazzale Aldo Moro 7, 00185 Roma, Italy.

V.N. Golosov, Department of Geography, Moscow State University, Vorob'evy Gory, Moscow 119899, Russia.

P. Grante, UMR 5600 – CNRS, 'Environnement – Ville – Société', 18 rue Chevreul, 69362 Lyon Cedex 07, France.

Florina Grecu, Department of Geomorphology, Faculty of Geography, University of Bucharest, 1 Bd. N. Balcescu, sect. 1, Bucharest, Romania.

David Higgitt, Department of Geography, University of Durham, Science Laboratories, Durham, DH1 3LE, UK.

N.N. Ivanova, Department of Geography, Moscow State University, Vorob'evy Gory, Moscow 119899, Russia.

M.K. Iwamoto, Department of Civil Engineering, Nishi-nippon Institute of Technology, Kanda-machi, Fukuoka 800-03, Japan.

Owen G. Kimber, Department of Geography, University of Durham, Science Laboratories, Durham, DH1 3LE, UK.

Hirohito Kojima, Science University of Tokyo, Remote Sensing Laboratory, Department of Civil Engineering, 2641 Yamazaki, Noda-City, Chiba 278, Japan.

Yoshimasa Kurashige, Graduate School of Environmental Earth Science, Hokkaido University, N10 W5, Kita-ku, Sapporo 060, Japan.

Adam Lajczak, Present address: School of Environmental Science, The University of Shiga Prefecture, 2500 Hassaka-cho, Hikone, Shiga 522-8533, Japan. Faculty of Earth Sciences, Silesian University, Sosnowiec, Poland.

Lucilla Laureti, Presidenza del Consiglio dei Ministri, Dipartimento per i Servizi Tecnici Nazionali, Gruppo di Lavoro 'Carta della Natura', Via Curtatone 3, 00185 Roma, Italy.

Marc R. Laverdière, Université Laval, Département des sols et de génie agroalimentaire, Sainte-Foy, Québec, G1K 7P4, Canada.

Angelo Lisi, Presidenza del Consiglio dei Ministri, Dipartimento per i Servizi Tecnici Nazionali, Gruppo di Lavoro 'Carta della Natura', Via Curtatone 3, 00185 Roma, Italy.

Francesca R. Lugeri, Presidenza del Consiglio dei Ministri, Dipartimento per i Servizi Tecnici Nazionali, Gruppo di Lavoro 'Carta della Natura', Via Curtatone 3, 00185 Roma, Italy.

Nicola Lugeri, Presidenza del Consiglio dei Ministri, Dipartimento per i Servizi Tecnici Nazionali, Gruppo di Lavoro 'Carta della Natura', Via Curtatone 3, 00185 Roma, Italy.

E. Lupia Palmieri, Dipartimento di Scienze della Terra, Università degli Studi di Roma 'La Sapienza', Piazzale Aldo Moro 5, 00185 Roma, Italy.

Lionel Mabit, Université Laval, Département des sols et de génie agroalimentaire, Sainte-Foy, Québec, G1K 7P4, Canada.

R. Marini, Dipartimento di Scienze della Terra, Università degli Studi di Roma 'La Sapienza', Piazzale Aldo Moro 5, 00185 Roma, Italy.

Fabio Matano, Dipartimento di Scienze della Terra, Università Federico II, Largo San Marcellino 10, Napoli 80138, Italy.

Sandro Moretti, Dipartimento di Scienze della Terra, Università di Firenze, Via La Pira 4, 1-50121 Firenze, Italy.

R. Morris, Central Coast District National Parks and Wildlife Service of N.S.W., PO Box 1393, Gosford, NSW 2250, Australia.

Shigeyuki Obayashi, Science University of Tokyo, Remote Sensing Laboratory, Department of Civil Engineering, 2641 Yamazaki, Noda-City, Chiba 278, Japan.

M.A. Ortiz-Pérez, Instituto de Geografía, Universidad Nacional Autónoma de México, Coyoacán 04510, D.F. Mexico.

J.L. Palacio-Prieto, Instituto de Geografía, Universidad Nacional Autónoma de México, Coyoacán 04510, D.F. Mexico.

Jean-Pierre Peulvast, UFR de Géographie, Université Paris Sorbonne, 191 rue Saint-Jacques, 75005 Paris, France and Orsay Terre, Université Paris Sud, Paris, France.

H. Piégay, UMR 5600 – CNRS, 'Environnement – Ville – Société', 18 rue Chevreul, 69362 Lyon Cedex 07, France.

G. Pizzaferri, Università di Parma, CNR, Compagnia Generale Ripreseaesee, Piazzale Aldo Moro 7, 00185 Roma, Italy.

R. Raffi, Dipartimento di Scienze della Terra, Università degli Studi di Roma 'La Sapienza', Piazzale Aldo Moro 5, 00185 Roma, Italy.

Denise J. Reed, Department of Geology and Geophysics, University of New Orleans, New Orleans, LA 70148, USA.

Paolo Russo, Dipartimento di Ingegneria, Università degli Studi di Ferrara, Via Saragat 1, 44100 Ferrara, Italy.

Yukiya Tanaka, Department of Geography, Faculty of Education, Fukui University, Fukui 910-8507, Japan.

P.R. Tecca, National Research Council, IRPI, Padua, Italy.

A. Urbani, Department of Earth Science, University of Rome, Italy.

Stanislas Wicherek, Centre de Biogéographie-Ecologie, UMR 180 CNRS-ENS, Le Parc 92211 Saint-Cloud, Paris, France.

PART 1   Slopes and Landslides

1 Stability Analysis of Prediction Models for Landslide Hazard Mapping

CHANG-JO F. CHUNG
Geological Survey of Canada, Spatial Data Analysis Laboratory, Ontario, Canada

HIROHITO KOJIMA
Science University of Tokyo, Remote Sensing Laboratory, Chiba, Japan

AND

ANDREA G. FABBRI
International Institute for Aerospace Surveys and Earth Sciences, ITC, Enschede, The Netherlands

ABSTRACT

This chapter discusses the influence of causal factors for landslide hazard mapping and stability of the prediction results. Among many quantitative models, we consider a model based on the theory of fuzzy sets with the algebraic sum operator. In the model, layers of geoscience maps represent the spatial information used for the prediction of areas in which the geomorphologic setting is similar to the ones in which a particular type of mass movement has taken place. One of the main challenges in the selection of these causal factors is that of how to compare two prediction models based on two different sets of casual factors. In the application discussed here, a study area in the Rio Chincina region of central Colombia is considered in which a spatial database was constructed for hazard mapping of 'rapid debris avalanches'. In the database eleven map-layer causal factors were selected, and the landslides of rapid debris avalanches were divided into two subsequent periods, PRE-1960 (prior to 1960) and POST-1960 (after 1960). For the prediction, the eleven layers of information were integrated as evidence toward a proposition that 'points in the study area will be affected by future mass movement'. The analysis discussed here assesses the stability of the predicted hazard map with respect to the introduction or removal of each causal factor, i.e. of each map layer, according to the following steps.

1. A prediction map using all eleven causal factors was first constructed. Eleven prediction maps were generated by subsequently eliminating different single factors. Difference or DIF-maps were computed between the eleven maps and the prediction map using all the eleven factors. All prediction maps were obtained using the PRE-1960 landslides only. A 'matching rate' was then defined as a quantitative indicator associated to the DIF-map.
2. Using the matching rates, the causal factors were divided into two groups: the 'influent factor group' and the 'non-influent factor group'. For each of the three prediction maps based on the influent factor group, non-influent factor group and all eleven factors, a prediction-rate curve was obtained by comparing the prediction map and the distribution of the POST-1960 landslides. The prediction power of the factors used in each prediction

was measured by the statistics from the comparison. The subsequent investigation was on the prediction differences between the two groups as fully described by the prediction-rate curve.

3. The influent factor group was also studied in terms of its relevance in the prediction.

The results of the study led to the following conclusions for the study area. (i) Although the division of two groups was rather artificial, only six causal factors were considered as 'influent factor group'. The prediction power of the prediction map based on these six factors is as good as that of the map using all eleven factors. The six factors are lithology, distance to valley head, aspect, slope angle, elevation and relief patterns. The first three factors have stronger influence in the prediction. (ii) The prediction power of the final map based on the six influent causal factors was illustrated by the corresponding prediction-rate curve. Comparing with the prediction map based on all eleven factors, the prediction map was stable enough to be used for land-use planning.

1.1   INTRODUCTION

About one-quarter of the natural disasters in the world appear to be directly or indirectly related to landslides, due to rainfall, local downpour, earthquakes and volcanic activities. Human interventions by constructing social infrastructure are often the trigger of the landslide phenomena. 'When', 'Where' and 'What scale' are important aspects of landslides in the prediction of geomorphologic settings and conditions in which landslides are likely to occur. The problem is critical in developing countries where warning and protection measures are particularly difficult to implement due to the limitation of economic conditions (Hansen 1984). The aim of this contribution is to predict where landslides may occur and analyse the stability of the predictions.

Many research activities have been carried out for landslide prediction, using various kinds of map data (e.g. Carrara 1983; Carrara *et al.* 1992; Kasa *et al.* 1991; Chung and Fabbri 1993; Chung and Leclerc 1994; Fabbri and Chung 1996). Recently, the analysis of satellite remotely sensed data has also been applied to the slope stability evaluation (Obayashi *et al.* 1995). Some of the difficulties were:

1. the selection of causal factors (usually specially compiled map data) for landslide prediction;
2. the analytical procedure to test the influence of each causal factor in a prediction;
3. the interpretation of the results from a prediction.

In this chapter, we plan to provide a systematic procedure to identify and evaluate the influence of causal factors on landslide prediction, in a study area. The difference or DIF-maps represent an initial approximation of spatial correspondence between two prediction patterns in a study area. While more complete comparisons can be easily computed for all the classes of predicted values between pairs of predictions, the four-class DIF-maps between two corresponding binary patterns used here facilitate the identification and visualization of discrete spatial patterns.

In Figure 1.1, three hypothetical binary predictions (A, B and C) are considered, and the accompanying three maps (D, E and F) show three DIF-maps. The prediction maps were based on three models and the pair-wise DIF-maps of the three prediction maps were generated from them. The black ellipses represent ten unknown 'future' landslides to be

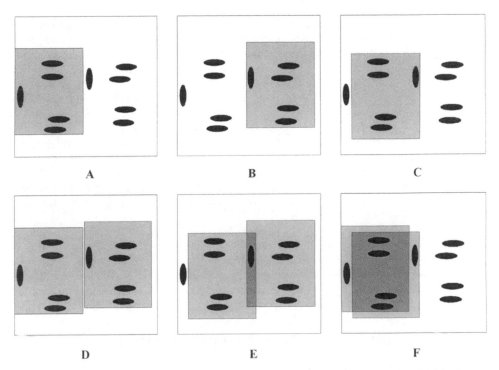

Figure 1.1 Prediction maps based on three models and the corresponding pair-wise difference or DIF-maps. The black ellipses represent ten 'future' landslides to be predicted by the prediction models. In each prediction map, the grey rectangle represents the hazardous area and it occupies 25% of the whole study area. The prediction powers of three prediction maps are identical, in that each map predicts five future landslides (50%) of ten landslides to come, i.e. each grey hazardous rectangle intersects five ellipses. (A) Prediction map from Model 1. (B) Prediction map from Model 2. (C) Prediction map from Model 3. (D) DIF-map between Model 1 and Model 2. (E) DIF-map between Model 2 and Model 3. (F) DIF-map between Model 1 and Model 3

predicted by the prediction models. In each prediction map, the grey rectangle represents a hazardous area and it occupies 25% of the whole study area. Each map predicts five future landslides (50%) of the ten landslides to come, i.e., each grey hazardous rectangle intersects five ellipses. It implies that the prediction powers of three prediction maps are identical. The DIF-map in Figure 1.1D shows no overlapping area and hence Model 1 (Figure 1.1A) and Model 2 (Figure 1.1B) are distinctly different. However the DIF-map in Figure 1.1F has much common area between the two models (Figure 1.1A and C). On one hand, if two models generate two prediction maps with the corresponding DIF-map shown in Figure 1.1D, then it would be very difficult to make a land-use map by combining two prediction maps. On the other hand, if two models generate the DIF-map shown in Figure 1.1F, it would be easy to combine two prediction maps because the two maps have a reasonably stable pattern. In this situation, we may be able to conclude that we can present a combined prediction from Model 1 and Model 3 with a certain confidence. As a quantitative indicator on a DIF-map, the following 'matching rate' was defined as:

Matching rate = common area (dark grey) / predicted hazardous area (light and dark grey)

If two maps match perfectly, then the matching rate is 1. If two maps do not match at all, then the matching rate is 0. Higher matching rates mean similar results in map 1 and map 2. A lower matching rate means that the two maps are very different. The matching rate for the DIF-map in Figure 1.1D is obviously 0, but the rate for DIF-map in (Figure 1.1F) is about 75%.

From an empirical study, assuming that two prediction maps have similar prediction powers, we came to a conclusion that the matching rate should be at least 75% to combine and subsequently interpret two prediction maps effectively. Otherwise two prediction maps are informing two different descriptions of the landslides. However, if the matching rate is lower than 75% and one prediction has much larger prediction power than the other map, then we may ignore the prediction map with the lower prediction power.

Neither the DIF-maps nor the associated matching rates themselves provide any valid measure of prediction power. It is only through the corresponding 'prediction rates curve' that we will provide that measure.

From the viewpoint of a land planner, the dark grey area in Figure 1.1E and F can be termed the 'stable hazardous' areas. On the other hand, the light grey areas in Figure 1.1E and F are termed 'non-stable hazardous' areas in the DIF-map. The 'non-stable hazardous' areas mean that we do not have much information on these pixels concerning the studied landslide hazard. According to Kasa et al. (1991, 1992) more supporting information is essential for decision-making in carrying out landslide prevention plans in the non-hazardous areas.

1.2 STUDY AREA, DATA SETS AND PREDICTION MODEL

1.2.1 Study Area and Causal Factors

The catchment of the Rio Chincina, located on the western slope of the central Andean mountain range (Cordillera Central) in Colombia, near the Nevado del Ruiz volcano, was used as a test for various landslide hazard zonation techniques. Van Westen (1993) made an extensive study of the region and constructed the database of the study area. Since then the database was made available as a case-study data set for many kinds of exercises and experiments on landslide hazard zoning by van Westen et al. (1993), with the name GISSIZ: training package of Geographic Information Systems on Slope Instability Zonation. It is on that data set that Chung et al. (1995) have developed several multivariate regression models for landslide hazard mapping.

The input data for landslide hazard mapping usually consist of several layers of map information. Each layer may be the result of map updating by experts, of field verification and of interpretation of aerial photographs. The resulting maps describe surficial and bedrock geology, soil type, slope, land use, geomorphology, mass movements, distance from active faults and other features, including man-made ones, that are relevant to slope instability. In addition, the identification of types and dates of landslide phenomena is critical to the application of predictive techniques.

Among many layers of spatial data constructed by van Westen (1993), he has suggested that the following seven data layers are 'causal factors' and are significantly related to landslide hazard: (1) bedrock lithological map; (2) geomorphological map; (3) slope-angle map; (4) land-use map; and three maps containing distances from the nearest valley head (5), from roads (6) and faults (7). In particular, translational mass movements termed 'rapid debris avalanches' (or 'derrumbes' in Spanish) were studied and predicted using the above causal factors. The initial seven maps and four additional ones derived from the elevation map are described in Table 1.1.

Table 1.1 Causal factors used in the Rio Chincina area, Colombia

Lithological map	Geomorphologic zone map	Land-use map	Elevation range map (m)	Relief class map (m)	Slope-angle map (degree)	Aspect class map	Map of valley density	Distance from valley head map (m)	Distance from road map (m)	Distance from fault map (m)
Unmapped area	Unmapped area	Traditional farming system	1200–1250	0–2	0–5	North	0	0–25	0–25	0–25
Gneissic intrusives	Romeral fault zone	Technified farming system	1250–1300	2–4	5–10	Northeast	1	25–50	25–50	25–50
Schists	Western Hills	Modern–intermediate farming system	1300–1350	4–6	10–15	East	2	50–	50–	50–75
Volcanic and meta-sedimentary	Terrace	Grass and shrubs	1350–1400	6–8	15–20	Southeast	3	51		75–100
Alluvial sediments	Modern–intermediate	Other crops	1400–1450	8–10	20–25	South	4			100–
Flow materials, alluvial and ashes		Built-up areas	1450–1500	10–12	25–30	Southwest	5			
Lake deposits		Bare	1500–1550	12–14	30–35	West	6			
Lahar deposits		Forest	1550–1600	14–16	35–40	Northwest				
Pyroclastic flow deposits			1600–1650	16–18	40–	Flat				
Mix of pyroclastic and debris flow			1650–1700	18–						
Andesite intrusive			1700–1750							
Tertiary sediments			1750–1800							
Fluvio-volcanic sediments			1800–1850							
			1850–1900							
			1900–							

Elevation, relief class, slope angle and aspect were calculated from DEM.
Ranges: '25–50' means more than or equal to 25 and less than 50. '100–' means more than or equal to 100.

Only the scarps of the rapid debris avalanches were considered as the trigger areas of the landslides in this study. There are 450 scarps in the study area and they are shown in Figure 1.2A. Further work by Chung *et al.* (1995) led to modifications of the database with the subdivision of the mass movements into two successive time intervals: those that had occurred 'prior to 1960' and those which occurred 'after 1960' (i.e. during 1961–1988). These two groups are termed PRE-1960 landslides, consisting of 174 scarps with 5515 pixels, and POST-1960 landslides consisting of 276 scarps with 4589 pixels.

In the spatial database it was pretended that the time of the study was approximately the year 1960. All the spatial data available in 1960 were compiled and were used to construct prediction models by establishing quantitative relationships between the scarps of the PRE-1960 landslides and the remainder of the input spatial data set. The predictions based on those relationships were then evaluated by comparing the predicted hazard classes with the distribution of the scarps of the POST-1960 landslides.

Each map consists of 779×561 pixels (a total of 437 019 pixels), and each pixel corresponds to a $12.5\,m \times 12.5\,m$ square area on the ground. These types of maps were also defined as the 'causal factors' in the evaluation of the properties of the land for different land uses by Kojima and Obayashi (1991). Slope angle, elevation range, relief class and aspect class maps were obtained from the digital elevation model (DEM) with the Rika University Image Processing Analysis System (Obayashi *et al.* 1995).

1.2.2   Prediction Model, Visualization and Rate

In this study, the fuzzy set model with the algebraic sum operator (see Appendix) was selected based on an earlier comparative study on the performance of prediction models (Chung and Fabbri 1999). Figure 1.2B shows the prediction results obtained using the PRE-1960 landslides shown in grey polygons and the six causal factors based on the fuzzy algebraic sum model. The POST-1960 landslide occurrences shown in black polygons in Figure 1.2B were used to evaluate the prediction results.

To visualize this predicted pattern, the predicted pixel values ranging between 0 and 1 were sorted in descending order. Then the ordered pixel values were divided into 40 classes. Each class occupies 2.5% of the whole study area. The division was made as follows. The pixels with the highest 2.5% estimated values were classified as the 0–2.5% class, shown in purple; the pixels with the next highest 2.5% values were represented in light magenta, and were classified as the 2.5–5% class. The same procedure was repeated 38 more times, for classes 2.5% apart, and the resulting 40 classes were shown in the corresponding colours (Figure 1.2B).

Following a procedure used by Chung and Fabbri (1999, 2000) to evaluate the prediction results, the 40 classes of the prediction pattern in Figure 1.2B were compared with the distribution of the POST-1960 landslides shown in black polygons in Figure 1.2B and also shown in Figure 1.2A. When a class was intersected with a landslide scarp, the portion of the scarp could vary from one pixel to the whole scarp. When the portion of the intersection was greater than 1% of the scarp, it could be assumed that the landslide with the scarp has been predicted and this rate was shown in the last column in Table 1.2 and illustrated as the grey curve with squares in Figure 1.3. However, if we assumed that the landslide with the scarp has been predicted only if the portion of the intersection was greater than 20% of the scarp, then the rate would obviously be smaller than the previous 1% rate as illustrated by the black curve with circles in Figure 1.3 and the rates are also shown in the third column in Table 1.2. On the other hand, if we compare the classes in Figure 1.2B with the distribution of the 4589 POST-1960 landslide pixels directly, and we count the number of these 4589 pixels in each

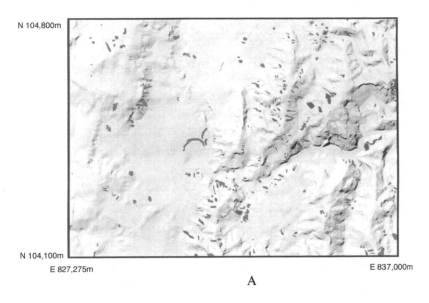

N 104,800m

N 104,100m

E 827,275m

E 837,000m

A

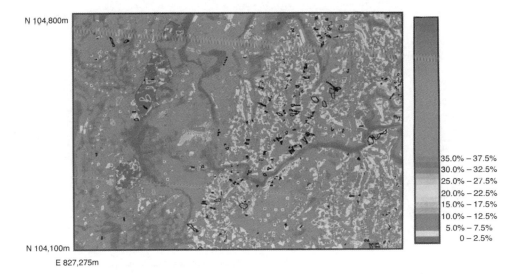

N 104,800m

N 104,100m

E 827,275m

35.0% − 37.5%
30.0% − 32.5%
25.0% − 27.5%
20.0% − 22.5%
15.0% − 17.5%
10.0% − 12.5%
5.0% − 7.5%
0 − 2.5%

B

Figure 1.2 (Plate 1) Landslide occurrences and hazard prediction in the Rio Chincina study area, Colombia. (A) Red identifies the distribution of the PRE-1960 and blue the distribution of the POST-1960 rapid debris avalanches. (B) The landslide hazard prediction is shown using the six influent causal factors and the fuzzy set algebraic sum operator

Table 1.2 Three prediction rates of the prediction map using the PRE-1960 landslides and the six influential causal factors based on fuzzy set algebraic sum operator (gamma = 1) in Equation 1.7 in the Appendix. Column A was obtained by comparing the prediction map (with classes 5% apart) with the 4589 pixels of POST-1960 landslides. Columns B and C were obtained by comparing the prediction map with the scarps of 276 POST-1960 landslides. For Column B, even if only 1% of the scarps of a landslide was contained in a class in the prediction map, the landslide was assumed to be predicted in that class. For the second column, if at least 20% of the scarps of a landslide were contained in a class in the prediction map, the landslide was assumed to be predicted in that class. Columns A–C are shown as three curves in Figure 1.3

Classes (%)	Proportion of 4589 pixels of POST-1960 landslides A	20% of the scarps of 287 POST-1960 landslides B	1% of the scarps of 276 POST-1960 landslides C
0–5	0.207	0.288	0.506
0–10	0.301	0.428	0.751
0–15	0.382	0.612	0.816
0–20	0.470	0.628	0.841
0–25	0.524	0.653	0.868
0–30	0.583	0.754	0.910
0–35	0.633	0.795	0.923
0–40	0.683	0.873	0.938
0–45	0.735	0.911	0.953
0–50	0.792	0.925	0.955

Equivalent curves in Figure 1.3: A, grey curve with squares; B, black curve with circles; C, grey curve with triangles.

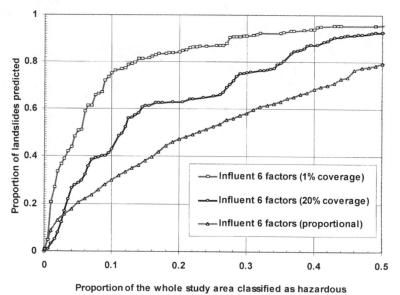

Figure 1.3 Prediction-rate curves for the six influent factors. 1% coverage (grey curve with squares) prediction-rate curve assumes that the landslides are predicted whenever 1% of the landslides are overlapped with the prediction pattern; 20% coverage (black curve with circles) curve assumes that the landslides are predicted whenever 20% of the landslides are overlapped with the prediction pattern. Proportional prediction-rate curve (grey curve with triangles) is based on the comparisons between the prediction pattern and the 4589 post-1960 landslide pixels directly

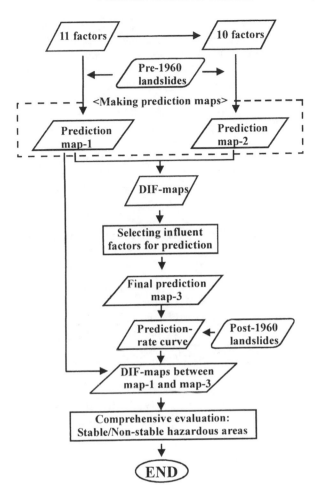

Figure 1.4 Operational flowcharts of the stability analysis for landslide hazard prediction

class as a rate, then we obtain another rate for the evaluation. These rates are shown in the second column in Table 1.2 and shown as a grey curve with triangles in Figure 1.3. The corresponding rates for the three curves in Figure 1.3 are shown in the columns in Table 1.2. If 20% of the scarp of a landslide was predicted as a hazardous area, then we thought that it was reasonable to assume that the landslide had been predicted. Because of this assumption, the only rates with 20% coverage were used in this chapter and were termed 'prediction rates'.

1.3 ANALYTICAL PROCEDURE

The stability analysis performed in this study corresponds to the computational steps described in the operational flowchart shown in Figure 1.4. The steps are as follows:

1. *Analysis for all factors.* Using all eleven factors, generate a prediction map. Successively compute eleven prediction maps by removing a single factor in turn from the set of eleven factors. Compute DIF-maps between those eleven maps and the prediction map generated by combining all eleven factors together. The corresponding matching rate as a quantitative indicator on the DIF-map was computed.
2. *Selection of the 'influent factors'.* For prediction, using matching rates and DIF-maps, divide the causal factors into two groups: the 'influent factor group' and the 'non-influent factor group'. Investigate the difference in the prediction by analysing these two groups, separately, in terms of prediction-rate curves.
3. *Construction of the final prediction map.* Using the influent factor group, we have constructed the final prediction map shown in Figure 1.2B. In addition, the prediction-rate curves represent the support of the set of layers to the patterns obtained.

1.4   INFLUENCE OF CAUSAL FACTORS ON PREDICTION

1.4.1   Influence of Single Factors

At first, a prediction map, termed map-1, was computed by combining all the eleven factors using the scarps of the PRE-1960 landslides. Then, to evaluate the influence of individual causal factors on the prediction, eleven prediction maps, termed map-2, were computed by successively eliminating each single factor. The pixels with the highest 10% estimated values among all pixels were considered as hazardous areas. In practice there are no rules on how to select an optimal threshold of hazard values, but the pixels with the highest 10% predicted values in map-1 cover over 70% of the 276 scarps of the POST-1960 landslides.

The highest 10% of the pixels were assigned the value of 1, while the other pixels were assigned the value of 0, to indicate non-hazardous areas. Based on the combination of results between pairs of map-1 and map-2, the pixels were classified into four groups and assigned four different values. The four values and the corresponding colour for graphic displays (red, yellow, green and blue) are shown in Table 1.3 and displayed on the DIF-maps in Figures 1.5 and 1.6. In three DIF-maps in Figure 1.1, the dark grey corresponds to red and the white corresponds to blue. The descriptions of these groups of pixels are as follows:

Group 1: the pixels identified as hazardous in both map-1 and map-2 are assigned the value of 3. On the DIF-map, these pixels are coloured in red.

Group 2: the pixels identified as hazardous in map-1, but as non-hazardous in map-2, are assigned the value of 2. On the DIF-map, these pixels are coloured in yellow.

Table 1.3 Combination of events, pixel values and colours in the difference maps (DIF-maps), considering the 10% highest predicted hazard values. Map-1 and map-2 were computed using the PRE-1960 and the POST-1960 training data, respectively

Combination of events		Map-2	
		Hazardous	Non-hazardous
Map-1	Hazardous	Red, 3, Group 1	Yellow, 2, Group 2
	Non-hazardous	Green, 1, Group 3	Blue, 0, Group 4

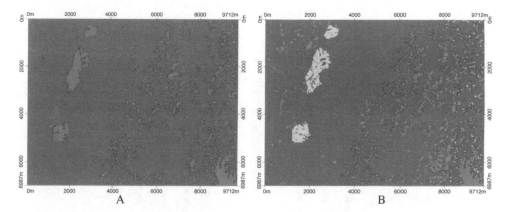

Figure 1.5 (Plate 2)   Two difference maps (DIF-maps) for stability analysis in the Rio Chincina study area, Colombia. (A) Map of all eleven causal factors (map-1) versus a ten-factor map (map-2), i.e. removing the 'distance from road' map: matching rate is 99.1%. (B) Same as in (A) but removing the 'lithology' map: matching rate is 58.1%

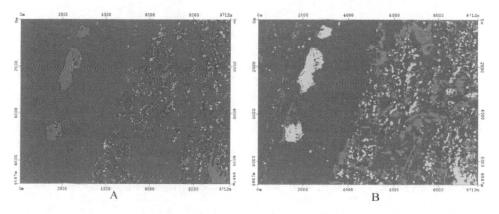

Figure 1.6 (Plate 3)   Two difference maps (DIF-maps) for stability analysis in the Rio Chinchina study area, Colombia. (A) Map of the six influent causal factors (map-1) versus that of all eleven-factor map (map-2): matching rate is 79.1%. (B) Map of the six influent causal factors (map-1) versus that of only the five non-influent factor map (map-2): matching rate is 6.1%

Group 3: the pixels identified as non-hazardous in map-1, but as hazardous in map-2, are
   assigned the value of 1. On the DIF-map, these pixels are coloured in green.
Group 4: the pixels identified as non-hazardous in both map-1 and map-2 are assigned the
   value of 0. On the DIF-map, these pixels are coloured in blue.

As defined earlier:

Matching rate = no. of red pixels / (no. of red pixels + no. of yellow pixels + no. of green pixels)
             = no. of red pixels / (no. of pixels in the whole area − no. of blue pixels)

Figure 1.7 shows matching rates as percentages for the single factor elimination. The lowest match rate is 58.1% for 'lithological map' factor, while the highest match rate is 99.1% for the

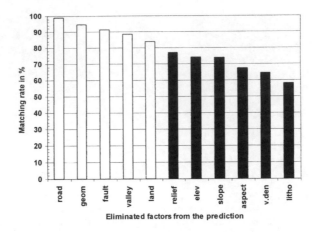

Figure 1.7 Matching rates after removing single causal factors in turns (litho, lithology; v. den, valley density; aspect; slope; elev, elevation; relief; land, land-use; valley, distance from valley heads; fault, distance from faults; geom, geomorphology; road, distance from roads)

factor 'distance from roads'. For these two extreme situations, the DIF-maps are shown in Figure 1.5A and B. It is clear that the 'lithological map' factor has great influence on the prediction results. In this study, for the situations indicating less than 80% in matching rate, six causal factors were identified as belonging to the 'influent factor group': (1) lithology, (2) distance from valley heads, (3) aspect class, (4) slope angle, (5) elevation range and (6) relief class. The remaining five factors were considered to belong to the 'non-influent factor group'. In Figure 1.7, black bars and white bars indicate the influent group and non-influent group respectively.

1.4.2 Analysis of 'Influent Factors' and 'Non-influent Factors'

To evaluate the differences in prediction by using 'influent factors' and 'non-influent factors', the following three situations were considered:

(A)   using all the eleven factors;
(B)   using the six 'influent factors' only;
(C)   using the remaining five 'non-influent factors' only.

The DIF-map, as shown in Figure 1.6A, was computed between situations (B) and (A). The DIF-map between (B) and (C) is shown in Figure 1.6B. Corresponding matching rates are 79.1% and 6.1%, respectively. In Figure 1.6A, there are many 'red' pixels. On the contrary, in Figure 1.6B there are few 'red' pixels and many 'yellow' and 'green' pixels. It is clear that the prediction patterns using the 'non-influent factors' are almost insignificant.

To evaluate the performance of the prediction models based on the above three situations, the corresponding prediction rates were calculated as shown in Figure 1.3. In Figure 1.3, situation (C), which uses only the non-influent factors, shows a poor prediction. This result corroborates what was displayed in the difference of DIF-maps in Figure 1.6A and B.

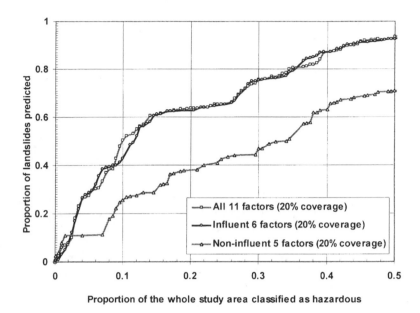

Proportion of landslides predicted

Proportion of the whole study area classified as hazardous

Legend:
- All 11 factors (20% coverage)
- Influent 6 factors (20% coverage)
- Non-influent 5 factors (20% coverage)

Figure 1.8 Prediction-rate curves for all eleven factors (grey curve with squares), the six influent factors (black curve with circles) and five non-influent factors (grey curve with triangles)

1.5   FINAL PREDICTION MAP AND ITS PREDICTION

Using the stability analysis (described in Figure 1.4), we can select the influent causal factors for prediction. As shown in Figure 1.8, the two prediction rates for the prediction maps using all the eleven factors and for those using only the six influent factors are nearly identical. The corresponding DIF-map of Figure 1.6A, however, shows that over 20% (100% − 79.1%) of the two prediction maps are different. This represents one of the difficulties in evaluating the prediction results. To evaluate the prediction results, we must obtain not only the prediction rate curves but also the corresponding DIF-map. While the stability through the DIF-map provides confidence in the prediction pattern, the prediction-rate curve gives us the necessary statistics to interpret the prediction results. In this example, the stability of the prediction pattern is a borderline case and it suggests that we may need 'better' data or/and 'better' models for the prediction.

The pixels with the red colour (or dark grey in Figure 1.1), i.e. value 3 in the DIF-maps, can be interpreted as having a higher likelihood of future landslide occurrence. Also the geomorphologic setting of the studied rapid debris avalanches (the particular 'landslide type' representing the most frequent mass movement in the study area) might be similar for these pixels. From the viewpoint of a land-use planner, this red area can be termed as the 'stable hazardous area', as shown in Table 1.4. On the other hand, the 'non-stable hazardous areas' are the areas covered by the yellow pixels and the green pixels in the DIF-map. The 'non-stable hazardous area' means that we do not have much information on these pixels concerning the studied landslide type.

From the results of the prediction map, hazardous or non-hazardous areas can be identified. When we have these classified areas delineated in land-use planning or civil engineering

Table 1.4 Interpretation of the difference maps (DIF-maps)

Prediction Map-1 using the six 'influent causal factors'	Prediction Map-2 using all eleven causal factors	Colour and pixel value in the DIF-map	Assessment
Hazardous	Hazardous	Red, 3, Group 1	Stable hazardous
	Non-hazardous	Yellow, 2, Group 2	Non-stable hazardous
Non-hazardous	Hazardous	Green, 1, Group 3	Non-stable hazardous
	Non-hazardous	Blue, 0, Group 4	Stable non-hazardous

evaluations (i.e. for human activities), the areas identified as hazardous will have very little activity on them because they are classified with high confidence in the prediction. Therefore they are of little interest and have been referred to as 'stable hazardous area' and 'stable non-hazardous area' from a decisional point of view by Kasa *et al.* (1992) and Obayashi *et al.* (1995), i.e. the areas are 'safe' from a decisional point of view. On the contrary the 'non-stable hazardous area' is decisional 'unsafe' because discrepancies exist concerning its classification as hazardous or non-hazardous.

1.6   CONCLUDING REMARKS

The results of this study are summarized as follows.

1. A stability analysis procedure is useful for comparing two prediction maps,. The analysis was used for evaluating (1) the influence of causal factors on landslide prediction, and (2) the stability of the prediction map. Through this procedure, it was found that six causal factors – (1) lithological map, (2) distance from the valley heads, (3) aspect class, (4) slope angle, (5) elevation range and (6) relief class – were significant for the prediction of rapid debris avalanches in the study area. In particular, the first three causal factors seem to have higher impacts on the prediction.
2. The prediction power is measured only by the prediction-rate curves, not by the matching rates. However, we must consider both statistics to interpret the prediction map with some confidence.
3. As a final product for land-use planners, we have divided the whole study area into two subareas, 'stable hazardous area and stable not hazardous area' and 'non-stable hazardous area' based on the DIF-maps. To decision makers and land-use planners, these divisions are useful and important not only for identifying the areas where landslides are not likely to occur in the future, but also for helping in the selection of areas in which field investigation is necessary to verify and validate the results, thus improving the cost-effectiveness of time-consuming surveys.

The stability analysis procedure, with the DIF-maps and the associated prediction-rate curves, contributes to the construction of a final product, for the prediction of landslide hazard. This holds true not only for the particular fuzzy set model used here, but also for any model based on map overlays. Pairing the DIF-maps with the respective prediction-rate curves allows capture of the changes in spatial patterns corresponding to the changes in prediction rates.

It is essential that the prediction results be interpreted in the field, describing the conditions in areas where landslides have not occurred yet and which are at some distance from where they have already occurred. In other words, particulars of the predicted maps that seem to have either low or high stability should be derived and validated in the field. In addition, based on the results of the stability analysis presented in this study, if the level of detail and the scale of a study permit it, further investigation should be made on the effectiveness of satellite remote-sensing data (Kojima and Obayashi 1991; Obayashi *et al.* 1995) for the landslide prediction.

ACKNOWLEDGEMENTS

We wish to thank David F. Garson of Geological Survey of Canada, in Ottawa, who read an earlier draft of the manuscript and provided many useful comments. We gratefully acknowledge Cees J. van Westen of the International Institute for Aerospace Surveys and Earth Sciences, The Netherlands, who provided the data set of the Rio Chincina study area.

APPENDIX: BASIC CONCEPTS IN PREDICTION MODELS

A.1   Favourability Function

Consider a pixel p in **A** denoting the whole study area and the following proposition:

$$\mathbf{T_p}: \text{`p will be affected by a future landslide of type } \mathbf{D}\text{'} \tag{1.1}$$

At pixel p in **A**, we have the information on the m map layers, $(c_1, \ldots, c_m)$. The problem is to define a function, at every pixel p, representing a degree of support that p is likely to be affected by a future landslide given the m evidences $(c_1, \ldots, c_m)$ at p. Let us define:

$$f(\mathbf{T_p} \mid \text{Given } m \text{ evidences, } c_1, \ldots, c_m) \text{ at every p in } \mathbf{A} \tag{1.2}$$

Chung and Fabbri (1993) have defined such a function as the 'favourability function' for the proposition $\mathbf{T_p}$ in 1.1. In this chapter, we have used only one type of favourability function in 1.2 which is the membership function of Zadeh's fuzzy set (Zadeh 1978) consisting of all pixels which are likely be affected by a future landslide of type **D** given the m evidences at a pixel. i.e.

$$f(\mathbf{T_p} \mid c_1, \ldots, c_m) = \mu\{p \mid c_1, \ldots, c_m\} \tag{1.3}$$

where $\mu\{p \mid c_1, \ldots, c_m\}$ is the membership function of a fuzzy set consisting of all pixels that are likely to be affected by a future landslide of type **D** given the m evidences at a pixel p.

A.2   Basic Idea of Zadeh's Fuzzy Set Model

The fuzzy set S consists of all the pixels $(p \in \mathbf{A}$: whole area) where the proposition in 1.1 is 'likely' to be true (Zadeh 1965, 1968). S is defined by a membership function μ_S:

$$\mu_S : \mathbf{A} \to [0, 1] \tag{1.4}$$

It represents a 'grade of membership', or a 'degree of compatibility'. We may also interpret the membership function as a 'possibility function' of p where the proposition, $\mathbf{T_p}$ is possibly true. Conversely, the fuzzy set S is uniquely defined by the membership function μ_S:

$$S = \{(p, \mu_S(p)), p \in \mathbf{A}\} \tag{1.5}$$

The definitions in 1.4 and 1.5 imply that the fuzzy set and the corresponding membership function are equivalent. Usually a fuzzy set is defined by a conceptual idea, and then the corresponding membership function is constructed to determine a 'degree of support' for each member. In practice, the membership function (also termed 'possibility function' by Zadeh (1978)) μ_S is difficult to construct. Our prediction model can be represented by constructing the membership function at each pixel in 1.4 after observing the m evidences ($c_1, \ldots, c_m$) at p in $\mathbf{A}$, as a function describing the support for the condition that p is likely to be affected by a future landslide.

A.3   Construction of Membership Function

We are going to construct the membership function; $\mu_S\{p|c_1, \ldots, c_m\}$ indirectly. Let us first look at an individual membership function based on information of a single layer, i.e. $\mu_{S_k}\{p \mid c_k\}$. Then it can be easily estimated by:

$$\mu_{S_k}\{p \mid c_k\} = \text{size of area } F \cap A_{kc_k} \text{ / size of area } A_{kc_k} \qquad (1.6)$$
$$S_k = \{(p, \mu_{S_k}\{p \mid c_k(p)\}), p \in \mathbf{A}\}$$

where A_{kc_k} denotes the area whose pixel has the same pixel value, c_k in the kth layer, and $F \cap A_{kc_k}$ denotes the area affected by the past landslides within A_{kc_k}.

Then using the algebraic sum operator in the theory of fuzzy set, we can obtain:

$$\mu_S\{p \mid c_1, \ldots, c_m\} = 1 - \prod_{k=1}^{m} (1 - \mu_{S_k} \{p \mid c_k\}) \qquad (1.7)$$

See Chung and Fabbri (2001) for a more comprehensive treatment of fuzzy sets in landslide hazard prediction.

REFERENCES

Carrara, A. 1983. Multivariate models for landslide hazard evaluation. *Mathematical Geology*, **15**(3), 403–427.

Carrara, A., Cardinali, M. and Guzzetti, F. 1992. Uncertainty in assessing landslide hazard and risk. *ITC Journal*, **1992-2**, 172–183.

Chung, C.F. and Fabbri, A.G. 1993. The representation of geoscience information for data integration. *Nonrenewable Resources*, **2**(2), 122–139.

Chung C.F. and Fabbri, A.G. 1999. Probabilistic prediction models for landslide hazard mapping. *Photogrammetric Engineering and Remote Sensing*, **65**(12), 1389–1399.

Chung, C.F. and Fabbri, A.G. 2001. Prediction model for landslide hazard using a fuzzy set approach. In Marchetti, M. and Rivas, V. (eds), *Geomorphology and Environmental Impact Asssessment*. A.A. Balkema, Rotterdam, 31–48.

Chung, C.F. and Leclerc, Y. 1994. A quantitative technique for zoning landslide hazard. *Papers and Extended Abstracts for Technical Programs of IAMG '94*. International Association for Mathematical Geology Annual Conference, Mont Tremblant, Quebec, Canada, October 3–5, 1994, 87–93.

Chung, C.F., Fabbri, A.G. and van Westen, C.J. 1995. Multivariate regression analysis for landslide hazard zonation. In Carrara, A. and Guzzetti, F. (eds), *Geographical Information Systems in Assessing Natural Hazards*. Kluwer Academic Publishers, Dordrecht, 107–133.

Fabbri, A.G. and Chung, C.F. 1996. Predictive spatial data analysis in the geosciences. In Fisher, M., Scholten, H.J. and Unwin, D. (eds), *Spatial Analytical Perspectives on GIS in the Environmental and Socio-Economic Sciences*. Taylor & Francis, London, GISDATA Series no. 4, 147–159.

Hansen, A. 1984. Landslide hazard analysis. In Brundsen, D. and Prior, D.B. (eds), *Slope Instability*. Wiley, New York, 523–602.

Kasa, H., Kurodai, M., Kojima, H. and Obayashi, S. 1991. Study on the landslide prediction model using satellite remote sensing data and geographical information. In Bell, D.H. (ed.), *International Landslide Symposium*. Balkema, Rotterdam, 983–988.

Kasa, H., Kojima, H. and Obayashi, S. 1992. Applicability of the slope failure prediction for various types of slope failure. *Journal of Japanese Society of Civil Engineers*, **444**/6–16, 11–20.

Kojima, H. and Obayashi, S. 1991. Development of the land use capability classification model applying the satellite multi-spectral scanner data. *Journal of Japan Society of Civil Engineers*, **427**/6–14, 65–74 (in Japanese).

Obayashi, S., Kojima, H. and Katushi, F. 1995. Improvement of the accuracy extracting areas in danger of landslide using satellite multispectral data. *Journal of Japan Society of Civil Engineers*, **534**/6–30, 173–184 (in Japanese).

van Westen, C.J. 1993. *Application of Geographical Information Systems to landslide hazard zonation*. PhD Thesis, Technical University of Delft, International Institute for Aerospace Survey and Earth Sciences, Enschede, The Netherlands, ITC Publication 15, Vol. 1, 245 pp.

van Westen, C.J., van Duren, H.M.G., Kruse, I. and Terlien, M.T.J. 1993. *GISSIZ: Training Package for Geographic Information Systems in Slope Instability Zonation*. ITC Publication 15, Enschede, The Netherlands.

Zadeh, L.A. 1965. Fuzzy sets. *IEEE Information and Control*, **8**, 338–353.

Zadeh, L.A. 1968. Probability measures of Fuzzy events. *Journal of Mathematical Analysis and Applications*, **10**, 421–427.

Zadeh, L.A. 1978. Fuzzy sets as a basis for a theory of possibility. *Fuzzy Sets and Systems*, **1**, 3–28.

2 Geomorphology, Stability Analyses and Stabilization Works on the Montepiano Travertinous Cliff (Central Italy)

L. D'ALESSANDRO
Department of Earth Science, University of Rome, Italy

R. GENEVOIS
Department of Geology, Paleontology and Geophysics, University of Padua, Italy

M. BERTI
Department of Earth Science, University of Bologna, Italy

A. URBANI
Department of Earth Science, University of Rome, Italy

AND

P.R. TECCA
National Research Council, IRPI, Padua, Italy

ABSTRACT

The Roccamontepiano area (central Italy) is characterized by the presence of a travertine plate resting on slightly overconsolidated Plio-Pleistocene silty clays. This plate forms a subvertical face about 30 m high and is marked by a sequence of arch-shaped features resulting from rotational sliding and lateral spreading phenomena. At present, the travertinous cliff shows the same morphological features as just before the last catastrophic event of 1765 that destroyed the Roccamontepiano village, claiming about 700 lives. The high risk associated with the active mass movement processes led the local administration to carry out a careful investigation plan consisting of field surveys, site investigations and laboratory tests. The data collected permitted a definition of the geological and hydrogeological characteristics of the area and the physical and mechanical properties of the materials. On this basis, finite difference analyses have been performed in order to evaluate the present stability conditions of the cliff and to verify the effectiveness of various stabilization works. These analyses confirm the critical stability conditions of the cliff and show the greater effectiveness of piezometric drawdown compared with slope regrading in improving slope stability.

2.1 INTRODUCTION

This work represents a widening of previous studies on the Montepiano landslide (D'Alessandro 1982; D'Alessandro *et al.* 1987, 1989). It is based on data collected during recent field survey and subsoil investigations which permitted a better definition of the geological and hydrogeological characteristics of the area and an evaluation of the physical

Applied Geomorphology: Theory and Practice. Edited by R.J. Allison.

and mechanical characteristics of the materials. A detailed geomorphological survey has also been carried out to relate the observed instability phenomena with the reconstructed geological model. On the basis of collected data, finite difference stability analyses have been performed in order to investigate the present stability conditions and to evaluate the effectiveness of the different types of remedial work.

2.2 GEOLOGICAL AND GEOMORPHOLOGICAL CHARACTERISTICS

The Roccamontepiano area is located next to the northeast edge of the Majella structure, between the River Foro and the River Alento (Abruzzo region, central Italy), and it is characterized by the presence of a pseudo-rectangular-shaped travertine plateau, extending in a NW–SE direction for about 2.3 km, lying on a marine clay formation (Figure 2.1).

The plateau, located between 610 and 650 m above sea level (a.s.l.), deepens slightly towards the NE, with an average gradient of about 5%. It is characterized by the presence of small valleys running SW–NE, and by dolines and other pseudo-karstic features along its northern margin. Generally, these depressions are aligned and elongated in the SW–NE direction, related to the presence of fractures, probably of tectonic origin, that facilitated dissolution processes. Red soils cover the plain continuously and uniformly except for the southern area, where the major elevations of the plateau (630–650 m a.s.l.) are reached. The margins of the plateau consist of vertical cliffs with height ranging from a few metres in the western part to over 30 m on the eastern side. The travertine plate rim is characterized by a set of arcuate features, more pronounced upstream of Montepiano. A close set of large-scale fractures more than 1 km in length and up to several metres in width is present along the northern border of the travertine plateau and along part of the eastern border. Fractures having major length and depth derive from tension effects and are subparallel to the cliff; typically they are gashes and affect the entire thickness of the travertine. The largest fracture, 350 m long, is situated 10 to 50 m from the cliff edge, and isolates a travertine block of about 200 000 m^3. Other variously oriented discontinuities (both large-scale and small-scale) derive from the unhomogenous response of the travertinous mass to deformations induced by gravitational effects. A large (about 4 km^2) and fairly continuous landslide debris belt, with high slope gradients (seldom less than 30%), is located at the cliff toe. This belt shows high variability in width and thickness, as found from field and airphoto interpretation and from the examination of some borehole logs carried out for different purposes in the area. The maximum width of more than 2 km can be found downstream of the Ripa Rossa (Figure 2.1), whilst the maximum thickness of more than 20 m is located immediately above Roccamontepiano village. Isolated blocks, up to several cubic metres in volume, can be found near Terranova and as far as the River Alento, about 2.5 km from the cliff. Most of these blocks can be related to the major historic landslide which occurred on 24 June 1765 (Del Re 1835; De Virgiliis 1837: Almagià 1910a, b). The blocks lying in the northern part of the area probably derive from different events. Southwards, the debris belt narrows to a few hundreds of metres in width and along the western margin it exists only discontinuously and with a maximum width of a few tens of metres (Figure 2.1).

Lithotypes outcropping in the area (Figure 2.1) consist of marine and continental deposits. They are briefly described in the following sections, from bottom to top, in chronological order (see D'Alessandro (1982) for more details).

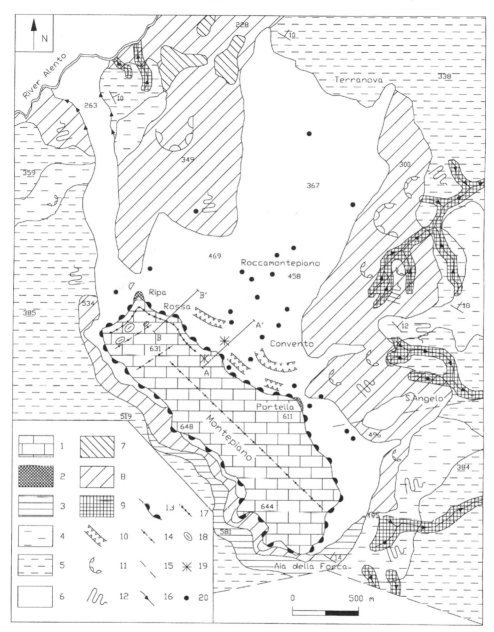

Figure 2.1 Geomorphological map (elevation in m a.s.l.). (1) Travertine (Lower Pleistocene?); (2) Sand and Conglomerate (Lower Pleistocene); Bottom Clay (Upper Pliocene); (3) arenaceous-pelitic association; (4) pelitic-arenaceous association; (5) pelite dominant; (6) landslide deposits; (7) alluvial terraces; (8) eluvial-colluvial and weathering deposits; (9) badlands area; (10) trench; (11) landslide scarp; (12) area affected by superficial creep; (13) landslide scarp; (14) fracture of probable tectonic origin; (15) supposed fault; (16) gully; (17) tension crack; (18) doline; (19) area investigated by subsoil survey and location of first remedial works (see text for explanation); (20) small spring

2.2.1 Bottom Clay (Upper Pliocene)

This soil consists mainly of marine clays, grey-blue or yellowish due to weathering, with a generally indistinct stratification. Parallel lamination is frequent in the upper portion of the formation, sometimes outlined by thin silt layers. Small-scale discontinuities were frequently observed in the clay mass. They are typically planar or gently curved, with a decimetric spacing and a matt surface, and form a network of apparently randomly distributed fissures. This formation is typically affected by weathering processes in its shallowest part (up to a few metres in depth) and outcrops mainly in the badlands around Montepiano plateau.

Palaeontological analyses showed a faunal association typical of Plio-Pleistocene transition in neritic environments.

Three different associations, characterized by a progressive increase of sand content, have been recognized:

1. pelite dominant: massive grey-blue clay with rare layers of laminated silty clay;
2. pelitic-arenaceous association: alternation of clay and silty clay in thin layers, sometimes laminated, grey and light brown, with rare yellowish fine and layers;
3. arenaceous-pelitic association: well stratified sandstone, sometimes cross-bedded, with grey sandy clay layers. This association outcrops in the southeast margin of the plateau (near Aia della Forca; Figure 2.1) showing a maximum outcropping thickness of about 20–25 m.

2.2.2 Sand and Conglomerate (Lower Pleistocene)

This formation comprises fine sands with silts, occasionally clayey and closely laminated and conglomerates with rounded calcareous pebbles ranging from a few millimetres to over 10 cm in diameter. The conglomerate has a sandy matrix and shows a generally low degree of cementation. Highly cemented layers with a calcareous sandy matrix can also be found. The formation has an average thickness of about 3–4 m. Micropalaeontological analyses show a low residue, with a faunal association characterized by a few species comparable with those of the Bottom Clay.

The formation outcrops in limited areas at the toe of the northwestern and southeastern margin of the plateau.

2.2.3 Travertine (Lower Pleistocene?)

This formation consists of different lithotypes (encrustation travertine, debris travertine, travertinous sand) with frequent facies transitions (visible at decimetric scale). Porosity is quite high and the whole mass is characterized by fractures and leaching processes. The overall thickness detected by drilling in the Ripa Rossa area is about 40 m and decreases slightly towards the southeast. The transition to the sand conglomerate layer below is sharp and slightly unconformable.

At the top of the travertine plateau and at the bottom of the dolina depressions, red silt and silty-sand deposits with pebbles and headers (*terra rossa*) are widely present, showing a maximum thickness of several metres.

2.2.4 Colluvium (Holocene–present)

These mainly consist of travertine and subordinately sand–conglomerate blocks embedded in a sandy-silty or sandy-clayey matrix. They derive from the accumulation of many landslide events and form a continuous belt of considerable thickness (locally more than 20 m) extending as far as the River Alento (Figure 2.1).

2.2.5 Structural Features

Some disjunctive structures interrupt, if only slightly, the otherwise markedly planar travertine plateau. These can be related to neotectonic displacements and are particularly evident in the northern area, where NE–SW fractures cut a major fracture parallel to the main axis of the plain (NW–SE direction). This latter fracture is located along the centre line of the plateau in correspondence with a visible break in the slope gradient.

Upper Pliocene deposits generally appear not to be disturbed by tectonics and are approximately monoclinal, with an average dip of 10–15° towards the northeast. Effects of tension tectonics activity may be observed in the area of S. Angelo (Figure 2.1), where a high-angle fault with a throw of about 20 m is present. No other evidence of dislocations occurs in the rest of the area.

2.3 INSTABILITY PHENOMENA

Instability phenomena with different kinematics, extent and state of activity occur along the entire Montepiano plateau involving the Travertine, the Sand and Conglomerate formation and the upper portion of the Bottom Clay.

Rock falls and toppling of minor importance characterize the northern part of the plateau and its southwestern margin. In these areas the volume of the detached blocks is generally small but extremely variable (from 1–2 m³ to a few cubic metres) as a consequence of both the physical characteristics of the travertine and the extent, orientation and distribution of the discontinuities. The runout distances range from several metres to a few tens of metres from the cliff toe depending on local morphology and soil characteristics.

On the other hand the NE margin of the plateau shows the presence of larger instability phenomena, and different morphological features can be detected SE and NW of the Convento area (Figure 2.1).

Narrow depressions, stretching parallel to the Montepiano cliff, characterize the area SE of the Convento, almost as far as Portella. They occur on the debris belt and in some places they probably reach the clay below, as is indicated by the occurrence of some springs. This morphological configuration, together with the occurrence of travertinous clints (parallel to the Montepiano cliff), is the result of instability processes (mainly lateral spreads) that affected the Travertine, whose blocks were broken during displacement and subsequently partially eroded.

At present, geomorphological evidence of the lateral spreads is partly obliterated and only residual features can be observed. Tensile fractures do not occur on the examined cliff edge, although a wide depression can be followed 200 m inside the plateau margin. Large topplings and falls of big blocks of travertine occur instead in the area NW of the Convento and along Ripa Rossa. In this area, tension fractures are present along the cliff (Figure 2.1), isolating big travertine blocks, some of which show clear rotational displacement; the occurrence of this kind of movement indicates the involvement of the clay below in the instability process.

With reference to this deformative mechanism, a major landslide occurred on 24 June 1765 (Del Re 1835; De Virgiliis 1837: Almagià 1910a, b). The slide started in the area immediately NW of the Convento and destroyed the Roccamontepiano village, causing about 700 deaths. At present, the travertinous cliff shows almost the same morphological features as just before the historic landslide. A picture of the time (Figure 2.2a) shows in fact that a set of fractures

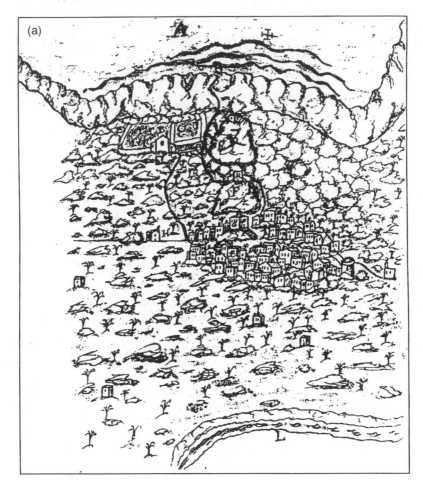

Figure 2.2   (a) Roccamontepiano before (a) and after (b) the landslide of 24 June 1765 in pictures of the time (from manuscript of unknown author)

parallel to the cliff, similar to that actually present in the same area, characterized the Travertine plateau before the landslide. Apparently, the main scarp of the 1765 landslide coincided with one of these major fractures (Figure 2.2b).

According to the field surveys and to the documents collected, the instability phenomena occurring on Montepiano can be considered quiescent in the area SE of the Convento and active in the NW area (Ripa Rossa).

2.4 SUBSURFACE INVESTIGATIONS

In 1993, an investigation was carried out in the Roccamontepiano area. The survey consisted of 22 boreholes (up to 50 m deep), ten of them equipped with open-pipe piezometers, two with Casagrande-type piezometers, and one with an inclinometer tube.

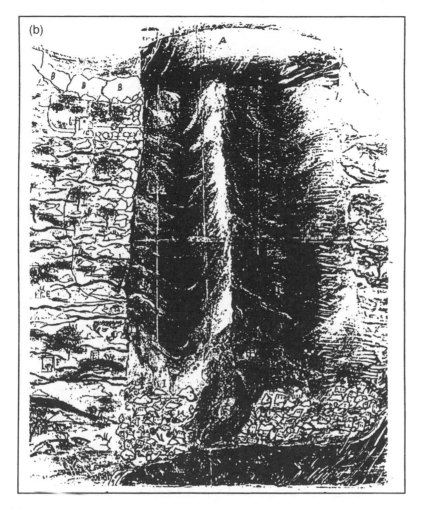

Figure 2.2 (*cont.*)

The data obtained from the survey on the plateau (610 m and 615 m in elevation) showed that the Travertine has a thickness of about 40 m and that it is characterized by marked lithological and structural heterogeneity due to the presence of sandy layers and to variously oriented small-scale fractures. In this area the Travertine is covered by sandy-silty and silty-clayey deposits (*terra rossa*) ranging about from 4 to 7 m in thickness. The transition between the Travertine and the Bottom Clay (about 565 m in elevation) is marked by a sandy-gravel layer with a thickness of about 3–4 m.

In correspondence with the historic landslide body, the top of the Bottom Clay deepens to 18–21 m from the ground surface, showing that the overall thickness of the travertinous landslide deposits is about 20 m (depending on the morphological conditions, however, the thickness of these deposits can be locally higher). From the available data it is not possible to establish if the upper portion of the Bottom Clay was involved in the historic landslide.

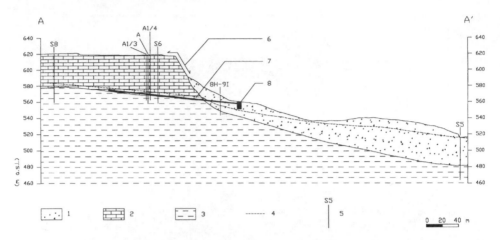

Figure 2.3 Schematic cross-section of Montepiano cliff showing the first completed remedial works: (1) landslide deposits; (2) Travertine; (3) Bottom Clay; (4) water table; (5) borehole; (6) slope regrading; (7) subhorizontal drain; (8) drainage well

Observations made on the piezometers installed in the boreholes reveal the presence of a groundwater table with phreatic levels ranging annually from 1 to 2 m above the top of the sandy-gravel layer. The water saturates only the basal part of the Travertine and flows gently northwards into the landslide deposits at the toe of the Travertine cliff. Moving downstream of the cliff, the water level becomes progressively nearer to the ground surface and several small springs appear depending on the position of the Bottom Clay, on the local morphology, and on the variations of the hydraulic conductivity.

Following the subsurface investigations, and considering the critical stability conditions of the cliff, first stabilization works were planned and completed in a small area located NW of the Convento. About 80 m downhill from the cliff edge, a drainage well with six subhorizontal drains about 150 m long has been excavated to reduce pore water pressure (Figure 2.3). Piezometric data collected before and after the installation of the drainage system showed a lowering of groundwater level of between 0.24 m and 1.26 m. A light regrading of the slope was also performed in order to reduce the weight of the unstable blocks and to remove the more fractured Travertine portions.

2.5   PHYSICAL AND MECHANICAL CHARACTERISTICS

A set of laboratory tests was performed on undisturbed and remoulded samples of the materials involved in the gravitational movements, with the particular aim of evaluating the mechanical characteristics of the silty-clayey formation (Bottom Clay). Minor attention was given to the geotechnical properties of the Travertine and landslide deposits, which are characterized by great lithological heterogeneity and are generally only passively involved in the deformative processes. The physical and mechanical properties are summarized in Table 2.1.

The Bottom Clay formation is classified as 'silt with clay' containing a small percentage of sand and belonging to the CL class (inorganic clays with low to medium plasticity) of the

Table 2.1 Physical and mechanical properties of samples

	WL	PI	IC	γ	c'	φ'	φ'_r	φ'_{fs}	σ_c
Bottom Clay	40–45	18–25	0.9–1.1	20–21	40–80	24–29	12–19	21–25	0.2–0.5
Travertine				18–22	0	30–33			4–15
Colluvium				18–20			25–29	27–30	

WL, liquid limit (%); PI, plastic index (%); IC, consistency index; γ, unit weight (KN m^{-3}); c', (kPa); φ', φ'_r, φ'_{fs} (°); uniaxial compressive strength, σ_c (MPa)

USCS classification (US Army Engineer Waterways Experiment Station 1960). Oedometric tests indicate that the clay is slightly overconsolidated, with an OCR ranging from 1.1 to 1.5.

As a consequence of the high porosity and of the presence of microfractures and small-scale discontinuities, the Travertine shows low values of uniaxial compressive strength and is classifiable as 'weak rock' (Table 2.1). The mechanical and hydraulic properties of the Travertine typically vary as a function of the orientation and of the size of the samples and extensive large-scale tests should be performed to obtain representative values of the rock-mass mechanical properties. These properties have been estimated on the basis of the intact rock strength and of the discontinuity geometry and characteristics using the RMR classification proposed by Bieniawski (1989). RMR indexes range from 25 to 35 and the inferred representative values of rock-mass shear strength and elastic parameters are reported in the next section.

Owing to the difficulty of collecting undisturbed samples on the Sand and Conglomerate formation, only direct shear tests on remoulded samples were done. The shear strength values showed in Table 2.1 are, therefore, not representative of the field peak strength but only of the available shear resistance at very high deformations.

The colluvium shows high variability of grain size and it is generally constituted by clasts ranging from a few centimetres to some metres in size, dipped in a sandy-silty/sandy-clayey matrix. The physical and mechanical properties summarized in Table 2.1 refer to the matrix only and the shear strength parameters were obtained by direct shear tests on remoulded samples.

2.6 STABILITY ANALYSES

In geotechnical terms, the stratigraphical sequence can be schematically represented as the superimposition of a rigid body (Travertine) over a plastic material (Bottom Clay) with the interposition of a thin layer of granular material (Sand and Conglomerate). In this arrangement, the tensile stress generated in the rigid body by the progressive deformation of the silty clays can easily exceed the tensile strength of the Travertine. Tension cracks develop along the border of the cliff causing blocks to come apart from the rocky mass. The stability conditions of these detached blocks are critical considering that their entire weight is loaded on the clay below.

The fissured nature of the basal clay causes the bulk strength of the material, that is the soil strength determined on samples large enough to be representative of the fissure pattern, to be considerably lower than the strength determined on small intact samples (Bishop and Little 1967; Marsland and Butler 1967; Lo 1970; Petley 1984; Chandler 1984a). Moreover, time-dependent processes such as swelling, softening, progressive failure and local internal strains cause a loss of soil strength with time so that the shear strength effectively available in the field results is unpredictable on the basis of laboratory tests on small undisturbed samples

(Skempton 1964; Duncan and Dunlop 1969; Law and Lamb 1978; Chandler 1984a, b; Thomson and Kjartanson 1985; Yoshida *et al.* 1991).

When the shear stress induced by the weight of a detached travertine block equals the diminishing shear strength of the clay, a failure surface starts to develop below the travertine block causing the drained strength of the clay to move towards the residual state. The development of a continuous sliding surface by progressive failure advances downward leading to significant strain in the zone of shear failure and, consequently, to the counterslope rotation of the travertine block. This process of advancing deformation and failure in the Bottom Clay is essentially slow, as indicated by the long time elapsed since the last catastrophic event and by the low rate of displacement of the travertine blocks. Montepiano inhabitants indicate that the width of some of these fractures has increased by about 3 m in the last 30 years. Even considering the low hydraulic conductivity of the clay, therefore, the phenomenon can be considered fully drained. When the failure surface approaches the ground surface, however, the rate of deformation rapidly increases and the final stage of the phenomenon can be extremely fast. Chronicles of the 1765 landslide report that within a few hours after the first evidence of movement (a travertine block that fell from the cliff), many fractures appeared in the houses, the village started to move as a coherent block and finally a major part of the travertine cliff suddenly collapsed destroying the village downhill. High water levels in the tension crack behind the travertine block can significantly reduce the available effective strength of the Bottom Clay leading to an acceleration of the spread of the failure surface which acts as a triggering mechanism of the final collapse. Past and present instability phenomena are mainly related to this morphological evolution. At present, evidence of incipient failure has been observed in the Ripa Rossa area where a big isolated travertine block shows lowering and partial counterslope rotation. In order to evaluate the present stability conditions of the travertine cliff in the Ripa Rossa area and to investigate the effectiveness of the stabilization works, the classic limit equilibrium analysis method is not fully suitable. In the rigid-block mechanics, in fact, the effect of progressive deformation and failure spreading can only be taken into account considering stress redistribution following local failure on a prefixed surface (Bjerrum 1967; Lo 1972).

Stability analyses were therefore performed using a two-dimensional explicit finite difference model (ITASCA 1995). The finite difference method is a numerical technique used for the solution of a set of differential equations, given initial values and shor boundary values (see, for example, Desai and Christin 1977). In the finite difference method, every derivative in the set of governing equations is replaced directly by an algebraic expression, written in terms of the field variables (e.g. stress or displacements) at discrete points in space; these variables are undefined anywhere else. The basic explicit calculation cycle invokes the equations of motion to derive new velocities and displacements from stresses and forces; then, the strain rates are derived from velocities and new stresses from strain rates. The model used simulates the behaviour of materials which may undergo plastic flow when their yield limit is reached and it is based on the discretization of the space in the elements, each of which behaves according to a prescribed stress–strain law.

The model allows fully coupled mechanical and groundwater calculation but the essentially drained behaviour of the slope before failure and the uncertainty of the mechanically induced pore pressures at failure lead to long-term effective stress analyses being an acceptable approximation of the real behaviour. The pore water pressure in the model has been considered simply as hydrostatically distributed below the water table and any groundwater flow or mechanically induced pore pressure variations have been taken into account. The analyses focused on the conditions leading to failure and cannot properly represent the slope evolution after a generalized failure is attained.

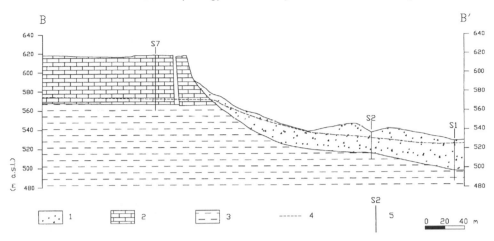

Figure 2.4 The analysed cross-section of Montepiano cliff: (1) landslide deposits; (2) Travertine; (3) Bottom Clay; (4) water table; (5) borehole

Table 2.2 Values of mechanical parameters used in analyses

	B	S	γ	c'	ϕ'
Bottom Clay	0.1	0.03	20	5	20
Travertine	15	5	22	100	30
Landslide deposits	0.08	0.02	20	2	30

B, bulk modulus (GPa); S, shear modulus (GPa)

The analysed cross-section (Figure 2.4) has been chosen taking into account both the maximum volume of the isolated travertine block and the distribution of the available data. The lithological model has been simplified combining the sand-gravel layer with the Travertine above. This simplification accounts for the limited thickness of the layer and for the fact that the sand-gravel material may be considered only passively involved in the deformative processes. The internal subdivisions in the Bottom Clay have also been neglected because of the mechanical prevalence of the clay even in the arenaceous-pelitic association.

The values of the mechanical parameters used in the analyses are listed in Table 2.2. As regard the silty clays, a sort of 'calibration' of the model has been done: starting from the peak shear strength parameters of the intact material, the effective friction angle and cohesion were progressively lowered in order to obtain in the modelled slope a steady-state condition of plastic deformation representative of the incipient instability phenomena observed in the field. The resulting limit values reported in Table 2.2 approximate the fully softened parameters determined through laboratory tests on remoulded samples. The fact that first-time slides in fissured plastic clays correspond to strength well below the peak is quite surprising but has been pointed out by many authors (Skempton 1964, 1970, 1985; Cancelli 1981; Chandler 1984a, b). These studies revealed that in these formations the mobilized shear strength in the field is very close to the fully softened strength of the clay or lies between the fully softened and the residual strength. The reasons given to explain such behaviour are different (presence of discontinuities, progressive failure, softening, time-dependent limit states) but the study of the prevailing factors in the analysed case is beyond the scope of this chapter.

For the Travertine, the available strength can be considered close to the peak strength of the rock mass while it can be assumed purely frictional along the large-scale discontinuities bonding the blocks. The mechanical parameters of the Travertine rock mass (Table 2.2) have been derived from the geomechanical classification (RMR; Bieniewski 1989) while the friction angle of the large-scale fracture (35°) has been inferred on the basis of unpublished data and of considerations on the surface roughness and wall strength (Barton and Choubey 1977). All materials have been modelled as elastic–perfectly plastic media with Mohr-Coulomb failure criteria.

Nine different initial conditions have been analysed by combining the following cases of groundwater level and slope geometry.

Groundwater level:
(A)   14 m above the base of the Travertine (natural condition with rain of high intensity and duration);
(B)   2 m above the base of the Travertine (light draining);
(C)   15 m below the base of the Travertine (heavy draining).

Slope geometry:
(1)   no regrading (natural profile of the slope);
(2)   7% of the volume of the unstable block removed (light regrading);
(3)   16% of the volume of the unstable block removed (heavy regrading).

The analysis results show that, in its natural condition (case A-1), when the field shear strength of the Bottom Clay gets near the fully softening values the isolated travertine block becomes unstable: the horizontal displacements monitored at the toe and at the top of the block (points P1 and P2 respectively; Figure 2.5) constantly increase according to the calculation cycles. The opposite directions of these displacements indicate a rotational-type failure mechanism, also evident from the overall distribution of the displacement vectors (Figure 2.6). The yielding zone is subcircular in shape and develops in the silty clays up to 70–80 m below the travertine block in depth and up to 180 m downhill from the cliff toe in width (Figure 2.7). Once the movement fully develops, the instability condition extends to the block lying behind.

In all the other analysed cases, the displacements reach a constant value and the slope stays in a stable condition. The computed values of displacements at stabilization do not have a real physical meaning but, depending on the initial groundwater level and slope geometry, give useful indications of the effectiveness of the remedial measures under consideration. For example, lowering the groundwater level to 2 m above the Travertine base, the maximum horizontal displacement becomes 107 cm without regrading (case B-1; Figure 2.8a) and 94 cm with a light regrading (case B-2; Figure 2.8b). In both cases, the silty clays remain in the elastic field; the yield zone reduces to the shallow portion of the colluvium and also the area subject to tensile failure in the Travertine becomes considerably smaller (Figure 2.9).

An overall picture of the results is shown in Figure 2.10, where the maximum horizontal displacements at stabilization for the different initial conditions are summarized. The diagram does not show the point relative to case A-1, where the displacements do not reach a finite value. The diagram underlines clearly the greater effectiveness of piezometric drawdown compared with slope regrading. In the case of limited draining, however, the displacements at stabilization would be quite high and could lead to a notable strength loss in the silty clays below the moving block, reducing the available strength towards the residual values.

With reference to the analysed failure mechanism, effective stabilization works could be the

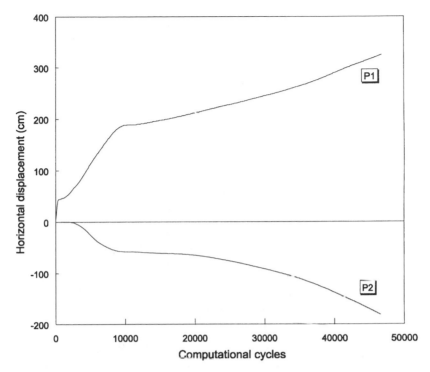

Figure 2.5 Finite difference analysis, case A-1: horizontal displacement computed in two points located at the toe (P1) and at the top (P2) of the travertine block (negative values indicate uphill displacement)

joining of slope regrading with a greater piezometric drawdown, which lowers the water table to within the upper portion of the Bottom Clay.

The slope regrading could also be of limited extent but the high degree of fragmentation of the travertine blocks makes the occurrence of topplings and rock falls very likely, which would represent in any case a serious hazard for the village below.

2.7 CONCLUSIONS

The detailed study of the geological and geomorphological characteristics of the Rocca-montepiano area and the results of the stability analyses carried out for different conditions allow us to draw the following conclusions.

1. The superimposition of a rigid body (Travertine) over a plastic material (Bottom Clay) causes the progressive horizontal deformation of the clay and, consequently, the tensile failure of the Travertine.
2. The deformation-induced strength loss of the clay and the pore pressure increases due to the saturation (even partially) of the tension cracks lead the isolated travertine blocks to critical stability conditions.
3. The instability process involves both the travertine block and the clay below causing an uphill extension of the instability conditions.

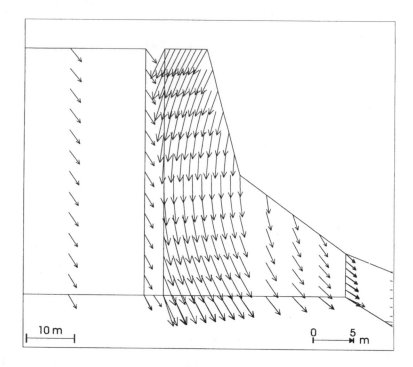

Figure 2.6   Finite difference analysis, case A-1: detail of displacement vectors in the travertine block

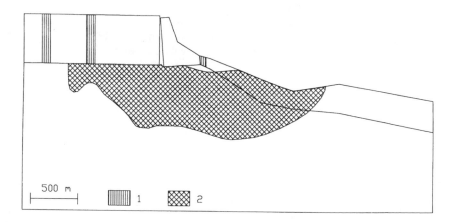

Figure 2.7   Finite difference analysis, case A-1: yielding zones (1 = at yield in tension; 2 = at yield in shear)

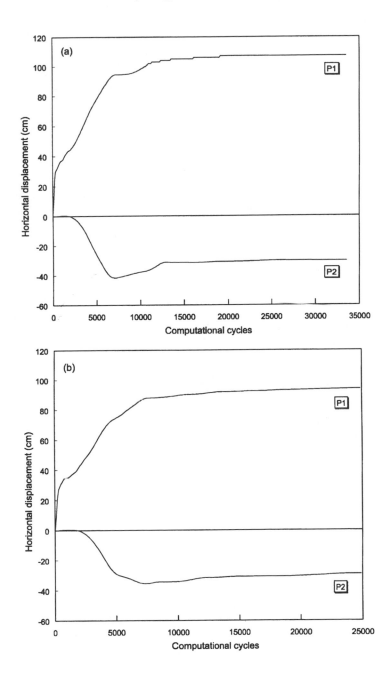

Figure 2.8 Finite difference analysis, cases B-1 (a) and B-2 (b): horizontal displacement computed in two points located at the toe (P1) and at the top (P2) of the travertine block (negative values indicate uphill displacement)

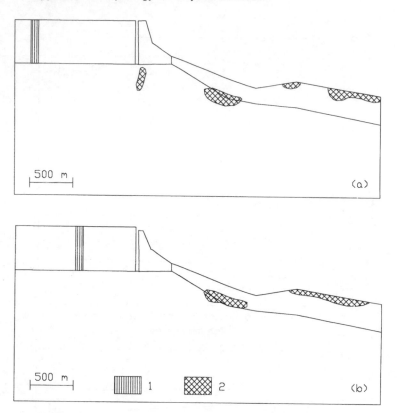

Figure 2.9 Finite difference analysis, cases B-1 (a) and B-2 (b): yielding zones (1 = at yield in tension; 2 = at yield in shear)

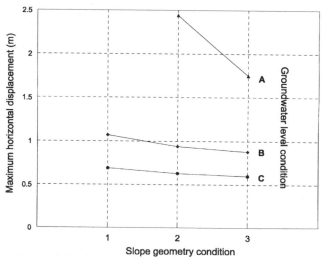

Figure 2.10 Maximum horizontal displacement at stabilization computed by finite difference analyses in different cases (see text for explanation)

4. The natural stability conditions of the travertine cliffs are almost critical: the finite difference analyses show that if high pore pressure values are associated with shear strength decreases to the fully softened values in the Bottom Clay, slope collapse is predicted.
5. Considering the morphological and hydrogeological conditions and the characteristics of the materials involved, the only feasible remedial work seems to be piezometric drawdown and slope regrading.
6. With reference to the rotational sliding, the modelling results show the greater effectiveness of piezometric drawdown compared to slope regrading in improving stability conditions.
7. Considering that the high degree of fragmentation of the travertine blocks makes the occurrence of topplings and rock falls very likely, future remedial works geared to protect the Roccamontepiano village should consist of both a very large piezometric drawdown and the removal of most of the fractured travertine portions.

ACKNOWLEDGEMENTS

Funding for this study was provided by MURST 1995 Professor R. Genevois. The manuscript benefited greatly from two anonymous reviewers who provided helpful criticisms and suggestions.

REFERENCES

Almagià, R. 1910a. Studi geografici sulle frane in Italia: l'Appennino centrale e meridionale. Conclusioni generali. *Memorie della Societè Geografica Italiana*, **14**, 435 pp.

Almagià, R. 1910b. La grande frana di Roccamontepiano (prov. di Chieti) (24 giugno 1765). *Rivista Abruzzese*, **25**, Teramo.

Barton, N. and Choubey, V. 1977. The shear strength of rock joints in theory and practice. *Rock Mechanics*, **12**, 1–54.

Bieniawski, Z.T. 1989. *Engineering Rock Mass Classification*. Wiley, New York, 251 pp.

Bishop, A.W. and Little, A.L. 1967. The influence of the size and orientation of the sample on the apparent strength of the London Clay at Maldon, Essex. *Proceedings of the Geotechnical Conference Oslo*, **1**, 89–96.

Bjerrum, L. 1967. Progressive failure in overconsolidated plastic clay and clay shales. *International Journal of the Soil Mechanics and Foundation Division*, ASCE, **93** (SM5), 1–50.

Cancelli, A. 1981. Evolution of slopes in over-consolidated clays. *Proceedings of 10th International Conference on Soil Mechanics and Foundation Engineering*, **3**, 377–380.

Chandler, R.J. 1984a. Delayed failure and observed strengths of first-time slides in stiff clays: a review. *Proceedings IV International Symposium on Landslides*, Toronto, 19–25.

Chandler, R.J. 1984b. Recent European experience of landslides in over-consolidated clays and soft rocks. *Proceedings IV International Symposium on Landslides*, Toronto, 61–81.

D'Alessandro, L. 1982. *La frana di Montepiano: indagini e studi – Relazione generale*. C.M. «Maielletta». Pennapiedimonte, Chieti, 86 pp.

D'Alessandro, L., Crescenti, U. and Genevois, R. 1987. La Ripa di Montepiano (Abruzzo): un primo esame delle caratteristiche geomorfologiche in rapporto alla stabilità. *Memorie della Società Geologica Italiana*, **37**, 775–787.

D'Alessandro, L., Genevois, R. and Romeo, R. 1989. A preliminary report on the Montepiano (Central Italy) landslide stabilization using regrading and drainage. In: *Stabilization of Landslides in Europe*.

International Society for Soil Mechanics and Foundation Engineering, European Regional Sub-Committee, Bogaziçi University, Istambul.

De Virgiliis, P. 1837. La Maiella, viaggio sentimentale. *Giornale Abruzzese di scienze, lettere e arti*, **7**, Chieti.

Del Re, G. 1835. *Descrizione topografica-fisica-economica-politica de' Reali domini al di qua del Faro nel Regno delle due Sicilie.* Tipografia Turchini, Napoli, 507 pp.

Desai, C.S. and Christin, J.T. 1977. *Numerical Methods in Geomechanics.* McGraw-Hill, New York.

Duncan, J.M. and Dunlop, P. 1969. Slopes in stiff-fissured clays and shales. *International Journal of the Soil Mechanics and Foundation Division*, ASCE, **95** (SM2), 467–493.

ITASCA 1995. *FLAC – Fast Lagrangian Analysis of Continua.* ITASCA Consulting Group, Inc. Minneapolis.

Law, K.T. and Lumb P. 1978. A limit equilibrium analysis of progressive failure in the stability of slopes. *Canadian Geotechnical Journal*, **15**, 113–122.

Lo, K.Y. 1970. The operational strength of fissured clays. *Geotechnique*, **20**, 57–74.

Lo, K.Y. 1972. An approach to the problem of progressive failure. *Canadian Geotechnical Journal*, **9**(4), 407–429.

Marsland, A. and Butler, M.E. 1967. Strength measurements on stiff fissured Barton Clay from Fawley (Hampshire). *Proceedings of the Geotechnical Conference Oslo*, **1**, 139–145.

Petley, D.J. 1984. Shear strength of over-consolidated fissured clay. *Proceedings IV International Symposium on Landslides*, Toronto, 167–172.

Skempton, A.W. 1964. Long-term stability of clay slopes. *Geotechnique*, **14**, 77–102.

Skempton, A.W. 1970. First-time slides in overconsolidated clay. *Geotechnique*, **20**, 320–324.

Skempton, A.W. 1985. Residual strength of clays in landslides, folded strata and the laboratory. *Geotechnique*, **35**, 3–8.

Thomson, S. and Kjartanson, B.H. 1985. A study of delayed failure in cut slope in stiff clay. *Canadian Geotechnical Journal*, **22**, 286–297.

US Army Engineer Waterways Experiment Station 1960. *The Unified Soil Classification System.* Technical Memorandum, 3–357.

Yoshida, N., Morgenstern, N.R. and Chan, D.H. 1991. Finite-element analysis of softening effects in fissured, overconsolidated clays and mudstones. *Canadian Geotechnical Journal*, **28**, 51–61.

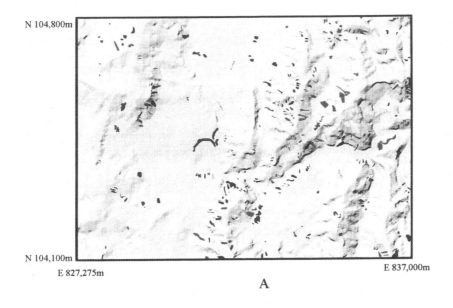

A

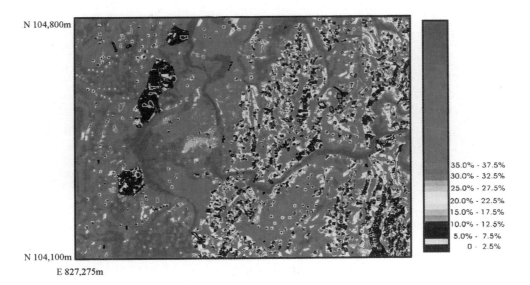

	35.0% - 37.5%
	30.0% - 32.5%
	25.0% - 27.5%
	20.0% - 22.5%
	15.0% - 17.5%
	10.0% - 12.5%
	5.0% - 7.5%
	0 - 2.5%

B

Plate 1 Landslide occurrences and hazard prediction in the Rio Chincina study area, Colombia. (A) Red identifies the distribution of the PRE-1960 and blue the distribution of the POST-1960 rapid debris avalanches. (B) The landslide hazard prediction is shown using the six influent causal factors and the fuzzy set algebraic sum operator (see also **Figure 1.2,** page 9)

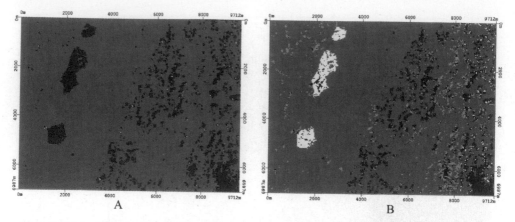

Plate 2 Two different maps (DIF-maps) for stability analysis in the Rio Chincina study area, Colombia. (A) Map of all eleven causal factors (map-1) versus a ten-factor map (map-2), i.e. removing the 'distance from road' map: matching rate is 99.1%. (B) Same as in (A) but removing the 'lithology' map: matching rate is 58.1% (see also **Figure 1.5,** page 13)

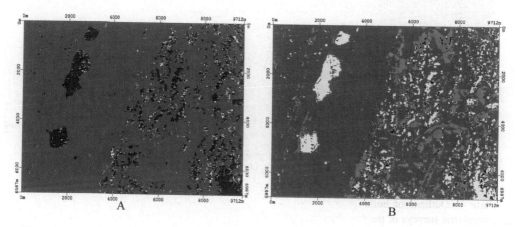

Plate 3 Two difference maps (DIF-maps) for stability analysis in the Rio Chincina study area, Colombia. (A) Map of the six influent causal factors (map-1) versus that of all eleven-factor map (map 2): matching rate is 79.1%. (B) Map of the six influent causal factors (map-1) versus that of only the five non-influent factor map (map-2): matching rate is 6.1% (see also **Figure 1.6,** page 13)

Plate 4 Outlines of buildings from December 1994 (red) and July 1996 (green) aerial photogrammetric surveys of the Corniglio slide. These features are overlain in the ortophoto of the July survey (see also **Figure 3.2,** page 45)

■	-30 - -20
■	-20 - -10
■	-10 - -0.5
■	0.5 - 10
■	10 - 20
■	20 - 30

0   500   1000   1500 Meters

Plate 5   Difference between elevation models of the December 1994 and the July 1996 data. The differences are shown as shaded relief, coloured reddish for the positive elevations (erosion), and bluish for the negative ones (accumulation) superimposed on the July 1996 digital ortophoto (see also **Figure 3.4,** page 47)

3 Photogrammetric Techniques for the Investigation of the Corniglio Landslide

G. GENTILI, E. GIUSTI AND G. PIZZAFERRI
CISIG, Parma, Italy

ABSTRACT

Methods and techniques of aerial photogrammetry have been applied to the Corniglio landslide, one of the largest in Europe. The directions and magnitudes of the horizontal movements of the slide have been defined by measuring the displacements of building corner outlines identified on large-scale orthophotos of the slide area prepared from aerial photos taken at different times. If there are no man-made features it is suggested that the centroids of landscape patches such as open fields in wooded areas or vegetation crowns may be used.

Where only aerial photography after the event is available, a similar technique can be applied in areas with man-made structures by comparing orthorectified photos with cadastral maps.

Differences between the 5 m grid digital terrain models have been used to document the vertical movements of the grid points within the slide area. The many patches, internal to the slide, showing no differences in elevation suggest that, contrary to what is reported in the literature, the slide does not move as a whole body but rather represents an area where material is eroded at the head of the crown and is accumulated at its foot over a stable base.

3.1 INTRODUCTION

Often techniques or products of one scientific discipline are useful to the investigations of other disciplines. The techniques of photogrammetry are applicable to all areal investigations and indeed offer the most useful means of extending point information to an area. In the well-known publication of Schuster and Krizek (1978) on landslides, some space is dedicated to the use of aerial photography but little mention is made of photogrammetric techniques as tools in geologic hazard studies.

When the Corniglio slide became active in 1994 the Regione Emilia Romagna, through its Servizio Difesa del Suolo (soil conservation service) of Parma headed by Ing. Larini, commissioned the Compagnia Generale Ripreseaeree of Parma to execute a series of photographic and other remote sensing aerial surveys over the slide area.

3.2 THE CORNIGLIO LANDSLIDE

The still-growing mountain chain of the Appennines, the backbone of Italy, is plagued with landslides. The landslide of Corniglio in the northern Appennines, is the largest of these; indeed it is one of the largest in Europe. The essential data concerning the slide are presented in Table 3.1 while its geographic location and geologic background are illustrated in Figure 3.1.

Applied Geomorphology: Theory and Practice. Edited by R.J. Allison.
© 2002 John Wiley & Sons, Ltd.

Table 3.1   Data on the Corniglio landslide

Average slope	8–23°
Elevation crown	1150 m
Elevation of bed of Parma River	550 m
Relief	600 m
Length of slide area	3000 m
Maximum width of slide	1000 m
Area of slide	$3 \times 19^6 \text{ m}^2$
Depth to sliding surface	30 to 120 m
Estimated volume of slide	$200 \times 10^6 \text{ m}^3$

Adapted from CNR (1996)

Three tectonic units are recognized in the area (Tellini and Verna 1996).

(1)   The Monte Caio unit is the oldest (Mezozoic), but is the uppermost unit in position, and is made up of a flysch sequence of limestones, marls and mudstones with a basal argillaceous chaotic complex. Within the immediate area of the slide, on Mount Aguzzo, it occupies the highest elevation while its stratigraphic sequence includes elements of the argillaceous complex at its base.

(2)   The Canetolo unit, formed in a sub-Ligurian basin, comprises a number of subunits which from the bottom upward include: (a) the subunit of Groppo Sovrano (Eocene) made up of the multicoloured argillites and calcilutites of the Riana Member (Paleocene?) followed by the thick sandstone sequences of the Groppo Sovrano proper; (b) the Petrignacola subunit, characterized by arenaceous and turbiditic sequences with a volcanic sedimentary component and a medium to thick stratification, locally very thick and conglomeratic; mixed with the Petrignacola Sandstones member (Lower Oligocene) are argillaceous limestone sequences; (c) the Bratica subunit which comprises basal sandstones (Isola di Palanzano member, analogous to the Ostia Sandstones, followed by Clays and Carbonates (Upper Cretaceous–Middle Eocene) with interspersed masses of Eocene carbonates of the flysch of Groppo del Vescovo and of Staiola. Conformably overlying these sediments are the T. Bratica Sandstones (Oligocene), a thickly stratified arenaceous-pelitic flysch interspersed with slumps and olistostromes of clayey-calcareous breccias. In the area of the Corniglio slide a lithologically heterogeneous unit has been located though its areal extension and time frame have yet to be defined. This chaotic unit has been called Melange di Lago. It contains lenses that have been interpreted as tectonic fragments from the Clays and Carbonates member limestones from the Groppo del Vescovo or Flysch of Staiola, Sandstones of Pte. Bratica, and olistostromes of these subnunits.

(3)   Within the Appenines the Tuscan–Umbrian units occupy a lower position and are represented by the Unità Cervarola (Lower Miocene) which in the slide area can be subdivided into (a) the Marra member, detached from the underlying Pracchiola Sandstones, which is represented by massive grey silty marls outcropping as a tectonic window near Marra, and (b) the Pracchiola member, characterized by thin to medium turbiditic sequences of pelitic sandstones which outcrop at the slide area as a small fenster on the Parma River. Within the immediate slide area the head of the crown is part of the flysch of Mount Caio which dominates the southern ridge of Mount Aguzzo. It consists of a strongly fractured marly-calcareous-pelitic formation about 100 m thick and strongly faulted from west to east. Its clayey base is only a few metres thick from tectonic lamination.

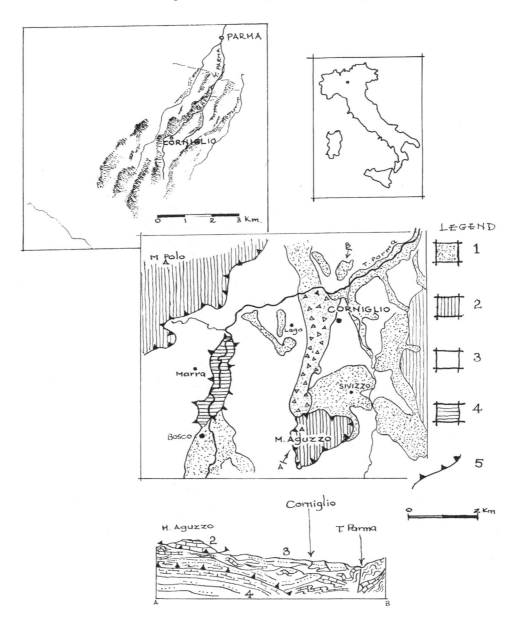

Figure 3.1 Geographic location and geologic background for the Corniglio landslide. Legend: 1 = Quarternary clastics (alluvium, detritus, slides, moraines, etc.). The Corniglio slide is indicated by open triangles. 2 = Mezozoic (Campanian–Maestrichian) Mount Caio Unit of the Ligurian domain. Sedimentary flysch facies (limestones, marls, mudstones). 3 = Eocene–Oligocene Canetolo Unit of the Subligurian domain. Chaotic bodies rich in marls and limestones, clays and sandstones. 4 = Lower Miocene Carvarola Unit (Pracchiola and Marra elements) of the Tuscan domain. Grey silty marl and pelitic sandstones. 5 = Contact among the tectonic units. The triangles point to the overlying unit

On both sides of Mount Aguzzo, but especially on the left side of T. Bratica, one finds remnants of old landslides mixed with great fragments of flysch that have moved as a mass retaining their structural appearance and providing evidence of a long sliding gravitative movement all around the ridge.

The western flank of the slide comprises mainly a chaotic argillaceous-calcareous mass tectonically thrusted over the sandstones of Pte. Bratica which has been named Unità di Lago. This unit is also found on the left side of Parma River in front of the slide. The eastern flank is represented mainly by the sandstones of Pte. Bratica which make up the ridge on which is located the town of Corniglio.

3.3 SLIDE CHRONOLOGY

The Corniglio slide of 1994–1996 is but the last of a long series of earth movements that go far back in time. An ancient legend of medieval origin has it that Saints Lucio and Amanzio, in whose names the main town square is named, invoked the occurrence of an earth slide to repay the town inhabitants for the poor hospitality shown them during their passage through Corniglio.

Almagià (1907) reports on all the historical activities of the slide, particularly those of 1612 and 1740.

We are dealing thus with an intermittent slide that in four centuries has been active at least four times with time intervals between 92 and 162 years. It is in this context that the great slide of November 1994 took place.

It began in the middle of the month after a period of intense rainfall, 1070 mm, of which 700 mm fell during September. A new scarp was formed at the head of the slide and rotational and translational movements took place throughout the slide area with velocities as high a 150 m per day. The slide was monitored by land surveys and with inclinometers down to a depth of 80 m. During 1995 the slide was quiescent and several engineering works were undertaken to drain it. On 31 December 1995 at 23:45 hours an earthquake of 3.3 magnitude was registered, with its epicentre at about 15 km southeast of the town of Corniglio, and the slide was reactivated. A new scarp was formed accompanied by fractures and earth flows. Another earth tremor of magnitude 2.2 was registered on 29 January 1996 and the earth movements that followed (30–80 cm per day) cracked some of the town's houses; the slide front began to impinge on the Parma River bringing fears of its closure and the formation of a lake. By April 1996 the channel width was reduced to 30 m or one-third of its original section. On 14 October 1996 a small earthquake, located about 90 km northeast of Corniglio, was registered and on 17 October, following long and intense rainfall, the slide began to move again starting from the crown area and propagating to the middle of the slide in a few days and to the Parma River within two weeks. Up to the end of November, when movements ceased, the slide moved about 7 m planimetrically along a sliding surface located between 80 and 150 m in depth (Larini et al. 1998).

3.4 AERIAL PHOTOGRAMMETRY AND GIS AS A TOOL FOR HYDROGEOLOGIC RISK STUDIES

The most popular use of aerial photogrammetry is the production of cartography and ortho-rectified imagery, a type of geographically referenced image (Greve 1996). While cartographic maps are the most common and widespread tools, orthoimages are fast becoming one of the most universally used mapping tools.

Scientists, cartographers and even the media are using orthoimages as cartographic background upon which thematic information is overlaid. The planimetric accuracy allows people to use orthoimages like maps for locating geographic features and for making direct measurements of the terrain: distances, angles and areas. On unrectified imagery such measurements can only be approximated because of image displacement and scale change caused by variations in local relief.

The recent popularity of geographic information systems (GIS) as a tool for combining diverse data sets to manage land resources and to model land uses, increases the demand for a computer version of a conventional orthophoto. In this context it is common to use a digital orthophoto as the basemap instead of a vector-drawn photogrammetric map because the orthophoto is less expensive while it offers all the benefits of photographic images. Because digital photogrammetry is in an early stage of evolution, it is appropriate to examine some practical relationships between detail requirements, photo scale, pixel size and data volume as related to resource mapping and inventory.

Geomorphic and environmental hazard studies involve the use of aerial photography at the scale of approximately 1:5000 to 1:40 000. Modern cameras and films can produce photos with resolution of 50 to 125 lines per mm and even higher. These values equate to ground resolutions in the range of approximately 4 to 80 cm and require pixel resolutions in the range of 8 to 20 μm to acquire all the photographic detail when scanning. However, for most land resources mapping tasks there is little advantage to scanning photos at the resolution pixel size of the original photograph. Therefore, a pixel size of 28 to 21 μm is used because it provides adequate clarity of image details and accuracy of measurement for a photographic scale of 1:40 000 or larger. At these pixel size values the scanning standard of a standard aerial photo of 23 × 23 cm produces data files of about 64 to 113 megabytes.

Various tools have been developed for image scanning by the photogrammetric and the printing industry. The rotating drum scanner has a high level resolution performance and a high range of density while the flatbed scanner has a more accurate geometry. As a consequence all photogrammetric scanners are flat and should meet the following requirements: geometric positioning accuracy with a precision of ±2 μm, radiometric resolution of up to 1024 grey levels, and a pixel resolution of 10 μm or better.

3.5 PHOTOGRAMMETRIC APPLICATIONS TO THE CORNIGLIO SLIDE

Soon after the occurrence of the Corniglio landslide, the Servizio Difesa del Suolo of Parma commissioned a series of aerial photography surveys over the slide area. Six low level (about 1:12 000 scale) aerial photo surveys were taken from December 1994 to July 1996 using a 152 mm Wild RC20 camera with panchromatic film.

All the man-made features (buildings outlines, roads), hydrographic boundaries (river edges, lake shores, coastal lines), and elevation contour lines were extracted in digital form through a Zeiss P1 analytical stereoplotter. In addition, the last flight of 4 July 1996 was chosen to produce a digital orthophoto of the slide area which required the collection of ground control points for the orientation of the photos, the creation of a digital elevation model to correct the displacement due to relief, and the scanning of the photos to obtain digital images to be orthogonally rectified.

Ackermann's analytical aerial triangulation program was used to correlate the field surveyed x,y,z coordinates of about 30 ground control points with their equivalent photographic features. The planimetric accuracy of these points is 38 cm (RMS) and the altimetric accuracy is ±42 cm (RMS). The program uses this correlation to interpolate the coordinates of other points. The resulting stereo model was entered in the analytical stereoplotter in order to map the features, contour lines and spot heights for the creation of the digital terrain model (DTM).

Contour lines with 2 m intervals, spot heights, and breaklines that depict abrupt changes in elevation and increase the accuracy of the DTMs were interpolated by the SCOP software for the generation of a DTM in the form of a regular grid. The aerial photographs were scanned with a Zeiss SCAI precision scanner with a pixel resolution of 28 μm that corresponds to a ground resolution of approximately 33 cm.

The ground control points, the points derived from the aerotriangulation computations, and the DTM with grid spacing of 5 m provided an adequate framework for image orientation and orthorectification which was carried out on a Silicon Graphics workstation using Zeiss PHODIS OP software. The orthophotos were subsequently mosaicked to create a seamless image on disk. These data provided the necessary input to evaluate the areal distribution of the planimetric and vertical displacement of the slide. For this purpose the vectors taken from the stereoplotter were entered into an ARC/INFO software package to create planimetric data that depict man-made features such as buildings and roads. In particular, two maps were made of buildings outlines from the December 1994 and the July 1996 data thus covering essentially the entire period of slide motion, at least for the lower part of the slide area. This is illustrated in Figure 3.2 where the two maps are superimposed on the orthophoto for the July 1996 flight.

From these data the actual vector displacements of buildings corners can be calculated as shown in Figure 3.3, where it is apparent that the scalar values of the displacements are not uniform but tend to increase in a downhill direction. So as to illustrate a technique possible when no aerial photography of before-the-event conditions are at hand, we made a similar evaluation of planimetric displacement of buildings by comparing a cadastral map of the area from the 1930s with the July 1996 image (Gentili 1998).

The results were even more striking though the number of buildings appearing on both maps were of course fewer. In these examples we have measured the displacements making reference to the same corner of buildings for simplicity. A more precise measurement would have used the centroid of the outline of the objects being displaced. This is of particular significance where no man-made structures exist and one has to make recourse to the outline of natural features that can be recognized in the before and after photographs such as open spaces within wooded areas, the outline of rock outcrops, and similar.

For the vertical displacement of the slide terrain we have used very simply the difference in land elevation of the DTMs, the digital elevation models of the orthophotos consisting of the elevations of points spaced 5 m apart. The data of the DTMs from the December 1994 and the July 1996 surveys, corresponding to the period of maximum displacement of the slide, were entered in the GRID module of ARC/INFO and subtracted resulting in an image of differences of elevations that ranged in value between −28 m in the upper part of the slide area where the terrain was eroded, to +17 m in the lower part where the soil accumulated. This is shown in Figure 3.4 as classes of relief areas superimposed on the July 1996 digital ortho-photo. Figure 3.4 provides a striking image of the slide from which, first of all, one can trace very accurately the boundary of the active slide area.

It should also be noted, however, that there are many patches throughout the middle part of the slide area, where no differences of elevation, to an approximation of about 1 m, were calculated. This would imply that the ground did not move, or that it moved less than could be registered by the DTM differences, or that it slid downhill along a plane parallel to the ground surface. The first of these conjectures is at odds with the generally accepted view that the entire area of the Corniglio slide moves as a whole body.

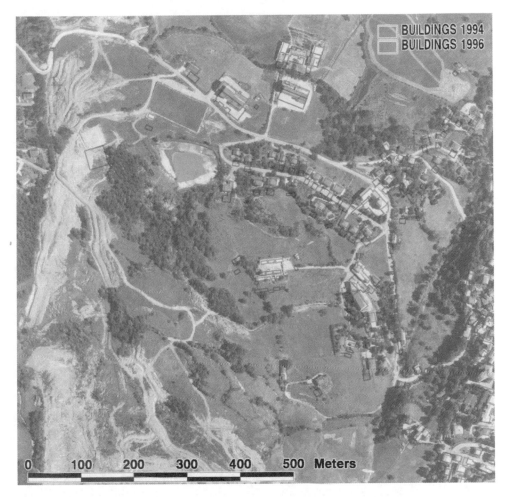

Figure 3.2 (Plate 4) Outlines of buildings from December 1994 (red) and July 1996 (green) aerial photogrammetric surveys of the Corniglio slide. These features are overlain in the ortophoto of the July survey

3.6 CONCLUSIONS

Techniques employed in photogrammetry, especially with the arrival on the market of the latest aerial films and cameras, provide a powerful means of measuring ground displacements of less than 1 m with low level aerial photography taken before and after a dynamic geologic event. If no aerial photography is taken before the event it may be possible to measure displacements with respect to the aerial photography existing in archives. In particular, where man-made structures are present, it may be possible to make use of cadastral maps.

An application of photogrammetry to one of the largest slides in Europe, the Corniglio slide in the northern Appennines, indicates that it is possible to measure planimetric displacements of buildings corners with an accuracy of about 40 cm. Nearly the same accuracy

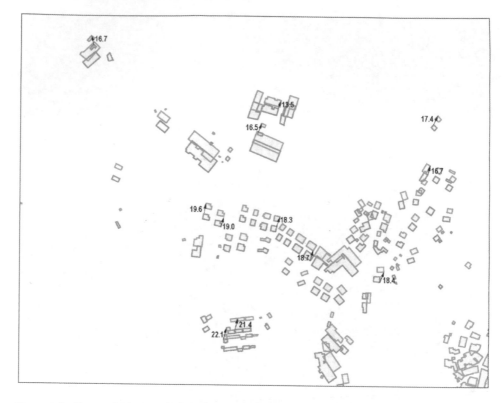

Figure 3.3 Vector displacement of selected buildings corners from the data of Figure 3.2. Note the general movements towards the northeast and the decrease of the displacement from more than 20 m to about 17 m

can be achieved in measuring vertical displacements. A map of differences in elevation of a net of grid points 5 m apart (DTM) over the slide area shows terrain erosion of about 28 m in the crown and an accumulation of about 17 m at the toe. The elevation difference map shows that many parts of the slide area, especially in its centre, underwent either erosion or deposition in a pattern of leopard skin-like patches. This suggests that, contrary to what is reported in the literature, the Corniglio slide may not move as a whole body but rather it may consist of erosional and depositional patches superimposed on a, at least in part, stable base. Clearly this point is of some importance to the land use planning of the area requiring some further studies concerning the overall mechanics of the slide.

ACKNOWLEDGEMENTS

We gratefully acknowledge permission of the Compagnia Generale Ripreseaeree of Parma to use data from their remote sensing aerial surveys. We also wish to thank Professor Tellini of Parma University for having made available information on the geology of the slide area before its publication.

	-30 - -20
	-20 - -10
	-10 - -0.5
	0.5 - 10
	10 - 20
	20 - 30

0　　　　　500　　　　　1000　　　　1500 Meters

Figure 3.4 (Plate 5)　Difference between elevation models of the December 1994 and the July 1996 data. The differences are shown as shaded relief, coloured reddish for the positive elevations (erosion), and bluish for the negative ones (accumulation) superimposed on the July 1996 digital ortophoto

REFERENCES

Almagià, R. 1907. Studi geografici sopra le frane in Italia vol. 1. *Mem. Soc. Geogr. It.*, 8, 95–101.

CNR (Consiglio Nazionale delle Ricerche) 1996. *La grande frana di Corniglio (Appennino Parmense).* Publication No. 1526 of Gr. Naz. Difesa Catastrofi Idrogeologiche, CNR, Modena.

Gentili, G. 1998. Integration of cartographic and cadastral data in a common information system. *Proceedings of ISPRS Symposium*, Comm. IV, Vol. XXXII, Part 4, 176–179.

Greve, C. (ed.) 1996. *Digital Photogrammetry – An addendum to the Manual of Photogrammetry.* American Society for Photogrammetry and Remote Sensing, Bethesda, Maryland.

Larini, G., Marchi, G., Pellegrini, M. and Tellini, C. 1998. La grande frana di Corniglio (Appennino Settentrionale, Provincia di Parma), rimobilizzata negli anni 1994–1996. *Atti Conv. Int. CNR: La prevenzione delle catastrofi idrogeologiche: Il contributo della ricerca scientifica*, Alba (Cuneo) 5–7 November 1996.

Schuster, R.L. and Krizek, R.J. 1978. *Landslides (Analysis and Control).* Special Report 176, Transport Research Board, National Research Council, National Academy of Sciences, Washington, DC.

Tellini, C. and Verna, L. 1996. *La frana di Corniglio.* L'Orsaro (CAI, Parma), Anno XVI, No. 2.

4 Risk-Prone Lands in Hilly Regions: Mapping Stages

FLORINA GRECU

Department of Geomorphology, University of Bucharest, Bucharest, Romania

ABSTRACT

The risk map (Figure 4.6) represents the final stage of a laborious analytical approach. For the hilly regions, this map expresses the vulnerability of the lands to degradation processes. The most important variables that influence the risk-proneness of hilly lands, under temperate climate, are lithology, landforms (geomorphic processes, slopes, drainage density) soil and vegetation. The methods used to draw up the map of risk-prone lands in a morphohydrographic basin in the Transylvanian Tableland are careful mapping of geomorphic processes, superposition of this map with the geological, soil and soil erosion maps and also with slopes, drainage density and vegetation maps. We then proceeded to regionalize the geomorphic processes and morphodynamic factors. We have also used indices for vertical erosion, areal erosion and landslides. The legend presents terraces and interfluves without present risk or with low risk; slope areas with low, medium, high, very high or excessive current risk (at degradation processes); floodplains with low, medium or high risk (at flooding); river beds with vertical or side erosion. The geomorphic risk map could be a better substitute for the general geomorphic map of some characteristics.

4.1 CONCEPTS

The map of risk-prone lands is the final phase of a painstaking analytical endeavour. The few maps worked out in Romania so far are the result of lasting investigations, the map itself emerging as a synthesis of the respective authors' geomorphological research (Balteanu *et al.* 1989; Cioacă *et al.* 1993; Schreiber 1980; Sandu 1994; Grecu 1997a).

Unless the regionalization of the phenomena is based on quantitative indicators and detailed mapping, and phenomena are ranked arbitrarily, one cannot claim a correct assessment of risk-prone lands, or provide comparisons between the studied regions.

Since the majority of land degradation prevention and control works, as well as management and territorial planning schemes, are focused on small drainage basins, an analysis of land risks in these territorial units is most appropriate. The basin–slope notion used in the specialist literature is translated as a relatively small drainage basin that functions in keeping with its drainage network and relief. Our analysis embraces these morphohydrographical concepts (Grecu 1992).

Establishing the basin size follows the drainage pattern of the Horton–Strahler system of establishing the hierarchy of the network (Zăvoianu 1985; Grecu 1992). Accordingly, elementary thalwegs are assigned a first-order rank. The stream completion index defines the state of the system.

Applied Geomorphology: Theory and Practice. Edited by R.J. Allison.
© 2002 John Wiley & Sons, Ltd.

The approach to natural risk phenomena is twofold: some consider disasters in terms of the number of casualties; others perceive extreme phenomena as a component of the normal evolution of the environment (Cotet 1978; Gueremy 1987; Chardon 1990; Gares *et al.* 1994; Ianos 1994; Rosenfeld 1994; Schheidegger 1994; Zăvoianu and Dragomirescu 1994; Grecu 1996, 1997b). 'Hazards are therefore simply part and parcel of the normal geomorphic evolution' (Scheidegger 1994, p. 24).

In this chapter we discuss the second approach. Evaluating hazards in terms of geographical system dynamics requires the establishment of the critical intervals beyond which certain factors may engender negative effects. Slope disequilibrium produced by extreme phenomena can be foreseen, prevented and controlled. In terms of protection measures, lands may change their degree of vulnerability.

The United Nations glossary of major terms used in disaster studies defines vulnerability as the 'Degree of loss (from 0% to 100%) resulting from a potentially damaging phenomenon' (United Nations 1992, p. 63).

In the temperate regions, with a hill and tableland relief, the notion of land vulnerability is partially synonymous with land degradation. We say partially because land degradation means destruction of the soil structure, and so sometimes loss of the subsoil. This kind of destruction may, or may not, impair geological strata bedding.

Both notions have economic significance: the first is more embracing, with casualties and material damage involved; the second is more restrictive, referring to the loss or reduction of the soil productive capacity.

Extreme geomorphic phenomena, like landslides (active), torrential precipitation, flooding etc., cause land degradation. The highly vulnerable degraded lands produce high material loss. They have both a direct, immediate impact on the population (casualties and loss of property) and an indirect, lasting one (fields destroyed, depletion of vegetal and animal output).

Risk phenomena studies are employed to determine the origin and evolution of various types of processes, locate (map) and regionalize them.

Risk means 'Expected losses (of lives, persons injured, property damaged and economic activity disrupted) due to a particular hazard for a given area and reference period. Based on mathematical calculations, risk is the product of hazard and vulnerability' (United Nations 1992, p. 51).

In this chapter, we use the morphohydrographical approach and geomorphic perspectives to demonstrate the stages needed to construct maps of risk-prone lands for hilly lands in the Calva Basin, Romania.

4.2 STAGES IN THE DEVELOPMENT AND CONSTRUCTION OF MAPS OF RISK-PRONE LANDS

Mapping such lands in the temperate zones of hills and tablelands requires successive mapping steps (inclusive of analytical maps).

If greater detail is required, methods should be adapted to reflect geological substrate and landform features.

The steps below are based on a geomorphological perspective (see also Section 4.2.3):

–   analysis of morphodynamic control factors (morphodynamic potential);
–   analysis of geomorphological processes, and their mapping;
–   regionalization of geomorphic processes and of morphohydrodynamic factors;

–   the construction of the map of risk-prone lands according to a legend structured in the previous stage (areas in which the intensity of the geomorphic processes and the morphodynamic factors are different).

4.2.1 Morphodynamic Control Factors

Assessing the morphodynamic potential of drainage basins must start from a set of concrete data, namely geographical position, morphometry, morphography, morphodynamic strength, conditioning or passive factors (rock, soil, vegetation, mean climatic values – freezing in the soil included – hydrological averages), dynamic or active factors (maximum precipitation, discharge, extreme air and soil temperature, maximum values of discharge and solid sediment transport, of water levels, human activity (also land use), seismicity, tectonics).

In a previous study we analysed all the morphodynamic factors in the Calva Basin (Grecu 1997a). Therefore, only the main quantitative morphometrical and morphographical data relating to our topic are given herein.

(i)   Morphometrical and morphographical analysis: reflects the state-of-the-art in the lasting evolution of the relief. It relies on some computed variables (for details see Munteanu 1991, Grecu 1992).

(ii)   Morphometrical data: maximum basin (and stream) altitude (H_{max}), minimum altitude (H_{min}), average altitude (H_{med}), length of mainstream (L), sinuosity coefficient (C_s), basin length (L_b), basin width (ρ: maximum ρ_{max}, minimum ρ_{min}, average ρ_{med}), surface (S), perimeter (P), Gravelius' coefficient (C_G), drainage density, relief amplitude, slope dip.

(iii) Morphographical data: slope transverse cross-section (convex, concave, mixed, straight), morphodynamic unit on slope, longitudinal profile of streams (profiles based on topographic maps), thalweg shape index for torrential bodies ($n = S_s/S_i$; $S_s =$ area above thalweg line; $S_i =$ area below thalweg line).

(iv)   Potential morphodynamic strength: depends on mean declivity of slope (average by I_v) order of magnitude; slope length L_v (by order of magnitude); categories of slope; length of mainstream (by sub-basins); mean declivity of mainstream slope (I_r average) by order of magnitude.

(v)   Exposition of slopes: by order of magnitude, referred to the stream network.

(vi)   Morphometrical models (a morphometrical and morphological synthesis): of the drainage net of the perimeter; of the surface area; of the slope.

The Calva Basin, intersected by parallel 46° N lat. and meridian 24°20' E long. lies in the Southern Transylvanian Tableland, in the northwest of the Hârtibaciu Tableland (Figure 4.1).

The River Calva is a right-hand tributary of the Visa which runs into the Târnava Mare, downstream from Copşa Mică town. Maximum basin altitude is 673 m (Hamba Hill) in the median sector, yet not in the source area. Minimum altitude is 298 m at its mouth, in the Şeica Mare settlement area.

The river springs from Zlagna Hill (625 m), crosses the Hârtibaciu Tableland from east to west along a distance of 29 km, at a level difference of 327 m. The basin area covers 179 km²; average altitude is 487 m.

Calculating basin area to basin length (27 km) yields a mean basin width of 6.6 km with top values of 12 km in its central part where it receives a c. 10 km long tributary (Valea Satului).

Looking at the drainage model (Figure 4.2B) these points follow.

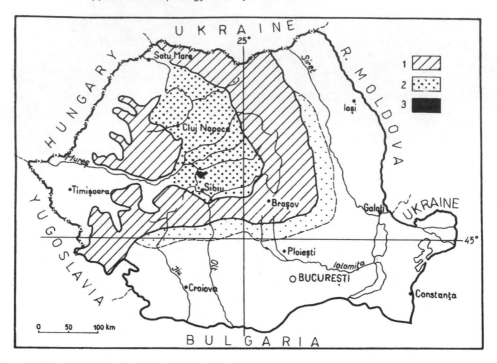

Figure 4.1 Geographical location within the Romanian territory. 1, The Carpathians; 2, The Transylvanian Tableland; 3, The Calva Basin

(a) The computed value of the number of stream segments in the highest order basin (S) being over unity, basin completion is above one (1.34), and so the junction ratio of 4.32 indicates a certain state of equilibrium throughout the basin. In this case, the calculated number of fourth-order river segments (5.78) exceeds the real, measured, value (4) (Table 4.1).

(b) The computed values of the sum of lengths and of average lengths for the highest-order basin (5) are far below the measured value, being oversized for the fourth-order segments. This is an indication that certain imbalances are entailed by fourth- and respectively third-order basin drainage models.

(c) Maps of the drainage net hierarchy of geomorphological processes, and of risk-prone areas reveal that second- and third-order basins are prone to headward and linear erosion.

(d) The same is revealed by analysing the first-order river segments, mainly involved in organizing run-off on slopes (graphical representation in semilog coordinates).

The outcropping deposits of the Calva Basin are remarkably uniform in terms of age and rock. Their physico-mechanical and chemical properties have been fairly well studied (Matei 1983; Grecu 1992) and the findings have revealed their proneness to landslides, sheet and linear erosion (Gârbacea and Grecu 1994).

The oldest surface deposits are of a Middle–Upper Miocene age. In general, the Transylvanian Basin Sarmatian consists of bluish-grey marls with intercalations of weakly cemented sands, several scores of metres thick. There are places in which rough sands alternate with large zones of packed marls, forming well-delineated complexes.

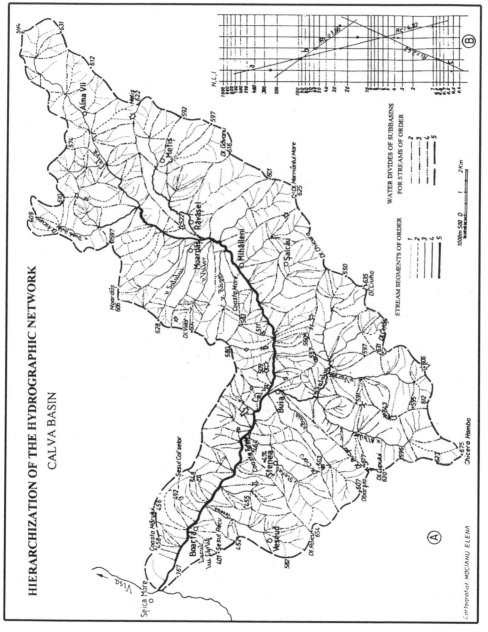

Figure 4.2 The hierarchy of the hydrographic network

Table 4.1 Drainage model data, Calva Basin

Parameter	Value*	Basin order					Ratio	Basin total
		1	2	3	4	5		
Number of stream	m	474	108	25	4	1	$R_c = 4.32$	$\sum N = 612$
segments (N)	c	466	108	25	5.78	1.34		
Length L	m	246	90	54	14	25	$R_L = 1.66$	$\sum L = 429$
(km)	c	149	90	54	32.5	19.6		
Average	m	0.52	0.83	2.02	3.5	25	$R_l = 2.43$	0.70
length l (km)	c	0.34	0.83	2.02	4.91	12		

* m = measured; c = computed

Other complexes are built up of a thick alternation of thin sands with sandy marls. This bedding succession is discontinued by dacitic tuffs, the latter being thinner in the upper section and thicker in the lower part of the Sarmatian. Most of the basin area is carved in Pontian deposits.

In a classification by lithological criteria, Pontian deposits are listed into the following horizons (Vancea 1960):

– lower sands;
– median marls;
– upper sands.

Each of these horizons contains numerous intercalations of marls or sands. The most recent formations are of Holocene age and consist of sands and muds.

The average annual rainfall amount is c. 591 mm (top quantities being registered in June – about 80 mm). Absolute rainfall values per 24 hours may reach some 100 mm (106 mm in 1941).

4.2.2 Present-day Geomorphic Processes

Contemporary geomorphic processes constitute major risk factors for land degradation. Therefore, large-scale mapping (1:25 000 and larger) and assessment of current dynamics of slopes (active or fixed) is an imperative step in any risk study of tableland regions.

As a rule, soil erosion studies group slope geomorphic processes according to their impact on the soil: sheet and linear erosion and landslides.

Indicators of the state of linear erosion, sheet erosion and landslides are based on the size of the area, or its length affected by the respective process (S_x) and the total surface area (S_t) studied (drainage basin), set in percentages (Figures 4.3 and 4.4).

$$E_x = \left(\frac{S_x}{S_t} \right) \times 100$$

Estimating the extent of the gully-covered area by the formula

$$E_r = \left(\frac{S_r}{S_t} \right) \times 100$$

(where S_r = gully area, km^2 and S_t = total area, km^2) is not very conclusive if field measurement data are omitted.

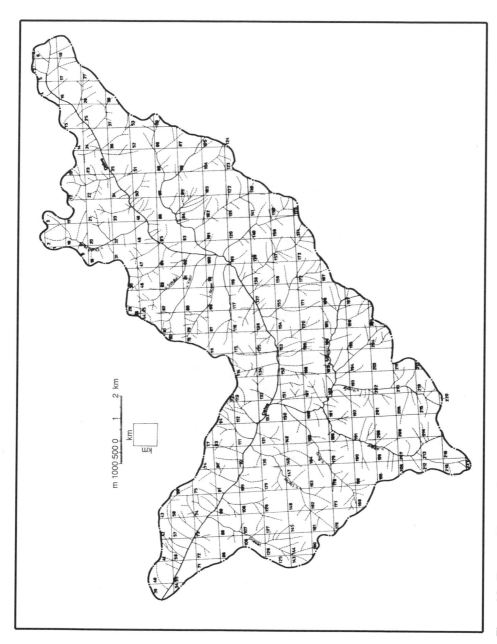

Figure 4.3 Topographic network

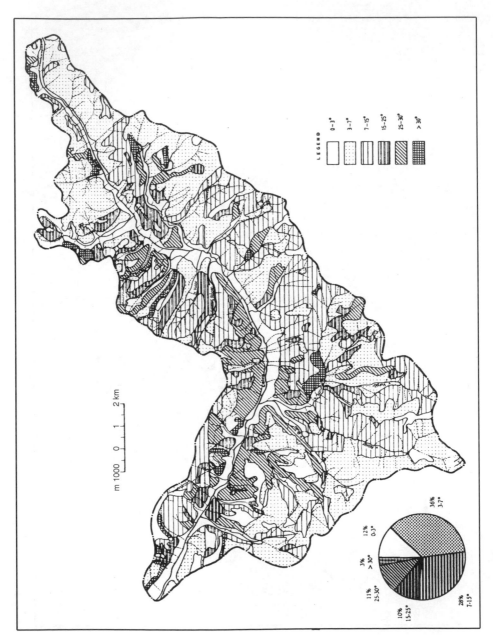

Figure 4.4 Slope map

Table 4.2 Drainage data-based risk parameter computation

	Formula	Unit
Occurrence frequency of segments	$\dfrac{\sum N_1}{S} = \dfrac{612}{179} = 3.41$	No. segments km^{-2}
Occurrence frequency of elementary thalwegs	$\dfrac{\sum N_1}{S} = \dfrac{474}{179} = 2.65$	N_1/km^{-2}
Incipient gully erosion	$\dfrac{\sum L_1}{S} = \dfrac{246}{179} = 1.37$	km km^{-2}
Full gully erosion	$\dfrac{\sum (L_1 + L_2)}{S} = \dfrac{246 + 90}{179} = \dfrac{336}{179} = 1.87$	km km^{-2}
Thalweg length density	$\dfrac{\sum L}{S} = \dfrac{429}{179} = 2.39$	km km^{-2}

In order to appraise how much of the area is subject to linear erosion, the following relation is used:

$$LE = \left(\frac{S_{LE}}{S_t}\right) \times 100$$

which is the ratio between the overall drainage network and total surface area.

Correlating the linear erosion coefficient with drainage network density ($L/S = $ km km^{-2}) yields relief dynamic in a linear perspective. However, since the linear erosion coefficient (LE) cannot easily be used even when measurements on the ground are made, length dimensions can be taken as a substitute, e.g. density of fragmentation, depth of fragmentation, as well as other calculations based on morphometrical models (the Horton–Strahler system). The degree of fragmentation can also be indirectly estimated by calculating the area covered by various slopes.

The extent of sheet erosion in a drainage basin includes the eroded soil area, in terms of the five soil erosion classes (Florea *et al.* 1987).

We mapped the primary contemporary processes.

The number of first-order segments (474) in the Calva Basin represents about 77% of the total stream segments measured (612). Referred to the overall basin area, the incidence of elementary thalwegs is 2.65 N_1 km^{-2}. This value comes fairly close to the occurrence frequency of the total segment number: 3.41 (Table 4.2).

Second- and third-order drainage basins record the highest density of elementary thalwegs. They have a high degree of linear and headward erosion.

Estimations of sheet erosion rely not only on rain-induced denudation, but also on the sheet slides of the herb cover, affecting all deforested slopes with higher than 15° declivity, i.e. the medium- to sharp-dipping ones (Moțoc 1975; Moțoc and Vatău 1992) (Figure 4.5).

Deep slides occur on very small areas, at the Miocene–Pliocene contact, in the south and southeast of the basin.

4.2.3 Quantitative Data Synthesis

Quantitative values of linear erosion, of relief energy and slopes were grouped by classes (Tables 4.3, 4.4, 4.5). Each class was assigned a number of points (from 1 to 5). The synthetic Table 4.6 presents the total number of points per square kilometre (see Figure 4.3). The risk classes correspond to the following points:

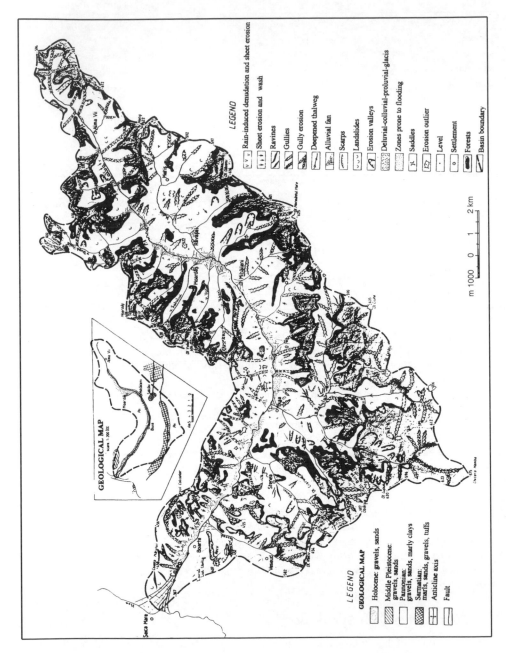

Figure 4.5 Present-day geomorphic processes

Table 4.3 Drainage density

Classes (km km^{-2})	Surface (km^2)	(%)	Value number	Relative frequency (%)	Cumulated frequency (%)	Points
0–1	23	12.90	51	23.06	23.06	*1*
1.1–2	60	33.30	66	29.84	52.90	*2*
2.1–3	64	36.10	70	31.65	84.55	*3*
3.1–4	25	13.70	30	13.56	98.11	*4*
4.1–5	7	3.90	4	1.89	100.00	*5*
Total	179	100	221	100		

Table 4.4 Relief energy

Classes (km km^{-2})	Surface (km^2)	(%)	Value number	Relative frequency (%)	Cumulated frequency (%)	Points
<50 m	6	3.30	17	7.69	7.69	*1–2*
51–100 m	50	27.70	65	29.41	37.10	*3*
101–150 m	106	59.50	115	52.05	89.15	*4*
>150 m	17	9.40	24	10.85	100	*5*
Total	179	100	221	100		

Table 4.5 Values of slopes

Categories of slope	0–3°	3–7°	7–15°	15–25°	25–30°	>30°	Total
Surface (km^2)	21.60	64.50	50.70	17.40	19.40	5.40	179.00
Surface ratio (%)	12.07	36.02	28.32	9.72	10.83	3.04	100.00
Value number	90	115	142	72	68	39	
Points	*1*	*2*	*3*	*4*	*5a*	*5b*	

0–2 points: no risk – code 1
3–4 points: low risk – code 2
5–6 points: moderate risk – code 3
7–8 points: high risk – code 4
9–10 points: very high risk – code 5a
more than 10 points: severe risk – code 5b

Each box contains the corresponding code of the respective risk category.

4.2.4 Regionalization of Risk Factors

Regionalization of risk factors, that is of geomorphic processes and morphodynamic potential, is a preliminary stage in drawing up a map of risk-prone lands. So far, such maps do not specifically reveal this operation, but it can be detected in the final editing of the map.

Given that the risk factors per square kilometre are varying, we used the method of map superposition (of geomorphic processes, vegetation, soil erosion and slope) in order to outline smaller areas at geomorphic risk. The quantities of precipitation with a major impact on relief dynamics do not vary with the basin. Therefore, they are not taken into calculation when outlining the areas at various degrees of risk.

Table 4.6 Synthetic table of quantitative data

No. points	RE		DD		Slope								Total
	Values	Points	Values	Points	0–3°	3–7°	7–15°	15–25°	25–30°	>30°	SS	Points	
1	40	1	0	1			25				25	3	5
2	140	4	1.75	2			60	35			60	4	10
3	100	3	1.1	2			15	15			15	4	9
⋮													
221	80	3	0.25	1		25					25	2	6

No. = box number (see Figure 4.3). RE = relief energy $H_{max} - H_{min}$ (m). SS = selected slope values. DD = drainage density $(\sum L/S)$ (km km^{-2})

Mapping the Calva Basin proceeded from the interfluves, slopes and major channels by superposing the respective maps. In this way, areas of little, medium and high risk have been outlined in terms of relief (slope), vegetation, and soil erosion.

4.3   RESULTS: THE MAP OF RISK-PRONE AREAS IN THE CALVA BASIN

From what has been discussed so far it follows that the impact of geomorphic processes on land in hilly zones with a relatively dense population and intense agricultural practices depletes soil productive capacity, more precisely the different degree of risk for their degradation or vulnerability.

The main risk variables in the hills of temperate regions, in our case the Calva Basin are: lithology, relief (geomorphic processes, slopes, density of drainage network), soil and vegetation. Therefore, from the bulk of analytical information of a great quantitative and qualitative diversity, we focused on the maps of geomorphic processes, soils and soil erosion, density of drainage network, relief amplitude, and slopes. The map of geomorphic processes forms the geomorphic basis for assigning risk. Since rocks are relatively uniform throughout the basin, this variable does not produce unusual space variations. Strata inclination is marked out on the slope map.

The degree of geomorphic risk is variable on three big surfaces: horizontal or semi-horizontal surface areas; slopes; and channels (major and minor) (Figure 4.6).

Surfaces of interfluves and terraces with slopes below 5° are, as a rule, risk-free at present. Very narrow summits are aggressed by the headward erosion of first-order streams. Their incidence is fairly high (2.65 segments km^{-2}). Such summits also occur in afforested areas, e.g. in the Steana and Valea Satului basins covering a total area of 27.0 km^2 (i.e. 14.09% of the basin area).

Slopes are dominant landforms in the Calva Basin.

(i)   Low-risk *slopes* (glacis) with a very mild (glacis-like) dip of 3–7° stretch on either side of the median and lower courses of the Calva, Valea Satului and Steana rivers, over a distance of 12 km^2 (6.7% of the Calva Basin area). Surfaces of interfluves and terraces and the glacis likewise, are not degraded, being used as cropland.

(ii)   Medium- to high-risk slopes of various declivity (intermediate to steep and very steep) are afforested, with pseudo-gleized brown and albian leached soils. They extend on the left side of Calva River and the upper course of the Calva River, corresponding largely to the afforested areas (55.9 km; 31.23% of the Calva Basin area).

(iii)   High-risk slopes with medium to sharp and very sharp dip, are deforested, with clay-illuvial brown rocks and eroded leached brown soils. Present-day geomorphic processes are affecting the herb layer and partially the soil, too. Sheet erosion and sheet slide are dominant phenomena (31.3 km^2; 17.49% of the Calva Basin area). Lands have low productivity, being used as pastures and hayfields. The onset of these processes has been facilitated by the unkept man-made terraces.

(iv)   Very high-risk slopes are steep and very steep, deforested, with erodisoils and eroded clay-illuvial brown soils. Both sheet and deep erosion are extremely active, lands are no longer used for agriculture (34.3 km^2; 19.6% of the Calva Basin area).

(v)   Severely eroded slopes are very steep, with outcropping rock and eroded soil, deforested, active gully erosion with sharp-dipping slopes down to 15–30 m (14 km^2; 7.82% of the Calva Basin area). Attempts at settling them with acacia species have been made over small areas.

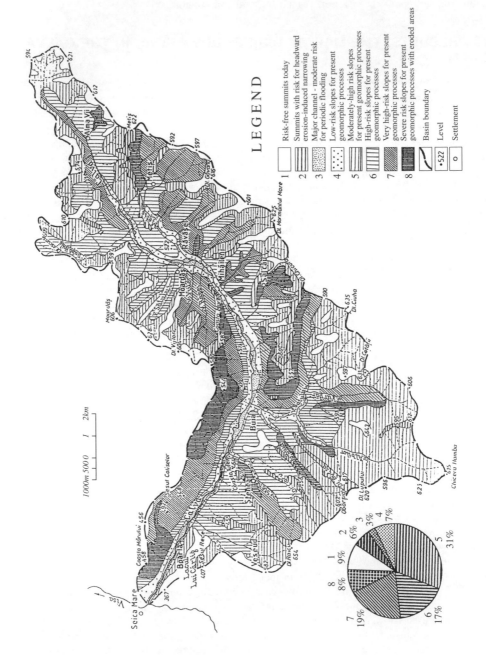

Figure 4.6 Map of risk-prone lands

Major channels are medium to high risk-prone due to flooding. The river thalwegs look like ditches deepened in their own alluvia; they have steep banks, formed of sands and muds. The minor channel is not embanked. The pattern of the minor channel and its relatively small depth, correlated with torrential precipitation, accounts for reduced flooding during the period May–August. The rain non-uniformity index in the Calva Basin is highest (15.3) in the months of May, June and August. The absolute maximum non-uniformity index rainfall value was in July (36.6 in 1985).

4.4 CONCLUSIONS

Some 50% of the Calva Basin area presents degraded lands with high, very high and severe risk situations, due primarily to deforestation.

The risk-prone map shows the present-day relief dynamics. The geomorphic risk map could be a better substitute for the general geomorphic map of hydrographic basins and could also be completed with landform characteristics. In conclusion, a general regional geomorphic map can be constructed.

REFERENCES

Balteanu, D., Dinu, M. and Cioacă, A. 1989. Hărţile de risc geomorfologic (Exemplificări din Subcarpaţii şi Podişul Getic). *Studii si cercetari de geologie, geofizica, geografie, Geografie*, **XXXVI**, 9–13.

Chardon, M. 1990 Quelques reflexions sur les catastrophes naturelles en montagne. *Revue de Geographie Alpine*, **LXVIII** (1-2-3), 193–213.

Cioacă, A., Bălteanu, D., Dinu, M. and Constantinescu, M. 1993. Studiul unor cazuri de risc geomorfologic nn Carpatii de la Curbură. *Studii si cercetari de geografie*, **XL**, 43–55.

Coteţ, P. 1978. O nouă categorie de hărţi – hărţi de risc – şi importanţa lor geografică. *Terra*, X(XXX), 3.

Florea, N., Bălăceanu, V., Răuţă, C. and Canarache, A. 1987. *Metodologia elaborării studiilor pedologice, Partea a III-a – Indicatorii ecopedologici*. ICPA, Bucureşti.

Gârbacea, V. and Grecu, F. 1994. Dealurile Şoalei. Caractere geomorfologice. *Studii si cercetari de geografie*, **XLI**, 49–58.

Gares, P., Sherman, D. and Nordstrom, K. 1994. Geomorphology and natural hazards. *Geomorphology*, 10, 1–18.

Grecu, F. 1992. *Bazinul Hârtibaciului. Elemente de morfohidrografie*. Academiei, Bucureşti.

Grecu, F. 1996. Expunerea la risc a terenurilor deluroase, în vol. Cercetări geografice în spaţiul carpato-danubian. *A II-a Conferinţă regională de Geografie*, Timişoara, 18–24.

Grecu, F. 1997a. Etapele întocmirii hărţii expunerii la risc geomorfologic a terenurilor din bazinele hidrografice de deal. *Memoriile Sectiilor Stiintifice, Academia Romana*, **XVII**, 1994, 307–323.

Grecu, F. 1997b. *Fenomene naturale de risc geologice şi geomorfologice*. Universităţii, Bucureşti.

Gueremy, P. 1987. Geomorphologie et risques naturels. Rapport introductif. *Revue de géomorphologie dynamique*, **3–4**, 98–107.

Ianoş, I. 1994. Riscul în sistemele geografice. *Studii si cercetari de geografie*, **XLI**, 19–25.

Matei, L. 1983. *Argilele panoniene din Transilvania*. Academiei, Bucureşti.

Motoc, M. 1975. *Eroziunea solului pe terenurile agricole şi combaterea ei*. Agrosilvică, Bucureşti.

Motoc, M. and Vatău, A. 1992. Indicatori privind eroziunea solului. *Mediul Inconjurator*, **III** (3).

Munteanu, S.A. 1991, 1993. *Amenajarea bazinelor hidrografice torenţiale prin lucrări silvice şi hidrotehnice*, **I**, **II**. Academiei, Bucureşti.

Rosenfeld, Ch. 1994. The geomorphological dimensions of natural disasters. *Geomorphology*, 10, 27–36.

Sandu, M. 1994. Harta de risc geomorfologic a culoarului depresionar Sibiu-Apold. *Lucrările Sesiunii Ştiinţifice anuale 1993*, Academia Română, Inst. de Geografie, Bucureşti, 44–49.

Scheidegger, A.E. 1994. Hazards: singularities in geomorphic system. *Geomorphology*, 10, 19–25.

Schreiber, W. 1980. Harta riscului intervenţiilor antropice în peisajul geografic al Munţilor Harghita. *Studii si cercetari de geologie, geofizica, geografie, Geografie*, **XXVII** (1).

United Nations 1992. *Internationally Agreed Glossary of Basic Terms Related to Disaster Management*. Department of Humanitarian Affairs, United Nations, Geneva.

Vancea, A. 1960. *Neogenul din bazinul Transilvaniei*. Academiei, Bucureşti.

Zăvoianu, I. 1985. *Morphometry of Drainage Basins*. Elsevier, Amsterdam.

Zăvoianu, I. and Dragomirescu, S. 1994. Asupra terminologiei folosite în studiul fenomenelor naturale extreme. *Studii si cercetari de geografie*, **XLI**, 59–65.

5 Rates and Mechanisms of Change of Hard Rock Steep Slopes on the Colorado Plateau, USA

OWEN G. KIMBER, ROBERT J. ALLISON AND NICHOLAS J. COX

Department of Geography, University of Durham, Durham, UK

ABSTRACT

The development of steep slopes in lithified jointed rock has been studied on the Colorado Plateau, USA. The rock mass material properties have been combined with geomorphological information in a distinct element computer program to develop models of cliff evolution. It is demonstrated in a numerical simulation study how the material properties in a rock mass, particularly the geometry of the discontinuities, affect the mechanisms of failure and therefore the development of steep slopes. The modelling has used geotechnical data from the Canyonlands area of the Colorado Plateau, Utah. High, vertical cliffs of horizontally bedded Jurassic sandstone are topped by a jointed cap-rock to form mesas, buttes and other spectacular rock landforms. While discontinuity orientation remains relatively constant spatially, spacing between joint sets varies, leading to differences in cliff development. Where the discontinuity spacing is greatest, buttes have become detached from the main cliff. The modelling demonstrates how the distinct element method can be used to elucidate landform evolution.

5.1 INTRODUCTION

Geomorphological studies of slope form and process are increasingly considering material characteristics in order to explain rates of development (Anderson and Richards 1987; Kirkby 1994; Ahnert 1996; Allison and Kimber 1998). However, the study of jointed rock masses has lagged behind work which has concentrated on soft, unlithified sediments (Allison 1994). In models of lithified steep rock slopes, the geomorphologist has to consider the properties of the discontinuities which cross-cut a rock mass as well as the intact blocks (Selby 1993). It is well understood that the joint pattern is a significant rock cliff control (Goodman 1980; Hencher 1987; Aydan and Kawamoto 1990) and this information has been incorporated in the development of computer programs written to examine the stability of rock slopes (International Society for Rock Mechanics 1988). The Distinct Element Method (DEM) is one example, combining a continuum and discontinuum approach to rock mechanics modelling.

It has been recognized for some time that the properties of the rock material in scarp faces help to explain rates and mechanisms of retreat (Schumm and Chorley 1966). A semi-quantitative approach which is regularly adopted is the rock mass strength (RMS) classification (Selby 1980; Abrahams and Parsons 1987), which combines rock mass mechanical and discontinuity properties into a single synthesized index. However, rock slope development has

Applied Geomorphology: Theory and Practice. Edited by R.J. Allison.
© 2002 John Wiley & Sons, Ltd.

been explained using links with rock mass morphometric (Schmidt 1994a), process-event (Hutchinson 1971) and strength characteristics (Nicholas and Dixon 1986). In the Jurassic Portland Limestone of the Isle of Purbeck, Dorset, variations of cliff retreat are not apparent (Allison 1989) and the effects of differences in rock failure mechanisms are countered somewhat by variations in rock strength (Allison and Kimber 1998).

The aim of the research presented here is to use Distinct Element Method computer software (UDEC) as a technique for improving the understanding of rock slope form in plan and profile in the Canyonlands Region of the Colorado Plateau, USA. There has been limited use of distinct element modelling in the examination of natural jointed rock mass landforms by the synthesis of geomorphological and rock mass geotechnical data. The advantage of UDEC is that it is a rigorous approach based on the principles of stress response within materials through time. Nevertheless a degree of simplicity is maintained, with easily interpreted output enabling the analysis of the effects of individual parameters upon slope development.

5.2   COLORADO PLATEAU AND FIELD DATA COLLECTION

The Colorado Plateau is one of the 15 United States geographical provinces covering $380\,000\,km^2$ in four states (Lohman 1981). The Plateau height ranges from 1500 m to 3000 m, with depths in the Grand Canyon as low as 600 m (Figure 5.1). Geomorphologically the Colorado Plateau has been influenced by the rapid incision of the Colorado River system about five million years ago (Graf *et al.* 1987). Virtually all of the rocks exposed on the Colorado Plateau are sedimentary rocks. They were uplifted at the end of the Mesozoic, during the Laramide Orogeny. Young (1985) suggests that scarp retreat during this period was rapid (1500 to 3800 m a^{-6}) when base levels were stable or rising. Since the end of the Orogeny, much slower rates (160 to 170 m a^{-6}) have occurred. The fact that the sedimentary layers have remained largely horizontal, resisting the forces which crumpled the surrounding Rocky Mountains, can be attributed to the normal faults inherited from Precambrian time. Added to this, climate limits the development of soils and vegetation, thus exposing the rock and making the Plateau an ideal location for geological study.

Geomorphologists studying cliff retreat on the Colorado Plateau have often based explanations upon the strength of the materials making up the slope. Koons (1955) describes how slope angles on the Colorado Plateau vary little, depending upon the angle of repose and the angle of sliding friction of the materials involved. Ahnert (1960) observed that scarps with the massive sandstone reaching the base of the cliff form a rounded profile, as opposed to the vertical profile of compound scarps. Schumm and Chorley (1966) attempt an understanding of the rates and mechanisms of scarp retreat. Explanation is given in terms of the rocks composing the scarp face and the structure of the scarp. Four key variables are the relative resistance of the cap-rock, the joint spacing of the cap-rock, the direction of dip, and the proportion of weaker rock exposed in the scarp. Oberlander (1977) suggests that cuesta-form landscapes seem to exemplify the equilibrium concept of slope development, in which each rock type is associated with a particular slope angle that equates erosional stress to surface resistance, resulting in efficient removal of weathering products. This leads to parallel rectilinear slope retreat which is common on the Colorado Plateau.

Young (1985) suggests that contemporary slope development on the Colorado Plateau is occurring by the parallel recession of scarps through the horizontal strata of varying resistance. Schmidt (1989) argues that the multi-level nature of scarp retreat on the different layers of rock on the Colorado Plateau makes the process almost inexhaustible. On a smaller

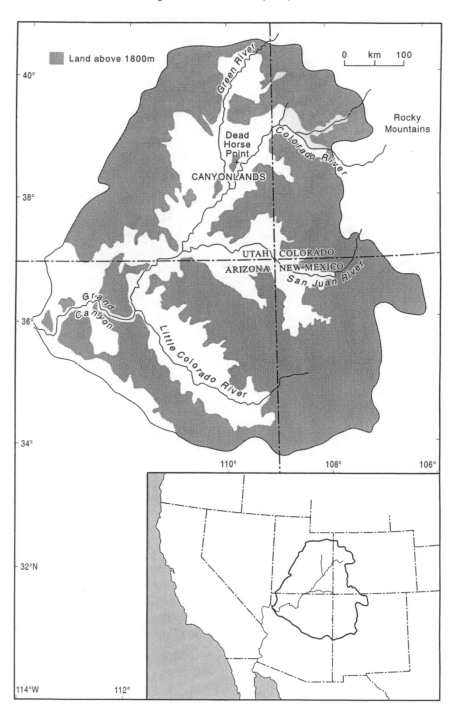

Figure 5.1   Location map

scale Schmidt (1991) demonstrates that the scarp back-slope is eroded by scarp recession. The length of the back-slope is controlled by the resistance and thickness of the scarp cap-rock unit, as well as the structural dip and the strength of the overlying rocks. Further morpho-metric analysis (Schmidt 1994b) demonstrates that Colorado Plateau scarps are less embayed in plan, with decreasing resistance and increasing thickness of the cap-rock and increasing structural dip. Nicholas and Dixon (1986) suggest that the rock fabric of the cap-rock, in terms of joint orientation and spacing, is the dominant control on scarp form and that the rock strength plays a minimal role. Cliff retreat is greater in embayments and headlands remain as resistant projections where joint spacing is greatest.

The field location for this study is Dead Horse Point State Park, which lies within the Canyonlands region at the centre of the Colorado Plateau (Figure 5.1). The region is typified by successions of cuesta-form compound scarps and surrounds the confluence of the Green and Colorado Rivers. The dominant feature of the Dead Horse Point State Park is the almost vertical cliffs of the mesas and buttes which are cut into horizontally bedded sandstone and are up to 400 m in height. The study of a geological cross-section of the Park leads to an elucidation of the geomorphological features (Figure 5.2). The soft red siltstone of the Chinle Formation and Triassic Moenkopi Formation forms the gently angled base of the large cliffs. The Wingate sandstone is the main cliff-forming unit in the Canyonlands region. This soft, fine-grained sandstone was formed during the Late Triassic as an aeolian deposit as deter-mined by the cross-bedding in the stone. The Wingate Sandstone is much less resistant than the overlying Kayenta. Without its protecting cover, the cliffs become more susceptible to weathering and there is a rapid recession of the cliff (Schmidt 1994a). The cap-rock of the cliffs and outlying buttes is the resistant Kayenta formation cemented with silica. This formation is a fluvial deposit comprising lenticular sandstone packages of reddish-brown arenite interbedded with minor reddish-brown siltstone/mudstone and carbonate conglom-erate (Luttrell 1987).

Much of the geomorphology of the Dead Horse Point State Park is controlled by variations in the Kayenta Formation cap-rock. The unit is well jointed, with horizontal bedding layers approximately 2 m thick and a complex pattern of nearly vertical joints which have an average spacing of about 5 m. As it is the joints within the Kayenta Formation which create the planes of weakness within the rock, it is thought that their geometrical pattern controls the development of the 400 m high cliffs. Similar suggestions have been made by Schmidt (1989) who demonstrated that cap-rock thickness is only related to the rate of scarp retreat when combined with rock resistance, which includes jointing.

Sites were selected at the top of the cliffs at Dead Horse Point State Park. The sample points reflect the varying extent of cliff development between the headlands and embayments along the cliff plan. At each sample point the height of the cliff, the mean slope angle of the free face, the angle of the basal unit, the orientation of the free face and other relevant morphometric characteristics were recorded. Joint data were collected for discontinuity strike, dip, spacing and persistence to establish the UDEC model mesh. In order to maintain consistency between sites, tapes were laid along perpendicular transects and 100 joint readings were taken for each site. Spacing between fractures was recorded along the transect. Data were trigonometrically converted to mean joint set spacing by relating the orientation of the transect line to the mean joint set strike. The study of jointed rock masses is a data-limited problem. Assumptions have to be made about the representation of the joint geometry within the rock mass (Starfield and Cundall 1988) but much can be concluded from simple representations of the cliff profiles.

Rock density, Young's modulus and Poisson's ratio are required materials parameters for UDEC input. Schmidt hammer hardness values were collected and correlated with other rock

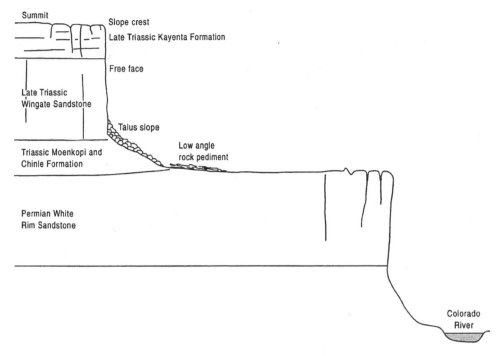

Figure 5.2 Geological and topographic cliff cross-section for the Canyonlands region

strength indices (Hucka 1965; Deere 1966) but there are many questions concerning the validity of this approach (Day and Goudie 1977; McCarroll 1987; Allison 1990; Kolati and Papadopoulus 1993). Due to limitations and constraints on field sampling, representative data were identified and used in the modelling exercise (Deere 1966; Dyke and Dobereiner 1991; Shakoor and Bonelli 1991; Ramamurthy and Arora 1994). The use of accurate, previously determined data is satisfactory as the methodology is designed to construct broadly representative models of the field sites as opposed to perfect representations of real-world conditions. Moreover, there is much discussion in the engineering literature on the accepted techniques for collecting rock strength data. It is acknowledged that the elastic deformation properties, including Young's modulus and Poisson's ratio, are the appropriate indicators of the response of intact rock under stress (Davis and Salvadurai 1996), but the commonly used approach of cutting a small rock sample in the field and determining the elastic properties from a stress/strain response laboratory situation is not particularly accurate (Herget 1988; Cristescu 1989) and the application to real-world situations inconsistent (Mohammad *et al.* 1997). It is pertinent to note that the analysis and modelling of rock slopes will always be a data-limited problem, but that strong conclusions can be made using simple approaches (Starfield and Cundall 1988).

5.3 MODELLING APPROACH

Advances in engineering, combined with the development of computer processing power, have led to the development of powerful models which can be used for the analysis of the response of earth materials under stress. Some early modelling attempts used physical hardware models and employed parameters such as the basal friction angle between blocks of rock. One example uses the drag of a belt moving underneath a model of bricks to simulate a jointed rock mass under gravity (Bray and Goodman 1981). The analogy between gravity and motion is the basis of the mathematical principle behind the limiting equilibrium analysis and associated Factor of Safety (Fs) analysis to determine slope stability (Goodman 1980). Limit equilibrium analysis approximates the slope as a series of columns with the failure mode for each block being determined by solving two statics problems: one assuming block sliding and the other rotation (Wyllie 1980; Warburton 1981; Chowdhury 1986; Nash 1987; Aydan et al. 1989; Scavia et al. 1990; Carter and Lajtai 1992). Limit equilibrium methods are restricted as the material properties are often not included and the mode of failure needs to be known for analysis. Pritchard and Savigny (1990) suggest that limitations restrict the method to analysis of small-scale toppling, where the process is limited to a planar failure surface and failure is facilitated by joint shear and separation.

More sophisticated modelling approaches use a finite element analysis (Zienkiewicz 1977; Hall 1996) which examines the interaction of an assemblage of blocks, but represents the rock mass as an implicit continuum. A rock mass is modelled by defining a mesh of blocks, which are connected at nodal points. An early version was demonstrated on rock topples at Hell's Gate Bluffs, British Columbia (Kakani and Piteau 1976). The analysis indicated zones of stresses in the rock mass and pivot points for topples. It was demonstrated that under high groundwater conditions the hinge points come within the zone of stresses, indicating that toppling is possible.

The most suitable modelling method for the analysis of failure mechanisms and slope development in jointed rock slopes is the distinct element method (DEM). It applies an explicit finite-difference method of solution to model large displacements and rotations of block systems (Cundall 1971). Stresses can be determined for each block individually, allowing the making and breaking of joint contacts, large displacements and rotations of blocks, as well as modelling the failure of block material (Pritchard and Savigny 1990). The main advantage of the DEM is that it applies a continuum modelling algorithm to the deformation response of individual rock blocks and a discontinuum approach to the inter-actions between blocks. The increase in detail over the finite element method requires greater computational power. The most commonly used DEM code is the Universal Distinct Element Code (UDEC) which is available for two- and three-dimensional analysis (Lemos et al. 1985; Cundall 1990; Itasca Consulting Group Inc. 1993). Originally developed for engineering applications, UDEC includes capabilities such as block deformation, a joint generator, fluid pressure and permeability. Pritchard and Savigny (1990, 1991) demonstrate that UDEC can accurately reproduce known limit equilibrium analysis and a base friction model.

There are many reports of UDEC being used for engineering applications. UDEC can represent the discontinuous deformation of jointed rock by thoroughly validating against analytical solutions and real-world examples (Brady et al. 1990; Choi and Coulthard 1990; Lemos 1990; Senseny and Simons 1994). Applications include the stability analysis of undermined cliffs lining the Loire valley (Homand-Etienne et al. 1990) and the analysis of toppling slopes to reveal failure surfaces (Ishida et al. 1987; Pritchard and Savigny 1991). The potential use for UDEC in nuclear waste isolation underground has been discussed by analysing the fluid flow capabilities in the model (Lemos and Lorig 1990). Block deformation was modelled in the Fjellinjen road tunnels under Oslo (Makurat et al. 1990) and UDEC has been used in monitoring an underground cavern in the Himalayas (Bhasin et al. 1996). However, the modelling approach does not appear to have been used before to examine the longer-term development of rock slopes in geomorphology.

UDEC is a command-driven program which is executed from an input file created using a text editor. The command input sequence commences by establishing the geometry of the jointed rock mass. The corners of each block are rounded at the recommended 1% of the block length to prevent a large build-up of stress at the corner point. Although it is possible to make fully deformable blocks, those defined in this study have been rigid because of their proximity to the ground surface, with a low gravitational stress. The intact material strength is high with slope failure occurring entirely along the discontinuities. There are seven built-in material models in UDEC of which the default elastic, isotropic model is used as it is most appropriate for the modelling situation. There are also four built-in joint material models and again use is made of the default joint area contact, Coulomb slip model, which represents closely packed blocks (Itasca Consulting Group Inc. 1993). By changing the built-in models, it is possible to examine softer deforming sediments under an overlying weight. After the material properties have been assigned, the initial problem conditions are set. Gravity is set at 9.81 m s^{-2} and stresses are set vertically through the model, at a gradient calculated from the material density, to simulate the weight of material above each point. Recent advances in science have demonstrated that in natural systems chaotic behaviour can occur among the components (Lorenz 1976; Gleick 1987). It is acknowledged that chaos can occur within the UDEC models (Itasca Consulting Group Inc. 1993) but several possible input scenarios were run with the data, all giving the same results.

The UDEC model is run by taking a series of calculational steps, linking block displacement and Newton's second law under loading. A variety of failure mechanisms may result from the modelled rock mass, with the system failing by the mode with the lowest stability. The first stage of the modelling is to fix the boundaries and set the defined mesh to reach equilibrium in a consolidation phase. The boundaries of the model can then be freed to allow displacement of blocks. The model is then left to run, with images of the mesh being saved every 2000 steps. Output from the model can be monitored by plotting the block geometry, block velocity, the history of displacement at locations or the history of unbalanced force in the model.

5.4 MECHANISMS OF SLOPE FAILURE

Goodman (1980) and Hoek and Bray (1981) classify movement of rock blocks into the three modes of plane sliding, wedge sliding and toppling, limited by the orientation of discontinuities within a rock slope. Plane sliding occurs only rarely on natural slopes when a failure plane daylights parallel to the slope face and the dip of the plane is less than the dip of the slope face. Wedge failure occurs when two discontinuities intersect to define a block with sliding occurring along the intersection. Toppling is the overturning of rock layers inclined steeply into the slope by bedding or flexural cracks. Field studies of toppling failure are common (Caine 1982; Holmes and Jarvis 1985; Woodward 1988; Aydan and Kawamoto 1992; Cruden et al. 1993; Glawe et al. 1993; Cruden and Hu 1994).

DeFreitas and Watters (1973) defined the kinematic failure of a single block upon the block geometrical parameters of the ratio between the base length b and height length h of the block, the angle of the base plane surface α and the angle of friction between the block and the base plane ϕ (Goodman and Bray 1976; Hoek and Bray 1981). The basic limiting conditions for block failure are listed in Table 5.1 and form a continuum between stability and instability.

Natural rock slope behaviour depends upon the dynamic interaction of numerous blocks with the displacement of a block being dependent upon the weight of overburden and stresses imposed by surrounding blocks. There is a need to understand the nature of jointed

Table 5.1 Boundary conditions for different types of rock block failure mechanism

Block failure mechanism	Boundary conditions
Stable	$\alpha < \phi$ and $b/h > \tan\phi$
Sliding	$\alpha > \phi$ and $b/h > \tan\phi$
Toppling	$\alpha < \phi$ and $b/h < \tan\phi$
Toppling and sliding	$\alpha > \phi$ and $b/h < \tan\phi$

α = base plane surface angle; ϕ = friction angle between block and base plane; b = block base length; h = block height length

rock mass failure as geomorphologists deal with landforms composed of numerous blocks and have had to assume the conditions of a single block failure in analysis (e.g. Caine 1982). The UDEC code was used to derive the same boundary conditions by varying the base plane angle (angle of bedding) and the b/h ratio (joint spacing) for a theoretical jointed rock mass (Figure 5.3) (Kimber 1996). The model was set up with fixed blocks of a height of 4 m and the joint friction angle set to 40°. It had the dimensions representing 30 m in height and 60 m in length. Both the block width and angle of the joint set controlling the base plane angle were varied for each model run, and the left-hand boundary was freed to allow failures to develop. In all, 112 meshes were defined in order to separate the geometric boundary

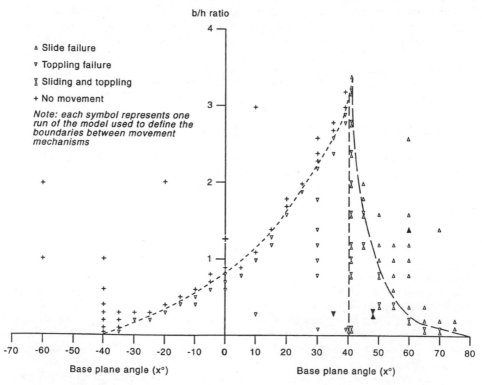

Figure 5.3 Geometry and limiting conditions for sliding, toppling and stable rock masses with a joint friction angle of 40° determined using the UDEC modelling approach

conditions which separate stable conditions, slides, topples and slide–topple complexes (Kimber *et al.* 1998).

The rock mass failure criteria are quite different from those for the failure of an individual block. The fact that at base plane angles greater than the angle of internal friction there is a small zone of toppling and sliding shows that the mechanism of sliding takes dominance, despite the toppling of individual blocks being kinematically possible. Another major difference is that toppling can occur where the base plane angle is horizontal, or even where it dips into the rock face, if there is a very low b/h ratio. This is a significant finding as kinematic analysis based upon an individual block would suggest stability under such circumstances. This is significant to the work presented here because it has implications for the consideration of failure mechanisms and slope development on the Colorado Plateau where jointed rock slopes are cut into horizontally bedded sedimentary rock. In many situations the base plane angle is close to horizontal, and given a low b/h ratio, a toppling failure mechanism will result. Those slopes which have a higher mean b/h ratio for blocks will be more stable and retreat at a slower rate. Natural rock slopes have a much greater form complexity, but the numerical simulation exercise has demonstrated facets of slope development behaviour, which greatly contributes to the understanding of jointed rock mass study on the Colorado Plateau.

5.5   THE CLIFFS AT DEAD HORSE POINT

In order to study the three-dimensional development of the rock cliffs at Dead Horse Point State Park, consideration needs to be made of the plan-form of the cliffs as well as the profile. By tracing the break in slope at the crest of the cliffs and plotting the joint geometry which cuts the horizontal plane for the field sites (Figure 5.4), interesting conclusions can be made. In a similar fashion to coastal cliffs, the scarp comprises headlands and embayments (Nicholas and Dixon 1986). The embayments have retreated further than the headland locations. Headland scarps may become detached from the main cliff to become isolated islands or buttes. In the Canyonlands region there are many buttes associated with headlands with various degrees of isolation. At the southernmost end of Dead Horse Point the headland is close to becoming isolated, with attachment being maintained at a 30 m wide neck. By considering the joint characteristics in the upper, ledge-forming Kayenta Formation, inferences can be made about the variations in cliff development.

Discontinuities measured in the field for the sites at Dead Horse Point State Park were nearly all close to vertical, with bedding being horizontal. In all but one instance there are two joint sets. The stereoplots for the data from the field locations indicate very strong pole concentrations (Figure 5.5), demonstrating a low sample bias in the joint readings because of consistently repeatable joint sets occurring in the Kayenta cap-rock. In comparing the joint geometry between sites it is apparent that the spacing between discontinuities is greater at the headland sites. The mean joint spacing at headland sites is 9.43 m and for the embayment sites, 4.27 m. It is interesting to note from the 1000 Schmidt rebound values measured at Dead Horse Point that the mean of the headland sites is 39.3, and the mean for the embayment sites is 42.9, but there is no statistical base to the difference. The differences in spacing for the whole data set can be confirmed by plotting a quantile plot for joint spacing readings for the headland and embayment sites (Figure 5.6). The fact that the points plot above the line of equality clearly shows that the joint spacing is typically greater at the headland sites, a result which has been observed elsewhere on the Colorado Plateau (Nicholas and Dixon 1986). A further insight can be gained from the cliff plan diagram. It appears that there are zones, or

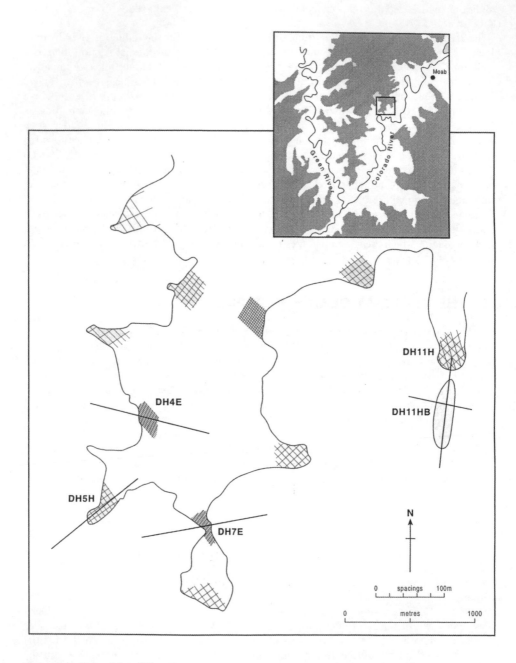

Figure 5.4 Plan of the cliffs at Dead Horse Point State Park showing the surface joint geometry for the field sites and cross-sections used for modelling

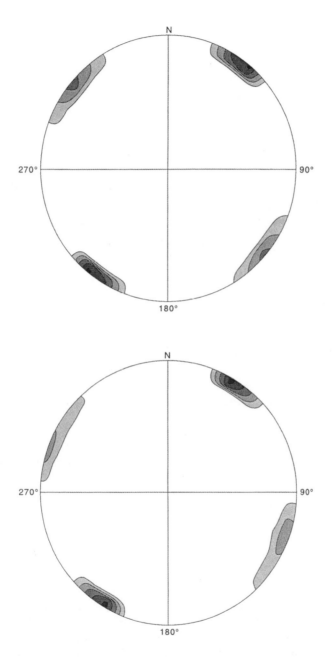

Figure 5.5 Equal area stereographic projections with poles contoured from site DH5H. The upper net represents headland sites and the lower net represents embayment sites

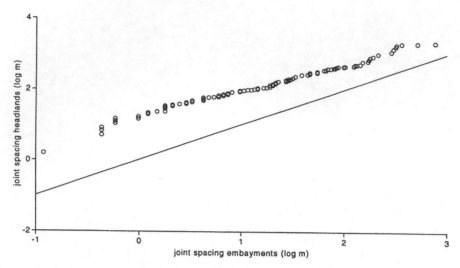

Figure 5.6 A quantile plot using joint spacing data from all headlands and all embayment sites at Dead Horse Point. Data are plotted successively and the line of equality is indicated

patches, of joint geometries which have similar characteristics. There is no general region of narrow spaced joints and a region of wider spacings.

It also appears from the cliff plan diagram (Figure 5.4) that the issue of rock cliff development at Dead Horse Point State Park is more complex than a case of being related to joint spacing differences. At each site there is one joint set which consistently strikes at 120° from north. The second joint set strikes at 20° from north at the embayment sites and nearer to 50° from north at headland sites. Both geometric situations result in the same rectangular shape of block but it is difficult to reason why a rock mass with blocks cut by 120° and 50° joint sets should be more resistant. A further observation from the diagram is that the strike of the scarp face at each site is consistent with the strike of one of the two joint sets. This is easier to explain in that the cliff face cuts along the line of one of the joint sets which acts as a line of weakness. A consideration of the joint geometric distribution for locations at Dead Horse Point State Park does give an indication of the b/h ratio which controls the stability of the cliffs. By considering the cliff profile and interaction of the joint sets at field sites using computer simulation, further insight can be gained into the nature and role of the failure process, and how the discontinuities control the resulting form.

5.6 RESULTS

After the data analysis was completed for the eleven field sites at Dead Horse Point State Park, characteristic sites were identified for the modelling exercise. Sites were selected for modelling a headland cliff profile, embayment cliff profile, a cross-section of the neck at Dead Horse Point, and the development of a butte. The sites are indicated on Figure 5.4. Embayment site DH4E was chosen as being typical for the category. It has a vertical cliff section with a height of 140 m, a free face bearing of 164°, and mean joint spacing of 2.18 m. The intact blocks at site DH4E have a compressive strength of 79 MPa as determined by Schmidt hardness readings. The adjacent headland, DH5H, has a wider joint spacing in the Kayenta

Table 5.2 Characteristics of the cliff profiles and fracture patterns at Dead Horse Point State Park and UDEC model input parameters

	Units	DH4E	DH5H	DH7E	DH11H butte
Free face height	m	140	120	128	165
Toe slope height	m	220	219	280	219
Toe slope gradient	x°	47	45	46/44	34
Free face bearing	x°	164	21/44	172	
UDEC mesh bearing	x°	74	33	082	011/101
Joint set A: dip	x°	88N	90	90	89N
Joint set A: strike	x°	125	130	130	120
Joint set A: spacing	m	2.43	6.81	1.99	
Joint set A: mesh angle	x°	−88	90	90	89/87
Joint set B: dip	x°	87E	87E	88W	89E
Joint set B: strike	x°	011	040	292	149
Joint set B: spacing	m	1.97	7.52	2.31	
Joint set B: mesh angle	x°	−86	081	−87	88/88
Joint set C: dip	x°				88W
Joint set C: strike	x°				032
Joint set C: spacing	m				
Joint set C: mesh angle	x°				−85/−88
Schmidt hammer value	r	42	42	45	35
Bulk density	kg m^{-3}	2500	2500	2500	2500
Bulk modulus	GPa	30	30	30	30
Shear modulus	GPa	15	15	15	15
Joint normal stiffness	GPa m^{-1}	10	10	10	10
Joint shear stiffness	GPa m^{-1}	10	10	10	10
Joint friction angle	x°	28	28	28	28

Formation cap-rock of 7.14 m and an overall cliff height of 408 m. The two clearly defined joint sets striking at 130° and 40° have spacings of 6.18 m and 7.52 m respectively. The neck at Dead Horse Point embayment site, DH7E, warrants specific examination because of its critical field location. There is a distance of 27 m across the neck between the two vertical cliff sections and the site has a joint spacing of 1.99 m. The UDEC model mesh for site DH7E was oriented at 062° and the intact rock has a compressive strength of 89 MPa. If failure of the cliffs occurs at DH7E to break the neck, the headland at DH8H to the south will become isolated as a butte. The butte which has become detached from the headland at DH11H was selected for modelling because it is a classic landform with wide, high cliffs. It is impossible to collect data from the summit of the butte so joint characteristics from the adjacent headland, DH11H, were assumed to be relevant for the modelling exercise. Here there are three joint sets occurring in the Kayenta cap-rock, striking at 120°, 149° and 032°. The cliffs have an overall height of 384 m, with the vertical section comprising 165 m.

Data collected for each site were formatted as appropriate for model input (Table 5.2). Individual blocks were defined as rigid units for simplicity, because the cliffs modelled contain hard rock with failure occurring along the discontinuities rather than as deformation of the intact material. UDEC model meshes were oriented perpendicular to the trend of each cliff face and the discontinuity strike, dip and spacing data were converted by computer program into the appropriate orientation for the mesh (Kimber 1996). Once the model mesh geometry and material properties were assigned, compressive stresses were set to act vertically through the model based on weight of the overburden. A horizontal compressive stress gradient was set at the recommended value of half of the vertical value (Herget 1988). The basal and vertical boundaries were fixed to nullify movement in order for the blocks to settle and consolidate. For each of the sites the model was run for 10 000 iterations for consolidation and to reach equilibrium. Following consolidation, the vertical boundaries of the model representing the free cliff face were released in order to allow for displacement of blocks in the cliff. Blocks which became detached from the cliff face were automatically deleted as they are of no consequence to the rate of cliff retreat. UDEC does not accurately model material in free fall. Blocks which settle on the angled slopes of the Chinle Formation disintegrate quickly and do not accumulate as talus.

Output from the UDEC modelling process was logged and saved every 5000 cycles. The results of model runs are illustrated by sequences of plots for each of the representative field sites at the same stage of the modelling process. The whole of the model mesh is shown on the first plot at each site. The subsequent plots focus upon the upper part of the cliff profile. The first image is taken at 10 000 steps, representing conditions once equilibrium had been established (Howard 1988). Further images were printed after approximately 100 000, 200 000 and 500 000 steps.

At the embayment site, DH4E (Figure 5.7), equilibrium is reached before 10 000 steps. The discontinuity pattern in the Kayenta Formation is formed by two joint sets and horizontal bedding. One joint set dips at 85° into the free face and the other is vertical. It is apparent from the velocity vectors in the second plot in the sequence (Figure 5.7) that a toppling failure mechanism is operating. The narrow spacing between the two joint sets gives a b/h ratio well below the value required for stability and the blocks fail relatively rapidly, as shown in the subsequent plots. Initial development of the cliffs is rapid, but activity has subsided by the third and fourth plots in the sequence. Activity is concentrated further back in the cliff profile, and there is little movement of the blocks at the cliff edge. The rate of retreat of the modelled cliff is slow compared with other rock cliffs (Allison and Kimber 1998). Attempts have been made to estimate the rate of retreat of cliffs in the Canyonlands region by cosmogenic nuclide dating (Nishiizumi et al. 1993). The exposure age of two sides of a large toppled block in the Wingate sandstone were measured to have a 10 000 year difference. At the moment the modelled time steps are unrelated to elapsed clock time. The approach used in the study presented here is to compare relative rates of retreat between models but there is further potential in dating surfaces, and calibrating with real-world time.

At DH4E, there is some movement in the large cliff-forming blocks of Wingate Sandstone but there is insufficient removal of Kayenta material from above for the top of the block to be exposed. The motion of the toppling blocks above the cliff face does not force the larger cliff blocks to topple. The piles of Kayenta Formation blocks which remain at the top of the cliffs in the second and third model plots are unrealistic in the real world. Weathering would promote further toppling movement. At the scale of this study, weathering effects can be assumed to be approximately the same for each site and conclusions can be made about the relative rate of cliff retreat between sites, despite the exclusion of a weathering parameter.

At the adjacent headland site, DH5H, the joint configuration in the Kayenta Formation leads to a much higher b/h ratio of blocks (Figure 5.8). When the free face is released after equilibrium, the initial movement vectors again indicate a toppling failure process. By the second plot in the sequence it is apparent that there has been some displacement of blocks in the Kayenta Formation cap-rock mass. Displacement vectors are consistently oriented out of the cliff face, and tension cracks have developed in the top of the profile. One such crack has occurred between blocks which are failing and blocks which are stable. Tension cracks have been observed in other field locations and are indicative of a toppling failure mechanism (Bovis and Evans 1996; Ishida et al. 1987). The tension crack also propagates into the massive Wingate Sandstone mass and again some small movement of the large blocks occurs under

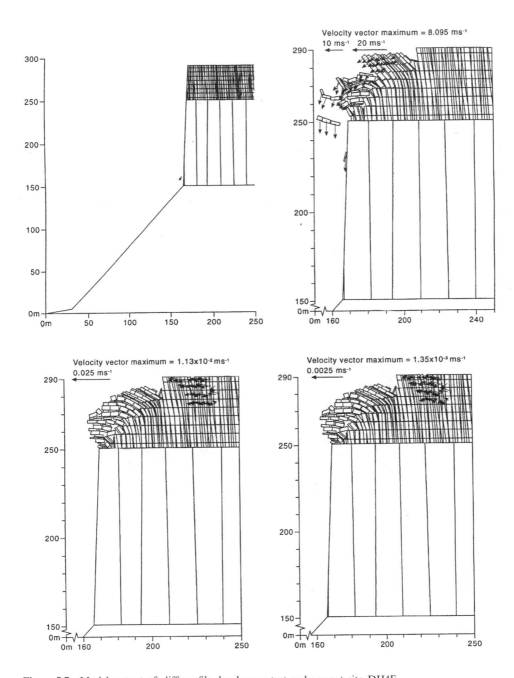

Figure 5.7   Model output of cliff profile development at embayment site DH4E

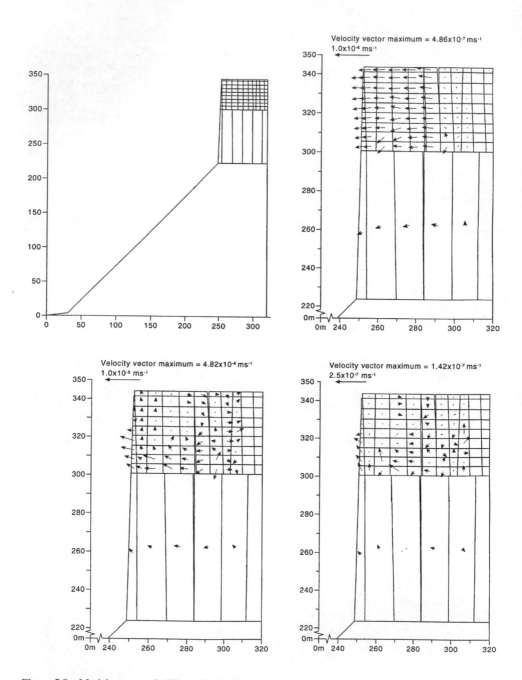

Figure 5.8   Model output of cliff profile development at headland site DH5H

the stress exerted by cap-rock movement. After the early phase of movement the blocks settle. By the third and fourth plots in the sequence the displacement vectors have a much smaller magnitude and are randomly oriented within the rock mass. This pattern of vectors indicates that the rock mass has stabilized again, occurring soon after 100 000 steps. Freeze–thaw weathering processes would possibly initiate further block displacement. The model mesh for site DH5H was bisected into the centre of the headland at a bearing of 33°. A model was also run for a second mesh at 90° to this, across the headland. Similar block movement occurs with initial toppling failure leading to tension cracks in the profile, before stability. As the plot appearances are similar, it has been decided not to include them in this chapter. Conclusions about the relative rate of retreat for headland sites have to account for failure occurring in several directions. The overall rate of retreat is less than that occurring at the embayment site.

If the plots for the embayment and the headland sites are compared after 100 000 steps (Figures 5.7 and 5.8), it can be seen that there is a large difference in the rates of cliff retreat. The scale of the displacement vectors indicates that a much greater rate of activity is occurring at the embayment sites and material has been removed by up to 45 m into the profile. A tension crack occurs at 60 m into the top of the profile for the embayment sites, whereas the crack occurs at 40 m into the profile at the headland site. Although quantitative differences between the development of cliffs at headland and embayment sites cannot be defined, it can be seen that there are clear differences in rates of cliff retreat at particular points during the modelling process.

When the profile of the narrow neck at Dead Horse Point State Park, DH7E, is considered (Figure 5.9), it can be seen that the UDEC-modelled cross-section reflects real-world conditions at the site. The bedding in the Kayenta Formation is horizontal and there are two close-to-vertical joint sets. One of the two sets dips slightly towards the west and hence the left-hand side of the modelled profile. The spacing between the discontinuities is typical for an embayment site at Dead Horse Point. Once equilibrium was reached in the model, both sides were released to allow failure. By the second plot in the sequence, taken at 100 000 steps, there is a rapid rate of cliff retreat by a toppling failure mechanism. At this point in the modelling process there has been less removal of material than had occurred at the embayment site, DH5E, although there is a much greater retreat than at the headland site, DH4H. It is clear in the third plot of the sequence that there is a much greater rate of retreat on the eastern side of the neck because of the geometric configuration of the joint sets. The fact that one joint set dips very steeply towards the west probably increases the stability on the western side of the neck. By the fourth plot, the few remaining blocks on the eastern side of the profile have stabilized, but there is still some slow activity on the western side. The activity is much greater at the neck than for the headland sites and the model suggests that the headland at Dead Horse Point will soon become isolated as a butte.

In order to gain further understanding of the three-dimensional nature of a butte, two model meshes were constructed, one oriented approximately north to south (Figure 5.10) and the second perpendicular to the first (Figure 5.11). The north–south profile reflects field conditions in that the butte is connected to the main cliff by a dormant neck at the base of the Wingate Sandstone. Discontinuity data were collected from the adjacent headland, DH11H, and assumed to be similar for the butte. Greater block displacement occurs on the narrower east–west model. Initial failures occur on the western side of the butte, with toppling blocks in the Kayenta Formation pulling over a large Wingate Sandstone block. The third plot in the sequence shows that large toppling failures are occurring on the eastern side of the model. By the last step in the sequence, the large block has been removed and there are few remaining blocks of the Kayenta Formation. The profile is similar to others of more disintegrated buttes

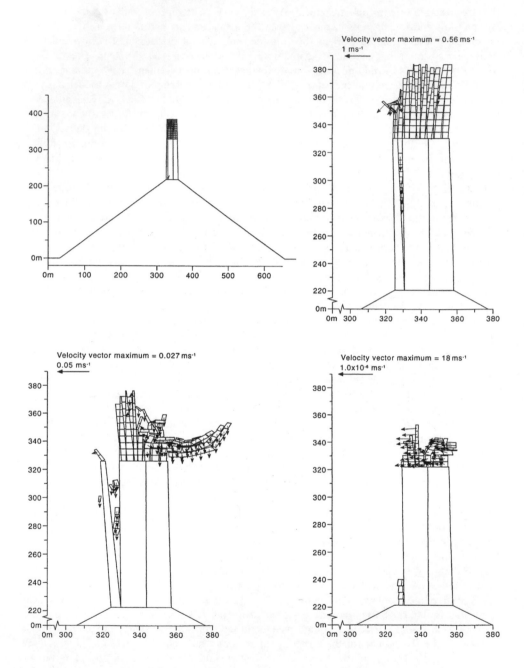

Figure 5.9 Model output of cliff profile development at the neck of Dead Horse Point

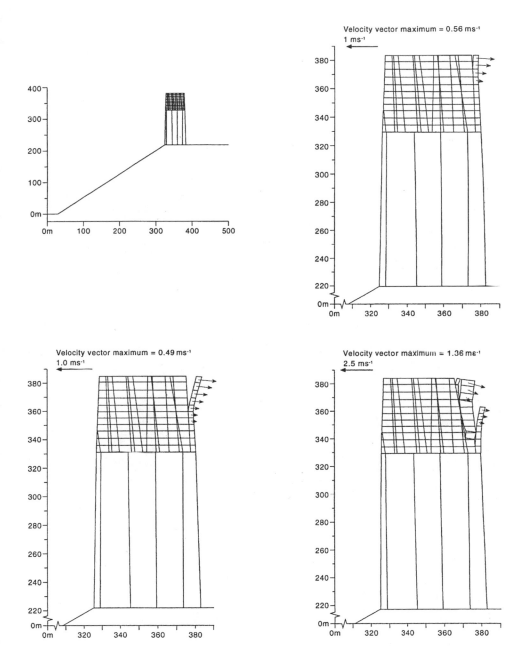

Figure 5.10 Model output of cliff profile development at butte DH11H, north to south axis

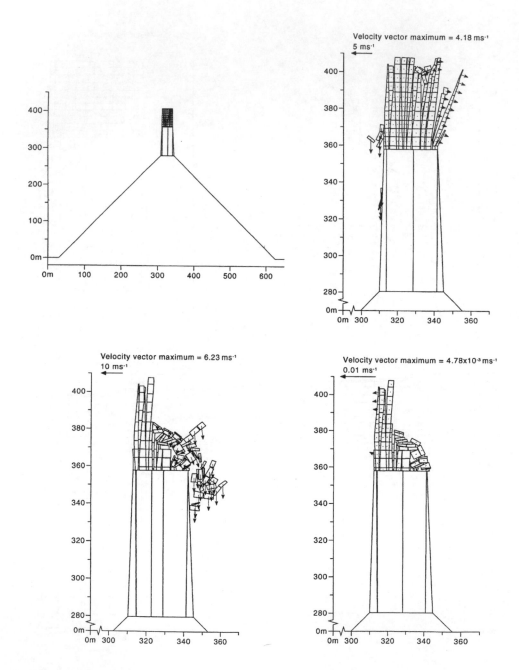

Figure 5.11 Model output of cliff profile development at butte DH11H, east to west axis

on the Colorado Plateau, with a reduction in height and width. At the same time there is a displacement of blocks in the north to south axis of the butte.

By comparing the size of the displacement vectors at equivalent stages in the modelling process and the number of failed blocks, it can be seen that the rate of cliff retreat is much greater for the east–west model. It is interesting to note that the joint configuration leads to greater failure at the northern end of the butte, close to the headland at DH11H (Figure 5.4). By the third and fourth plots of the sequence, columns of Kayenta Formation at the northern end of the butte are toppling but there is no movement at the southern end. At the southern end of the butte it may be assumed that there was once a headland connected to the main mesa as at Dead Horse Point and that the neck of the headland occurred to the north of the current butte. The greater displacement at the northern end could therefore be associated with the rock mass zone of weakness which occurred at the remnant neck. As the modelled butte had no variation of spacing in the cap-rock, it is the complex joint geometrical configuration that leads to variations in rates of cliff retreat in this case of an isolated butte. It is only by using the distinct element approach and constructing simple, representative butte landform models that such an insight into the processes of landform development can be attained.

5.7   CONCLUSION

By using powerful computer simulation software which combines data on material properties and morphology parameters, the opportunity exists to model slope development occurring on a time-scale that is difficult to record in the field. It is the control of the discontinuities in the cap-rock of Colorado Plateau rock masses which helps to explain the different rates of slope development between sites. In the Canyonlands Region at Dead Horse Point State Park, the plan of the compound scarps is divided into embayment and headland plan form features. It is the differences in the discontinuity pattern in the Kayenta Formation cap-rock which lead to the variations in the outcome of the modelling exercise, as there is a similarity between model results and real-world conditions at each site. At the embayment sites it is the narrow joint spacing, giving elongated blocks with a low b/h ratio, which causes the toppling failure mechanism of cap-rock material, and the relatively fast rate of retreat. Conversely at the headland site, the wider joint spacing leads to greater stability in the cliffs.

The geometry of the joint sets within the rock mass, as well as the spacing between joint sets, need to be considered for a full understanding of jointed rock mass development at Dead Horse Point. The other sites modelled highlight the benefits gained from using a distinct element approach. There is a continuum of rock mass landforms and understanding, initially gained from analysing the output of the neck model, to the models of the butte. The narrow joint spacing at the neck, DH7E, leads to the rapid modelled isolation of the Dead Horse Point headland as a butte. The geometry of the discontinuities indicates that the eastern side of the neck will fail more rapidly. Once a butte has become attached, as at DH11H, blocks continue to fail towards the neck at a greater rate, as determined by the joint geometry and not necessarily the spacing. The modelled butte disintegrates rapidly in all directions by a toppling failure mechanism.

The approach used in the study presented here can be used to elucidate rates and mechanisms of change of jointed rock cliffs on the Colorado Plateau. While the models are not perfect representations of cliff form and processes they are close approximations, which help to further explain the rates and mechanisms of rock slope change at Dead Horse Point.

ACKNOWLEDGEMENTS

Funding has been gratefully received from the Natural Environment Research Council and Durham University Research and Initiatives Fund. The authors would like to thank Chris Mullaney and Mark Scott for providing IT support. Field assistance was kindly provided by the staff at Dead Horse Point State Park, Utah, especially Lee Sjolberm, Assistant Manager, and Judi Lofland, Todd Overbye and Patrick Perrotti of the US National Park Service. Ian Harwood acted as a field assistant on many hot days and as a companion to share a beer with while enjoying Colorado Plateau sunsets.

REFERENCES

Abrahams, A.D. and Parsons, A.J. 1987. Identification of strength equilibrium rock slopes: further statistical considerations. *Earth Surface Processes and Landforms*, **12**, 631–635.

Ahnert, F. 1960. The influence of Pleistocene climates upon the morphology of cuesta scarps on the Colorado Plateau. *Annals of the American Association of Geographers*, **50**, 139–156.

Ahnert, F. 1996. The point of modelling geomorphological systems. In McCann, S.B. and Ford, D.C. (eds), *Geomorphology Sans Frontières*. Wiley, Chichester, 91–113.

Allison, R.J. 1989. Rates and mechanisms of change in hard rock coastal cliffs. *Zeitschrift für Geomorphologie*, **73**, 125–138.

Allison, R.J. 1990. Developments in a non-destructive method of determining rock strength. *Earth Surface Processes and Landforms*, **15**, 571–577.

Allison, R.J. 1994. Slopes and slope processes. *Progress in Physical Geography*, **18**, 423–433.

Allison, R.J. 1997. Geomorphological processes: rates, relationships and return intervals. In Marinos, P.G., Koukis, G.C., Tsiambaos, G.C. and Stournaras, G.C. (eds), *Engineering Geology and the Environment*. Balkema, Rotterdam, 2471–2490.

Allison, R.J. and Kimber, O.G. 1998. Failure mechanisms and rates of change in rock slopes and cliffs. *Earth Surface Processes and Landforms*, **23**, 731–750.

Anderson, M.G. and Richards, K.S. 1987. Modelling slope stability: the complementary nature of geotechnical and geomorphological approaches. In Anderson, M.G. and Richards, K.S. (eds), *Slope Stability*. Wiley, Chichester, 1–9.

Aydan, Ö. and Kawamoto, T. 1992. The stability of slopes and underground openings against flexural toppling and their stabilisation. *Rock Mechanics and Rock Engineering*, **25**, 143–165.

Aydan, Ö., Shimizu, Y. and Ichikawa, Y. 1989. The effective failure modes and stability of slopes in rock mass with two discontinuity sets. *Rock Mechanics and Rock Engineering*, **22**, 163–188.

Bhasin, R.K., Barton, N., Grimstad, E., Chryssanthakis, P. and Shende, F.P. 1996. Comparison of predicted and measured performance of a large cavern in the Himalayas. *International Journal of Rock Mechanics and Mining Science & Geomechanical Abstracts*, **33**, 607–626.

Bovis, M.J. and Evans, S.G. 1996. Extensive deformations of rock slopes in southern Coast Mountains, south-west British Columbia, Canada. *Engineering Geology*, **44**, 163–182.

Brady, B.H.G., Hsiung, S.H., Chowdhury, A.H. and Philip, J. 1990. Verification studies on the UDEC computational model of jointed rock. In Rossmanith, H.P. (ed.), *Mechanics of Jointed and Faulted Rock*. Balkema, Rotterdam, 551–558.

Bray, J.W. and Goodman, R.E. 1981. The theory of base friction models. *International Journal of Rock Mechanics and Mining Science & Geomechanical Abstracts*, **18**, 453–468.

Caine, N. 1982. Toppling failures from alpine cliffs on Ben Lomond, Tasmania. *Earth Surface Processes and Landforms*, **7**, 133–152.

Carter, B.J. and Lajtai, E.Z. 1992. Rock slope stability and distributed joint systems. *Canadian Geotechnical Journal*, **29**, 53–60.

Choi, S.K. and Coulthard, M.A. 1990. Modelling of jointed rock masses using the distinct element method. In Rossmanith, H.P. (ed.), *Mechanics of Jointed and Faulted Rock*. Balkema, Rotterdam, 471–477.

Chowdhury, R.N. 1986. Geomechanics risk model for multiple failures along rock discontinuities. *International Journal of Rock Mechanics and Mining Science & Geomechanical Abstracts*, **23**, 337–346.

Crïstescu, N. 1989. *Rock Rheology*. Kluwer Academic Publishers, Dordrecht.

Cruden, D.M. and Hu, X.-Q. 1994. Topples on underdip slopes in the Highwood Pass, Alberta, Canada. *Quarterly Journal of Engineering Geology*, **27**, 57–68.

Cruden, D.M., Hu, X.-Q. and Lu, Z. 1993. Rock topples in the highway cut west of Clairvoux Creek, Jasper, Alberta. *Canadian Geotechnical Journal*, **30**, 1016–1023.

Cundall, P.A. 1971. A computer model for simulating progressive, large-scale movements in blocky rock systems. In *Proceedings, International Symposium on Rock Fractures*, Nancy, France, paper II-8, 1–12.

Cundall, P.A. 1990. Numerical modelling of jointed and faulted rock. In Rossmanith, H.P. (ed.), *Mechanics of Jointed and Faulted Rock*. Balkema, Rotterdam, 11–18.

Davis, R.O. and Salvadurai, A.P.S. 1996. *Elasticity and Geomechanics*. Cambridge University Press, Cambridge.

Day, M.J. and Goudie, A.S. 1977. Field assessment of rock hardness using the Schmidt test hammer. *BGRG Technical Bulletin*, **18**, 19–23.

Deere, D.U. 1966. *Engineering Classification and Index Properties for Intact Rock*. Technical Report No. AFWL-TR-65-116, Air Force Weapons Laboratory, Research and Technology Division, Kirtland Air Force Base, New Mexico.

DeFreitas, M.H. and Watters, R.J. 1973. Some field examples of toppling failure. *Géotechnique*, **23**, 495–514.

Dyke, C.G. and Dobereiner, L. 1991. Evaluating the strength and deformability of sandstones. *Quarterly Journal of Engineering Geology*, **24**, 123–134.

Glawe, U., Zika, P., Zvelebil, J., Moser, M. and Rybar, J. 1993. Time prediction of a rock fall in the Carnic Alps. *Quarterly Journal of Engineering Geology*, **26**, 185–192.

Gleick, J. 1987. *Chaos: Making a New Science*. Viking, New York.

Goodman, R.E. 1980. *Introduction to Rock Mechanics*. Wiley, New York.

Goodman, R.E. and Bray, J.W. 1976. Toppling of rock slopes. In *Proceedings of the Speciality Conference on Rock Engineering for Foundations and Slopes*. ASCE (Boulder, Colorado), **2**, 201–234.

Graf, W.L., Hereford, R., Laity, J. and Young, R.A. 1987. Colorado Plateau. In Graf, W.L. (ed.), *Geomorphic Systems of North America*. The Geological Society of America, Boulder, Colorado, 259–302.

Hall, D.B. 1996. Modelling failure of natural rock columns. *Geomorphology*, **15**, 123–134.

Hencher, S.R. 1987. The implication of joints and structures for slope stability. In Anderson, M.G. and Richards, K.S. (eds), *Slope Stability: Geotechnical Engineering and Geomorphology*. Wiley, Chichester, 145–186.

Herget, G. 1988. *Stresses in Rock*. Balkema, Rotterdam.

Hoek, E. and Bray, J.W. 1981. *Rock Slope Engineering*. The Institution of Mining and Metallurgy, London.

Holmes, G. and Jarvis, J.J. 1985. Large-scale toppling within a sacküng type deformation at Ben Attow, Scotland. *Quarterly Journal of Engineering Geology*, **18**, 287–289.

Homand-Etienne, F., Rode, N. and Schwartzmann, R. 1990. Block modelling of jointed cliffs. In Rossmanith, H.P. (ed.), *Mechanics of Jointed and Faulted Rock*. Balkema, Rotterdam, 819–825.

Howard, A.D. 1988. Equilibrium models in geomorphology. In Anderson, M.G. (ed.), *Modelling Geomorphological Systems*. Wiley, Chichester, 49–72.

Hucka, V. 1965. A rapid method for determining the strength of rocks *in situ*. *International Journal of Rock Mechanics and Mining Science & Geomechanical Abstracts*, **2**, 127–134.

Hutchinson, J.N. 1971. Field and laboratory studies of a fall in Upper Chalk cliffs at Joss Bay, Isle of Thanet. In Parry, R.H.G. (ed.), *Stress Strain Behaviour of Soils*. Foulis, Henley-on-Thames, 692–706.

International Society for Rock Mechanics (ISRM) Commission on Computer Programs 1988. List of computer programs in rock mechanics. *International Journal of Rock Mechanics and Mining Science & Geomechanical Abstracts*, **25**, 183–252.

Ishida, T., Chigira, M. and Hibino, S. 1987. Application of the distinct element method for analysis of toppling observed on a fissured rock slope. *Rock Mechanics and Rock Engineering*, **20**, 277–283.

Itasca Consulting Group Inc. 1993. *Universal Distinct Element Code: Manual, Version ICG2.0.* Itasca Consulting Group Inc., Minneapolis.

Kakani, E.C. and Piteau, D.R. 1976. Finite element analysis of toppling failure at Hell's Gate Bluffs, British Columbia. *Bulletin of the Association of Engineering Geologists*, **13**, 315–327.

Kimber, O.G. 1996. *Rates and Mechanisms of Change in Hard Rock Steep Slopes.* Graduate Discussion Paper, Department of Geography, University of Durham, UK.

Kimber, O.G., Allison, R.J. and Cox, N.J. 1998. Mechanisms of failure and slope development in rock masses. *Transactions of the Institute of British Geographers*, **23**, 353–370.

Kirkby, M.J. 1994. Thresholds and instability in stream head hollows: a model of magnitude and frequency for wash processes. In Kirkby, M.J. (ed.), *Process Models and Theoretical Geomorphology.* Wiley, Chichester, 295–314.

Kolati, E. and Papadopoulos, Z. 1993. Evaluation of Schmidt rebound hammer testing: a critical approach. *International Association of Engineering Geology Bulletin*, **48**, 69–76.

Koons, D. 1955. Cliff retreat in the south-western United States. *American Journal of Science*, **253**, 44–52.

Lemos, J.V. 1990. A comparison of numerical and physical models of a blocky medium. In Rossmanith, H.P. (ed.), *Mechanics of Jointed and Faulted Rock*, Balkema, Rotterdam, 509–514.

Lemos, J.V. and Lorig, L.J. 1990. Hydromechanical modelling of jointed rock masses using the distinct element method. In Rossmanith, H.P. (ed.), *Mechanics of Jointed and Faulted Rock*. Balkema, Rotterdam, 509–514.

Lemos, J.V., Hart, R.D. and Cundall, P.A. 1985. A generalised distinct element program for modelling jointed rock mass. A keynote lecture. In *Proceedings of the International Symposium on Fundamentals of Rock Joints*. Björkliden, Sweden, 335–343.

Lohman, S.W. 1981. *The Geologic Story of Colorado National Monument.* US Geological Survey Bulletin 1508, US Department of the Interior.

Lorenz, E.N. 1976. Non-deterministic theories of climatic change. *Quaternary Research*, **6**, 495–506.

Luttrell, P.R. 1987. *Basin Analysis of the Kayenta Formation (Lower Jurassic), Central Portion Colorado Plateau.* MSc Thesis, Northern Arizona University.

Makurat, A., Barton, N.R., Vik, G., Chryssanthakis, P. and Monsen, K. 1990. Jointed rock mass modelling. In Barton, N.R. and Stephansson, O. (eds), *Rock Joints*. Balkema, Rotterdam, 647–656.

McCarroll, D. 1987. The Schmidt hammer in geomorphology: five sources of instrument error. *BGRG Technical Bulletin*, **36**, 16–27.

Mohajerani, A. 1989. Rock discontinuity spacing statistics: log-normal distribution versus exponential. *Australian Geomechanics*, **17**, 20–21.

Mohammad, N., Reddish, D.J. and Stace, L.R. 1997. The relation between *in situ* and laboratory rock properties used in numerical modelling. *International Journal of Rock Mechanics and Mining Science & Geomechanics Abstracts*, **34**, 289–297.

Nash, D.F.T. 1987. A comparative review of limit equilibrium methods of stability analysis. In Anderson, M.G. and Richards, K.S. (eds), *Slope Stability: Geotechnical Engineering and Geomorphology*. Wiley, Chichester, 11–75.

Nicholas, R.M. and Dixon, J.C. 1986. Sandstone scarp form and retreat in the Land of Standing Rocks, Canyonlands National Park, Utah. *Zeitschrift für Geomorphologie*, **30**, 167–187.

Nishiizumi, K., Kohl, C.P., Arnold, J.R., Dorn, R., Klein, J., Fink, D., Middleton, R. and Lal, D. 1993. Role of *in situ* cosmogenic nuclides ^{10}Be and ^{26}Al in the study of diverse geomorphic processes. *Earth Surface Processes and Landforms*, **18**, 407–425.

Oberlander, T.M. 1977. Origin of segmented cliffs in massive sandstones of south-eastern Utah. In Doehring, D.O. (ed.), *Geomorphology in Arid Regions*. Publications in Geomorphology, Binghamton, NY, 79–114.

Pritchard, M.A. and Savigny, K.W. 1990. Numerical modelling of toppling. *Canadian Geotechnical Journal*, **27**, 823–834.

Pritchard, M.A. and Savigny, K.W. 1991. The Heather Hill landslide: an example of a large scale toppling failure in a natural slope. *Canadian Geotechnical Journal*, **28**, 410–422.

Ramamurthy, T. and Arora, U.K. 1994. Strength predictions for jointed rocks in confined and unconfined states. *International Journal of Rock Mechanics and Mining Science & Geomechanical Abstracts*, **31**, 9–22.

Scavia, C., Barla, G. and Bernuado, V. 1990. Probabilistic stability analysis of block toppling failure in rock slopes. *International Journal of Rock Mechanics and Mining Science & Geomechanical Abstracts*, **27**, 465–478.

Schmidt, K-H. 1989. The significance of scarp retreat for Cenoxoic landform evolution on the Colorado Plateau, USA. *Earth Surface Processes and Landforms*, **14**, 93–105.

Schmidt, K-H. 1991. Lithological differentiation of structural landforms on the Colorado Plateau, USA. *Zeitschrift für Geomorphologie* S-B, **82**, 153–161.

Schmidt, K-H. 1994a. Hillslopes as evidence of climatic change. In Abrahams, A.D. and Parsons, A.J. (eds), *Geomorphology of Desert Environments*. Chapman & Hall, London, 553–570.

Schmidt, K-H. 1994b. The groundplan of cuesta scarps in dry regions as controlled by lithology and structure. In Robinson, D.A. and Williams, R. (eds), *Rock Weathering and Landform Evolution*. Wiley, Chichester.

Schumm, S.A. and Chorley, R.J. 1966. Talus weathering and scarp recession in the Colorado Plateaus. *Zeitschrift für Geomorphologie*, **10**, 11–36.

Selby, M.J. 1980. A rock mass strength classification for geomorphic purposes: with tests from Antarctica and New Zealand. *Zeitschrift für Geomorphologie*, **24**, 31–51.

Selby, M.J. 1993. *Hillslope Materials and Processes* (2nd edition). Oxford University Press, Oxford.

Senseny, P.E. and Simons, D.A. 1994. Comparison of calculational approaches for structural deformation in jointed rock. *International Journal for Numerical and Analytical Methods in Geomechanics*, **18**, 327–344.

Shakoor, A. and Bonelli, R.E. 1991. Relationship between petrographic characteristics, engineering index properties, and mechanical properties of selected sandstones. *Bulletin of the Association of Engineering Geology*, **28**, 55–71.

Starfield, A.M. and Cundall, P.A. 1988. Towards a methodology for rock mechanics modelling. *International Journal of Rock Mechanics and Mining Science & Geomechanical Abstracts*, **25**, 99–106.

Warburton, P.M. 1981. Vector stability analysis of an arbitrary polyhedral block with any number of free faces. *International Journal of Rock Mechanics and Mining Science & Geomechanical Abstracts*, **18**, 415–427.

Woodward, R.C. 1988. The investigation of toppling slope failures in welded ash flow tuff at Glennies Creek Dam, New South Wales. *Quarterly Journal of Engineering Geology*, **21**, 289–298.

Wyllie, D.C. 1980. Toppling rock slope failures: examples of analysis and stabilisation. *Rock Mechanics and Rock Engineering*, **13**, 89–98.

Young, R.A. 1985. Geomorphic evolution of the Colorado Plateau margin in west–central Arizona: a tectonic model to distinguish between the causes of rapid, symmetrical scarp retreat and scarp dissection. In Morisawa, M. and Hack, J.T. (eds), *Tectonic Geomorphology*. Allen and Unwin, Winchester, Mass., 261–278.

Zienkiewicz, O.C. 1977. *The Finite Element Method* (3rd edition). McGraw-Hill, London.

6 Slope Remodelling in Areas Exploited by Skiers: Case Study of the Northern Flysch Slope of Pilsko Mountain, Polish Carpathian Mountains

ADAM LAJCZAK

Faculty of Earth Sciences, Silesian University, Sosnowiec, Poland

ABSTRACT

The development of ski resorts in the mountains, associated with the increasing number of hikers in the summer season, has activitated slope degradation and soil erosion. An effect of this is increasing rates and extension of all morphogenetic processes modelling the slopes. Typical microforms occurring in areas impacted by skiing, hiking and sheep grazing have been investigated. This chapter explains the amount of slope degradation and its mode of remodelling due to the above forms of anthropopression. The study focuses on the northern flysch slope of Pilsko Mountain (Polish Western Carpathians) as one of the most important ski resorts in Poland.

6.1 INTRODUCTION

Skiing is one of the most recent forms of anthropopression in the mountains. Development of ski resorts in recent years associated with an increasing number of hikers in summer and autumn has activitated the degradation of vegetation and accelerated soil erosion. The result is quick local slope remodelling due to much more effective morphogenetic processes. In the literature on the subject ski-trail degradation appears much less of a problem than slope degradation due to other forms of anthropopression (Candela 1982; Behan 1983; Watson 1985; Cernuska 1986; Haimaier 1989; Etlicher 1990; Veyret *et al.* 1990; Tsuyuzaki 1994; Lajczak 1996a, b). The last few years have seen a growing number of ski trails in mountains in southern Poland, most of them situated on terrains with insufficient thickness of snow cover for skiing (Lajczak 1996a). Since the time when sheep grazing was banned, skiing and hiking have become the principal causes of slope degradation in highly elevated areas in the Polish Carpathian Mountains.

The aim of the examinations conducted on the northern flysch slope of Pilsko Mountain (Polish Western Carpathian Mountains), as one of the areas with the longest ski trails in Poland, is to establish the amount of slope degradation and its mode of remodelling due to skiing and hiking, with regard to bedrock resistance, climate and the period of time in service. The study area is presented in Figure 6.1.

Applied Geomorphology: Theory and Practice. Edited by R.J. Allison.

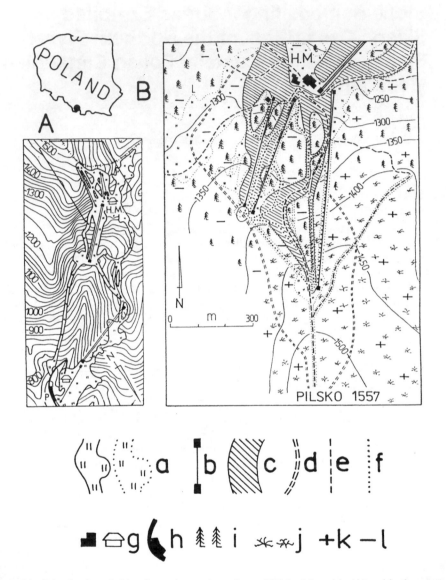

Figure 6.1   Distribution of ski trails on the northern slope of Pilsko Mountain (A) and in the upper part of the slope (B). a, Ski trails and glades; b, ski lifts; c, part of ski trails subjected to snow-levelling machines; d, limit of area used by skiers in the summit part of Pilsko Mountain; e, tourist paths; f, other unmarked paths; g, shelters and other buildings; h, roads and parking; i, upper mountain spruce forest; j, dwarf pine in subalpine zone; k, area underlain by hard sandstone; l, area underlain by soft sandstone and shale; H.M, Hala Miziowa glade

6.2 INVESTIGATION METHODS

Ski trails in the study areas have been divided into the following groups: those with preserved vegetation; those with devastated vegetation but with roots; those with bare soil; and those with easily erodible soil. Degraded areas are influenced by surface run-off, aeolian erosion, solifluction and fibrous ice effects. The expansion of erosion on the slopes has been examined by means of aerial photos since the 1970s. The rate of erosion and local soil deposition on the Pilsko northern slope have been examined by means of a repeated series of surveys in numerous cross-profiles on the ski trails and paths, made every 20 days on average in the period of May to October in the years 1993–1996. The eroded soil has also been examined from the point of view of its physical, especially water properties. The ski trails and paths on the top of Pilsko Mountain have been mapped geomorphologically at the scale 1:2000. The degree of erosion has been correlated with that of precipitation measured on the spot. The distribution of the variously degraded areas has been correlated with the thickness of snow cover on the ski trails measured regularly in numerous profiles every two weeks since the winter 1992/1993.

6.3 STUDY AREA

The ski trails on the northern Pilsko slope run through territory of 700–1480 m altitude (Figure 6.1). The gradient of the slopes varies between 10 and 45°. The bedrock is made of resistant sandstone covered with rubble waste which alternates with shale covered with silty waste. A large, easily erodible area is situated above the Hala Miziowa glade. For the last few years the snow conditions have been favourable for skiing only in the part of the slope above 1100 m altitude (Figure 6.2). During the winter season a growing contrast in the thickness of snow cover is observed among convex and concave sections of the slope on the ski trails. Snow conditions on the slope above 1270 m altitude are very favourable for skiing. The snow cover with thickness over 50 cm stays there for 100–140 days during the season (even to the end of April), and only on convex sections of slope with thin snow cover falls to 40–60 days (Lajczak 1996c).

The whole period of snow cover formation has been divided into three phases (Figure 6.3).

(1) Preliminary period of snow cover formation up to 50 cm in thickness on the ski trails, on average (for the last few years it has lasted until January). There are few skiers but the runs get destroyed intensively.
(2) The thickening of the snow cover. The snow-levelling machines and high number of skiers destroy the plants and soil in fixed spots, where the snow cover is not greater than 50 cm. The thickness of the snow cover quickly changes on the ski trails during every weekend period (Figure 6.2D).
(3) The snow cover vanishes in May, when skiers often continue skiing on shrinking patches of snow. The plant degradation and soil erosion increases locally and the morphological effectiveness of all slope processes is intensified then.

In the summer–autumn period the slope is used by hikers, for whom a few paths had been organized before the ski trails were built. Numerous additional unmarked paths have been developed since the 1980s and now they run along every ski trail and ski lift (Figure 6.1B). Sheep grazing, which was intensive in the past, is now disappearing.

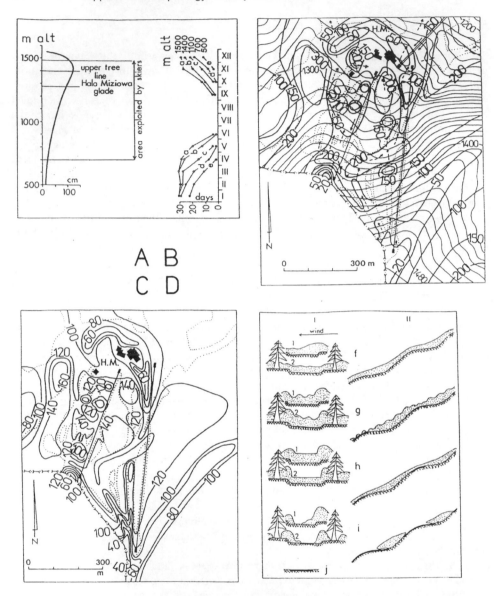

Figure 6.2 Snow conditions of the northern slope of Pilsko Mountain. (A) Mean thickness of snow cover at the end of March and number of days with snow cover in successive months of a year in altudinal profile of the slope. Elevation: a, 1500 m; b, 1400 m; c, 1100 m; d, 700 m; e, 500 m. (B) Spatial differentiation of the mean thickness (cm) of snow cover at the end of March in the upper part of the slope. Dotted line presents a border of ski trails and glades. (C) Mean time of occurrence of snow cover with thickness >50 cm (days per year). (D) Development of snow cover on ski trails during a weekend period: I, cross-profile of ski trail; II, longitudinal profile of ski trail; f, situation before a weekend; g, after two days; h, smooth snow cover on ski trail made by snow-levelling machine; i, beginning of snow melting. Stretches of ski trails; 1, wide; 2, narrow; j, degraded convex slope

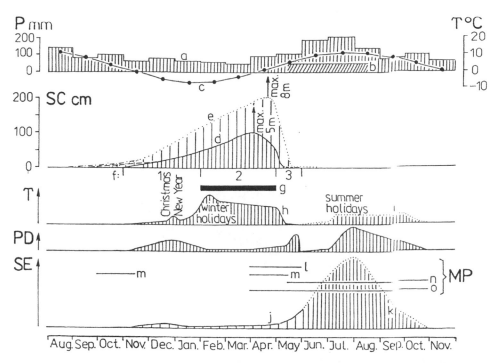

Figure 6.3 Seasonal variability of main factors influencing erosion rates on the slope: a, mean monthly sum of precipitation (P mm); b, period with downpours; c, mean monthly temperature (T°C). Thickness of snow cover (SC cm) on the Hala Miziowa glade during the last years (d) and before the 1980s (e). f, Phases of snow cover formation (for explanation of 1–3, see the text). g, Period when snow-levelling machines operate. Relative changes of number of tourists (T) visiting the Hala Miziowa glade: h, skiers; i, hikers. Intensity of plant degradation (PD) during a year due to skiing and hiking (relative rates). Relative changes of soil erosion rates (SE) on ski trails due to skiing (j) and hiking (k). Occurrence of morphological processes (MP): 1, solifluction; m, fibrous ice effects; n, aeolian erosion; o, surface run-off

6.4 RESULTS

6.4.1 Slope Degradation

Ski trails become degraded because of skiers, hikers and sheep grazing (Figure 6.4). Each of these ways of exploiting the slope causes the development of specific microforms, and each of them has a different impact on erosion.

As a result of sheep grazing there appear very narrow terraces, most of them parallel to each other, which cause minimal degradation of the ski trails and the surrounding areas. On some terraces used by hikers a minimal mass movement of pressed soil with turf can occur.

The ski trails exploited solely by skiers become degraded according to the scheme shown in Figure 6.5. The central parts of the trail contain areas with devastated turf; the convex spots of the slope devastated in a permanent way by the snow-levelling machines contain uncovered soil. This causes the immediate formation of small isolated hollows of up to $1\,m^2$ in area. When the hollows are more 10 cm deep, on average, boulders are revealed, exposed by the snow-levelling machines. They slow down the slope degradation which is lowest in areas of

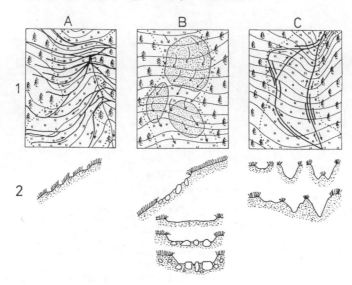

Figure 6.4   Slope degradation by sheep grazing (A), skiing (B) and hiking (C). 1, Plan of degraded ski trails; 2, longitudinal and cross-profiles of degraded areas with typical erosive microforms

resistant bedrock, and highest on the shale slopes. On the studied shale slope the areas of uncovered and eroded soil make up as much as 20% of the ski-trail area.

   Some of the ski trails are degraded as a result of being trampled down by hikers in either a concentrated or a dissipated way. The concentrated trampling down causes linear erosion where slope gullies appear (Figure 6.6). The reconstructed development of the gullies confirms the general scheme of the development of microforms of the kind described by Leopold *et al.* (1964). Within the gullies cut in the soft waste there appear short-lived sills and erosive pot-holes. Where the slope is flat there appear miniature alluvial cones. The speed of their formation and their shape depend on the bedrock lithology. Erosion is quickest in the part of the slope built by shale bedrock, where gullies of 50 cm depth, on average, occur. The amount of linear erosion also depends on how long the path has been used by hikers. It is greatest on the path which has been in use for about 100 years. In the trampled-down areas the erosion rate depends on the soil density which on the path in the top 10 cm of soil strata can reach 1.5 g cm^{-3}. Outside the paths within the glades and in the spruce forest, and also in areas covered by dwarf pine, water infiltration is 10 to 1000 times quicker in the surface soil strata with a density of 0.2–0.6 g cm^{-3}. It is clearly shown that the greater the soil density, the slower the water sinks and the greater the surface run-off flow becomes on the paths. During downpours which occur mainly in May–July, the episodic flow in the gullies can reach a discharge of up to 20 l s^{-1}. The increase in soil density above 1.35 g cm^{-3} makes it impossible for seeds to shoot and for plants to develop (Hildebrand 1983; Lajczak 1996b). That is why plants seldom grow on intensively trampled down paths. Within a period of several years the living roots will no longer be there and these soil strata will be eroded quickly. Hikers begin to pass around the newly formed slope gullies and with time a new path develops. The process recurs and it leads to the degradation of a large slope zone. In such places, especially in the subalpine zone, the effectiveness of all morpho-genetic processes increases. Where there is dissipated hiking traffic one can find large areas covered with rubble. These places, located above the upper tree line in areas built of hard sandstone, go through a process mainly of dissipated surface flow and deflation. The present

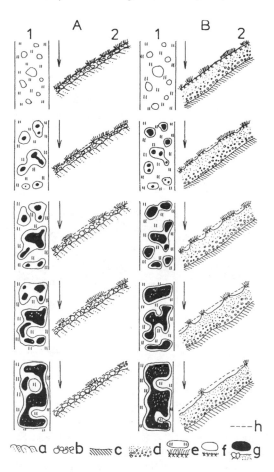

Figure 6.5 Scheme illustrating ski-trail degradation and microrelief development caused by skiing. Bedrock and slope waste: A, resistant; B, easily erodible. 1, Plan; 2, longitudinal profile. a, Sandstone; b, rubble waste; c, shale; d, silty waste; e, ski trail covered by turf; f, ski trail with devastated turf but with roots; g, ski trail with bare soil and rubble; h, initial surface of the slope

stage of slope gully development shows that linear erosion ceases when bedrock that is resistant to degradation becomes exposed. It has already come to this situation in the uppermost part of the slope, built of sandstone. It is only the presence of boulders 50 cm in diameter and of the solid rock that slows down the anthropogenic slope degradation. Slope gullies on shale bedrock are continuously deepened; many of these microforms are still at the initial phase. The potential volume of soil which may be eroded in this part of the slope is considerable, even in the neighbourhood of tourist trails used for a long time.

6.4.2 Erosion–accumulation Balance and Assessment of the Erosion Size During Research

Where there is turf on the paths the denudation balance is positive; where there is no turf the balance is negative. During the time of this research the gullies have never been more than

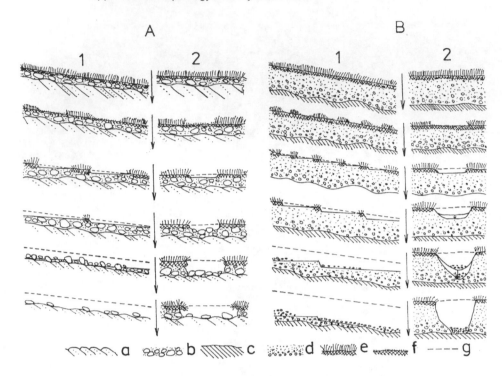

Figure 6.6 Scheme illustrating ski-trail degradation and microrelief development caused by hiking. 1, Longitudinal profile; 2, cross-profile; g, initial surface of the slope. Other symbols as in Figure 6.5

10 cm deep, although at times they have been 50 cm deep, on average. Short-lived erosive pot-holes have been getting silted and they have also been degraded by hikers. The average accumulation on flat places has been estimated to be 1 cm per year. Eighty per cent of the material has been deposited during the May–July period, that is the period of greatest linear erosion. During the research the erosion hollows caused by skiing and snow-levelling machines increased in diameter from 50 cm to 70 cm and in mean depth from 3 cm to 5 cm. Where there is no interference by skiers, snow-levelling machines and hikers, deposited soil is stabilized by plants.

The estimation of erosion and deposition rates covers a period of time when the snow cover has been thinner and the precipitation sum and the frequency of downpours smaller than during many previous years (Lajczak 1996b). One also has to consider that there will be more hikers in the years following the opening of planned tourist investments on Pilsko Mountain and surroundings. One may expect increased erosion on the paths leading through the silty waste. If a large portion of the ski trails becomes degraded, then downpours can cause rubble and mud flows, which do not occur there as yet.

Skiing contributes 30% and hiking 70% to the ski-trail degradation in the upper part of the northern Pilsko slope. As far as the volume of eroded soil is concerned, 10% is due to skiing and 90% to hiking (Lajczak 1996a, b). The role of sheep grazing in slope degradation has been negligible for the last few years. In the next few years skiing will be increasingly responsible for the surface degradation of ski trails, and hiking for the volume of eroded soil.

6.4.3 Morphological Consequences of Slope Degradation

Earthworks connected with slope profiling for ski trails and alternating eroded and overbuilt zones lead to smoothing of the longitudinal slope profile. In areas used exclusively by skiers and additionally smoothed by snow-levelling machines, the linear erosion of slopes does not occur. However, in ski areas penetrated by hikers, morphological effects produced by skiing are intensified by effects diversifying microrelief, which result from rill erosion. When erosion cuts reach the bedrock, their further development takes place by broadening. With time rubble, and next bare bedrock become exposed on large areas. Further degradation of these areas by hikers may effectively prevent their stabilization by plant cover. The described tendencies of slope development are fixed by the activation in such places, and particularly in the subalpine zone, of all morphogenetic processes, the range and intensity of which have increased since tourists appeared in this area.

6.5 CONCLUSIONS

Skiing has a moderate direct influence on ski-trail degradation. As a result of the combined impact of skiing and hiking the slope morphology changes and the effectiveness of all morphological processes increases. The volume of soil eroded from ski trails depends on the number of hikers using particular paths, mainly during their descent down the hill. The present rate of erosion depends on the developmental phase of a slope gully and on the lithology of bedrock. However, it is impossible to determine the effect of slope inclination on the extent of erosion in the scale of the whole investigated area. This effect is visible only in small areas of uniform granulometric composition of slope waste. In areas degraded exclusively by skiing no such effect was found. In order to diminish the erosion of the degraded area on the study slopes technical, biological and organizational solutions have been put into practice (Lajczak et al. 1996).

REFERENCES

Behan M.J. 1983. The suitability of commercially available grass species for revegetation of Montana ski areas. *Journal of Range Management*, 38.

Candela R.M. 1982. Pistes de ski et erosion anthropogenique dans les Alpes du Sud. *Mediterranee*, 3–4.

Cernuska A. 1986. Okologische Auswirkungen des Baues und Bertiebes von Schipisten und Empfelungen zur Reduktion der Umweltschaden. Council of Europe, Strasbourg. *Nature and Environment*, ser. 33.

Etlicher B. 1990. Observations sur la dynamique actuelle des versants dans best du Massif Central. *La terre et les hommes*, nouvelle serie, fasc. 32.

Haimayer P. 1989. Glacier-skiing areas in Austria. A sociopolitical perspective. *Mountain Resource Development*, 9.

Hildebrand E.E. 1983. Der Eifluss der Bodenverdichtung auf die Bodenfunktionen im forstlichen Standort. *Forstwis. Cbl.*, 2.

Lajczak A. 1996a. Present morphological development of the slopes intensively exploited by skiers: case study of the northern slope of Pilsko Mt., Polish Western Carpathians. *Proceedings of the International Conference INTERPRAEVENT 1996*, Garmisch-Partenkirchen, Band 2.

Lajczak A. 1996b. Impact of skiing and hiking on soil erosion of the northern slope of Pilsko. In Lajczak, A., Michalik, S. and Witkowski, Z. (eds), *The impact of skiing and hiking on the nature of the Pilsko Massif, Western Carpathians*. Studia Naturae, 41, Cracow.

Lajczak A. 1996c. The influence of skiing on snow cover at the top of the Pilsko Massif. In Lajczak, A.,

Michalik, S. and Witkowski, Z. (eds), *The impact of skiing and hiking on the nature of the Pilsko Massif, Western Carpathians*. Studia Naturae, 41, Cracow.

Lajczak A., Michalik S. and Witkowski Z. 1996. Conflict between skiers and conservationists and an example of its solution: the Pilsko Mountain case study (Polish Carpathians). In Nelson, J.G. and Serafin, R. (eds), *National Parks and Protected Areas*. NATO ASI Series, Subseries G 'Ecological Sciences', G 40, Springer-Verlag, Berlin.

Leopold L.B., Wolman M.G. and Miller J. 1964. *Fluvial Processes in Geomorphology*. Freeman, San Francisco.

Tsuyuzaki S. 1994. Environmental deterioration resulting from ski-resort construction in Japan. *Environmental Conservation*, 2.

Veyret Y., Hotyat M. and Bouchot B. 1990. L'erosion d'origine anthropique dans un milieu de moyenne montagne, le Massif Montdorien. *La terre et les hommes*, nouvelle serie, fasc. 32.

Watson A. 1985. Soil erosion and vegetation damage near ski lifts at Cairn Gorm. *Biological Conservation*, 33.

7 Effects of Run-off Characteristics on the Frequency of Slope Failures in Soya Hills, Northern Japan

YUKIYA TANAKA
Department of Geography, Fukui University, Fukui, Japan
AND
YASHUSHI AGATA
Department of Geography, University of Tokyo, Tokyo, Japan

ABSTRACT

The differential erosional topography is well developed on Soya Hills, Japan. The hills composed of Miocene hard shale and mudstone have a low drainage density, whereas those composed of Pliocene unconsolidated siltstone and Pleistocene unconsolidated sandstone have a high drainage density. These differences in drainage density were examined from the hydrogeomorphological point of view. The geomorphological and hydrological measurements were performed in four drainage basins underlain by hard shale, mudstone, unconsolidated siltstone and unconsolidated sandstone, respectively. The hydrological analysis shows clear differences in run-off characteristics between the drainage basin of muddy rocks and that of unconsolidated sandy rocks, i.e. the interflow component was dominant in the hard shale and mudstone drainage basins, and in contrast the quickflow component was dominant in the unconsolidated siltstone and unconsolidated sandstone ones. The differences in run-off characteristics resulted in a variety of erosional processes. The hillslopes of uncohesive sandy rock were eroded by surface flow under the condition of little forest cover, whereas those of consolidated muddy rock were dissected by slope failures under the condition of forest cover. This means that the drainage network of uncohesive sandstone basins was completed during the last glacial periods or at the Pleistocene–Holocene transition, and on the contrary that of consolidated muddy rocks basins had not been completed in the same period. Consequently, the variety of drainage density in the Soya Hills was explained by the difference in erosional phase due to the combined effects of the differences in the run-off characteristics and environmental changes.

7.1 INTRODUCTION

In the past two decades, quantitative research on differential erosional topography has been activated with the measurement of rock properties in the field and in the laboratory (e.g. Suzuki *et al.* 1985; Tanaka 1990). Recently, the differences in hydrological characteristics associated with rock types have also been examined (e.g. Freeze 1972; Sudarmadji *et al.* 1990; Hirose *et al.* 1994; Onda 1989, 1992, 1994; Tanaka *et al.* 1996; Tanaka and Agata 1997).

Applied Geomorphology: Theory and Practice. Edited by R.J. Allison.
© 2002 John Wiley & Sons, Ltd.

However, the relationships between rock types, the differences in erosional processes and the variety of hill morphology have not been well resolved yet from the hydrogeomorphological point of view in the Soya Hills, Japan. In addition, consideration of past landform development is necessary to find the reason why different rock types show such differences in topography. Such investigations, however, have not been performed in this area. In order to elucidate the relationships between the physical process of landform change and historical development of hill morphology, therefore, field measurements were performed.

7.2   STUDY AREA

The Soya Hills are located in the northern part of Hokkaido, Japan (Figure 7.1) and are lower than 250 m above sea level. The study areas are underlain by Neogene to Lower Pleistocene sedimentary rocks, which are divided into four rock types: i.e. (1) Miocene hard shale (Wakkanai Formation – Wk) with many joints of tectonic and weathering origin; (2) Miocene–Pliocene mudstone (Koetoi Formation – Kt) with fewer joints than Wk; (3) Pliocene unconsolidated fine-grained sandstone (Yuchi Formation – Yt); and (4) Pleistocene unconsolidated medium-grained sandstone (Sarabetsu Formation – Sa). These rocks are folded with dips of about 20° into north to northwest trending fold axes. The rocks are clearly different from each other in conepenetration hardness (P) and permeability coefficient (K), as shown in Table 7.1.

Differential erosional topography is clearly developed in the Soya Hills (Figures 7.1 and 7.2). Topographical characteristics such as drainage basin relief, stream length and drainage density are controlled by rock types as follows. The Wk hills show very coarse topographical texture with maximum basin relief and very few long and deep valleys, as compared with other rock hills. The Kt hills are topographically similar to the Wk hills, but have a lower basin relief than the Wk hills. The Sa and Yt hills show very rugged topographical texture with a low basin relief and a lot of short, well branching streams. The drainage density of second- or third-order streams is small (about 9 km km^{-2}) in the Wk and Kt hills, but large (about 15 km km^{-2}) in the Sa and Yt hills (Suzuki et al. 1985). The frequency of slope failure of the Wk and Kt hills is obviously larger than that of the Sa and Yt hills (Figures 7.2 and 7.3).

The valley side slopes of Sa and Yt hills are covered by a loam layer with a thickness of about 15 cm, contrasting with the cases of the Wk and Kt hills.

The climatic data are obtained from the AMeDAS (Automated Meteorological Data Acquisition System) station (located at Toyotomi, see Figure 7.3), Japanese Meteorological Agency, as follows. The mean annual temperature of this area is 5.6°C with a maximum temperature of 30°C and a minimum temperature of −23.9°C. The mean annual precipitation is 1100 mm. The hillslopes of this area are mostly covered with forest vegetation, which contains both coniferous and birch forests with some Sasa (bamboo) grass. Hourly precipitation data of the study area were obtained from the AMeDAS station.

7.3   METHODS

7.3.1   Setting of Experimental Drainage Basins

Four experimental drainage basins were chosen from the areas composed of each of four rock types, i.e. the Wk, Kt, Yt and Sa basins (Figure 7.3). They are 0.690 km^2, 0.242 km^2,

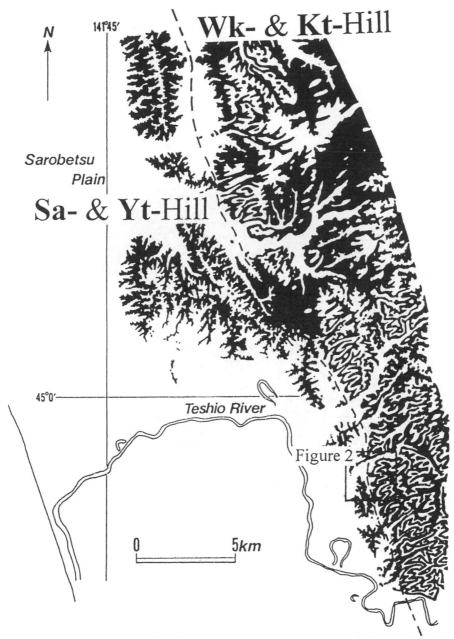

Figure 7.1 Zebra map of Soya Hill. This map was drawn by painting every alternate belt of 50 m in altitude jet-black on 1:25 000 topographic maps issued by Geographical Survey Institute of Japan (after Suzuki *et al.* 1985). Broken line shows the boundary between Wk and Kt hills and Sa and Yt hills. This boundary was compiled from the geological map (Nagao 1960; Hata and Ueda 1969). Sa, Sarabetsu formation (Lower Pleistocene unconsolidated medium-grained sandstone); Yt, Yuchi formation (Upper Pliocene unconsolidated fine-grained); Kt, Koetoi formation (Lower Pliocene mudstone); Wk, Wakkanai formation (Upper Miocene hard shale)

Table 7.1 Geological characteristics of the four rock formations in the studied area

	Wakkana fm	Koetoi fm	Yuchi fm	Sarabetsu fm
Geologic age	Miocene	Miocene–Pliocene	Pliocene	Pleistocene
Dominant lithology	Hard shale	Mudstone	Unconsolidated, fine-grained sandstone	Unconsolidated, medium-grained sandstone
P (kgf cm^{-1})*	1100	300	31	56
K (cm s^{-1})*	2×10^{-3}	1×10^{-7}	1×10^{-5}	1×10^{-3}

P, Cone-penetration hardness; K, permeability coefficient.
* After Suzuki *et al.* (1985)

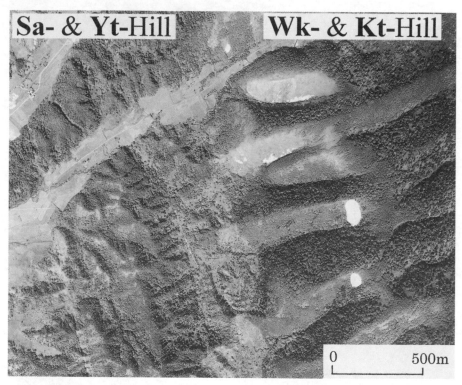

Figure 7.2 Aerial photograph of Soya Hills. Rock-controlled denudational hill morphology is clearly recognized. Location is shown in Figure 7.1

0.167 km^2 and 0.553 km^2 in area, respectively. These experimental drainage basins are located close to each other within a distance of less than 10 km. The climatic and vegetational conditions of these drainage basins are almost identical.

7.3.2 Hydrological Observation

Water discharge was measured continuously using a water level gauge and Parshall flumes. The observation sites are located at the mouth of each experimental drainage basin. The

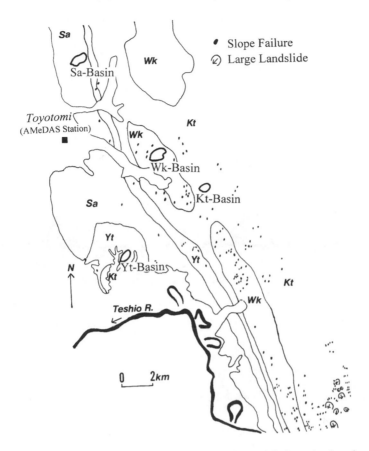

Figure 7.3 Distribution of slope failures around the experimental drainage basins of study area

measurement periods were from 12 June 1994 to 9 November 1994 and from 2 May 1995 to 27 October 1995. No snow cover was found during these measurement periods.

7.4 RUN-OFF CHARACTERISTICS

7.4.1 Recession Rate of Discharge

The hyetograph and hydrograph during the experimental periods are shown in Figure 7.4. The hydrograph shows that flood run-off occurs associated with every rainfall event for each drainage basin, and the recession rates of discharge of the Wk and Kt basins are smaller than those of the Sa and Yt basins.

7.4.2 Difference in Run-off Component Characteristics

Figure 7.5 shows the results of three components of the daily height of run-off, which were divided by the filter separation and AR method (Hino and Hasebe 1981, 1984). These three

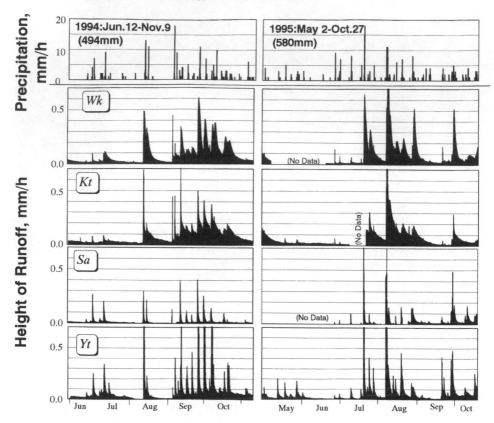

Figure 7.4 Hyetograph and hydrograph during the observation period of 1994 and 1995

components are supposed to correspond to quickflow, interflow and baseflow, as shown in Figure 7.5. This analysis was performed only for the data obtained in 1994, because of lack of data for 1995.

Differences in the run-off component ratio between drainage basins of the four rock types are shown in Figure 7.6. Based on the differences in run-off characteristics, the experimental drainage basins of four rock types are divided into two groups; first is the Wk and Kt group whose interflow component is relatively dominant; and second the Sa and Yt group whose quickflow component is relatively dominant.

7.5 DISCUSSION

Previous work shows that water plays an important part in hillslope processes, such as surface flow (e.g. Horton 1945; Schumm 1956; Tanaka 1957; Dunne and Dietrich 1980), seepage erosion (e.g. Dunne 1980; Laity and Malin 1985; Nash 1996) and landslide (e.g. Iida and Okunishi 1983; Dietrich *et al.* 1986).

Onda (1989) pointed out that the thick regolith zone has a larger water storage capacity than the thin one and hence the occurrence of slope failure is strongly influenced by the

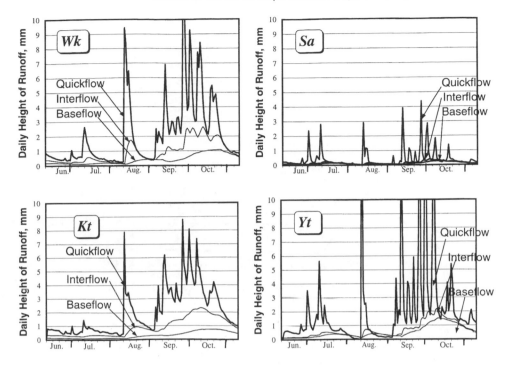

Figure 7.5 Results of filter separation and AR method for the data obtained from four experimental drainage basins in 1994

thickness of the regolith zone. In the Soya Hills, however, the thickness of the regolith zone does not play an important part in slope failure, because the thickness of the regolith is almost equal in each of the four rock types (Figure 7.7).

Onda (1994) also discussed the relation between run-off characteristic and slope processes in forested drainage basins in central Japan, and concluded that in the granite basins subsurface flow generates landslides, resulting in many slope failures, whereas in the basins of Palaeozoic sedimentary rocks surface flow is dominant and slope failure is rare. As a result, both the drainage density and slope failure are larger in granite basins with high subsurface flow and low surface flow than in the basins of the Palaeozoic sedimentary rocks with low subsurface flow and high surface flow. In the Soya Hills, on the contrary, the slope failures are distributed more densely in the Wk and Kt basins with low drainage density than in the Sa and Yt basins with high drainage density (Figure 7.3).

The differences in erosional processes associated with run-off characteristics between the Sa and Yt basins and the Wk and Kt basins, can be explained as follows. In the Sa and Yt basins, surface flow would cause severe gully erosion anywhere, when optimal conditions for gully erosion, such as destruction of forest, were provided. Figure 7.8 shows an example of a gully that occurred on the trail on the Sa hillslope. This landform implies that the Sa hills have the potential to form badland landforms with many rills and gullies. However, under the present forest-covered condition, slope failure is not frequent in the Sa and Yt basins (Figures 7.2 and 7.3). On the other hand, the authors discovered that there are still many gully-like small valleys in the Sa and Yt hills which are completely covered with thick vegetation and are not active now. Soil sections of the Sa and Yt hills show that there are loam layers which

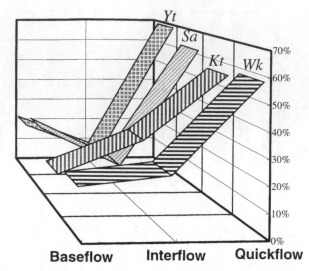

Figure 7.6 Difference in the run-off component characteristics between drainages basins in four rock types. Yt = silt; Sa = sand; Kt = mudstone; Wk = shale

are not the product of any erosional processes. These facts mean that erosional processes are inactive in the Sa and Yt basins at present and were surely active in the past.

Badland is formed on hillslopes composed of uncohesive deposits by gully erosion under conditions of poor vegetation cover (e.g. Schumm 1956; Bryan *et al.* 1978). Pollen analysis has revealed that grassland was dominant during 20 ka to 6 ka in the northern part of Hokkaido including the Soya Hills (Ono and Igarashi 1991). Igarashi and Yanai (1990) discovered a fossil gully in the Soya Hills and estimated that the gully was formed in 6 ka on the basis of ^{14}C dating and pollen analysis. The increased rainfall at the Pleistocene–Holocene transition generated hillslope incision in Japan (Oguchi 1988, 1996). Therefore, the drainage networks in the Sa and Yt basins are thought to have been formed before 6 ka when forest cover was poor and the rainfall amount started to increase. But after 6 ka, this area became forested (Ono and Igarashi 1991). As a result, gully erosion on the hillslopes became inactive.

Therefore, the landforms with high drainage density in the Sa and Yt hills are thought to be the relict topography of many gullies which were already well developed before 6 ka when the vegetation cover was sparse. At present, however, the development of valley landforms seems to be stopped by the surface vegetation cover.

On the Wk and Kt hills, gully erosion is not present but many shallow and broad surface failures are found. This leads to the recession of valley slopes. In these areas, subsurface flow would cause saturation of regolith with water or groundwater erosion. Therefore, many slope failures have been generated even under the present vegetation-covered conditions (Figures 7.2 and 7.3). At present, loam layers are not present in these areas and regolith zones are made from the colluvium of weathered material of the basement rocks. This is quite different from the Sa and Yt hills. These facts indicate that channel incisions into the hillslope and slope recessions of the Wk and Kt basins have been continuously active up to the present. It can be deduced that in the Wk and Kt basins, the optimal environments for erosion are completely different from those in the Sa and Yt basins, associated with the difference in run-off characteristics. Topographic change, especially the recession of valley-side slopes, is still active in the Wk and Kt hills, and has decreased the drainage density.

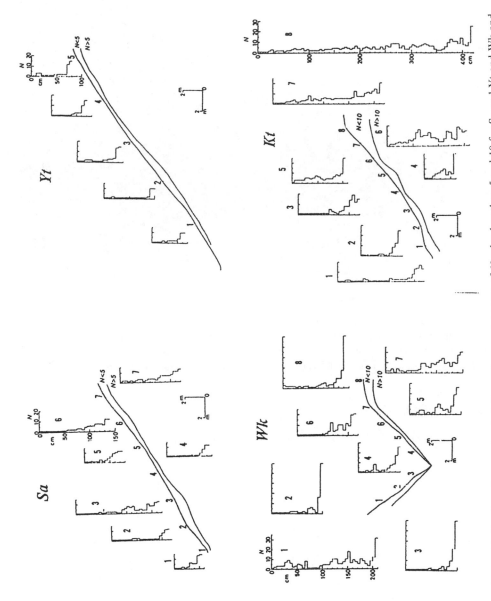

Figure 7.7 Results of penetration test. Regolith zones are zones of *N*-value less than 5 and 10 for Sa and Yt and Wk and Kt, respectively

Figure 7.8   Gully developed on a trail on Sa experimental drainage basin

Consequently, the combination of environmental change and variation in run-off characteristics caused the differences in the erosional processes between the Wk and Kt basins and the Sa and Yt basins. This would result in the difference in drainage density.

7.6   CONCLUSIONS

The present hydrological analysis revealed that the Wk and Kt basins are characterized by dominant subsurface flow which generates slope failure, whereas the Sa and Yt basins have dominant surface flow which generates gully erosion. Differences in the phase of erosional processes between the Wk and Kt basins and the Sa and Yt basins would play an important role in the formation of differences in hill morphology. In the Wk and Kt basins, slope failures are active at present, whereas in the Sa and Yt basins, incisions into hillslopes due mainly to gully erosion were generated only before 6 ka and have ceased at present. Therefore, the difference in the drainage density between the Wk and Kt basins and the Sa and Yt basins is thought to result from a combination of the differences in run-off characteristics and environmental change.

ACKNOWLEDGEMENTS

The authors would like to give special thanks to Professer Kaichiro Sasa and Professor Fuyuki Satoh of the Hokkaido University Forests, Faculty of Agriculture, Hokkaido

University, for their help throughout this study. Thanks are also due to Dr Yoshimasa Kurashige of the Graduate School of Environmental Earth Science, Hokkaido University, for fruitful discussion about the setting for experimental drainage basins in the field. The authors thank Dr Yuichi Onda of the School of Agricultural Sciences, Nagoya University, for helpful advice on hydrological observations.

REFERENCES

Bryan, R.B., Yair, A. and Hodges, W.K. 1978. Factors controlling the initiation of run-off and piping in Dinosaur Provincial Park badlands, Alberta, Canada. *Zeitschrift für Geomorphologie, Neue Folge, Supplement Band*, **29**, 151–168.

Dietrich, W.E., Wilson, C.J. and Reneau, S.L. 1986. Hollows, colluvium and landslides in soil-mantled landscapes. In Abrahams, A.D. (ed.), *Hillslope Processes*. Allen and Unwin, London, 31–53.

Dunne, T. 1980. Formation and controls of channel networks. *Progress in Physical Geography*, **4**, 211–239.

Dunne, T. and Dietrich, W.E. 1980. Experimental study of Horton overlandflow on tropical hillslopes: 1. Soil conditions, infiltration and frequency of run-off. *Zeitschrift für Geomorphology, Supplement Band*, **35**, 40–80.

Freeze, R.A. 1972. Role of subsurface flow in generating surface run-off 2. Upstream source areas. *Water Resources Research*, **8**, 1272–1283.

Hata, M. and Ueda, Y. 1969. *Geological map of 'Teshio'* (1:200,000). Geological Survey of Japan.

Hino, M. and Hasebe, M. 1981. Analysis of hydrologic characteristics from run-off data – a hydrologic inverse problem. *Journal of Hydrology*, **49**, 287–313.

Hino, M. and Hasebe, M. 1984. Identification and prediction of nonlinear hydrologic systems by the filter-separation autoregressive (AR) method: extension to hourly hydrologic data. In Stout, G.E. and Davis, G.H. (eds), Global Water: Science and Engineering – The Van Te Chow Memorial Volume, *Journal of Hydrology*, **68**, 181–210.

Hirose, T., Onda, Y. and Matsukura, Y. 1994. Run-off and solute characteristics in four small catchments with different bedrocks in the Abukuma Mountains, Japan. *Transactions Japanese Geomorphological Union*, **15A**, 31–48.

Horton, R.E. 1945. Erosional development of streams and their drainage basins: hydrophysical approach to quantitative geomorphology. *Geological Society of America Bulletin*, **56**, 275–370.

Igarashi, Y. and Yanai, S. 1990. Holocene fossil gully discovered from Sohya hill, northern Hokkaido, Japan. *Quaternary Research*, **28**, 413–417 (in Japanese with English abstract).

Iida, T. and Okunishi, K 1983. Development of hillslopes due to landslides. *Zeitschrift für Geomorphologie, Neue Folge Supplement Band*, **46**, 67–77.

Laity, J.E. and Malin, M.C. 1985. Sapping processes and the development of theater-headed valley networks on the Colorado Plateau. *Geological Society of America Bulletin*, **96**, 203–217.

Nagao, S. 1960. *Geological map and its explanatory text of 'Toyotomi'* (1:50,000). Geological Survey of Hokkaido.

Nash, D.J. 1996. Groundwater sapping and valley development in the Hackness hills, North Yorkshire, England. *Earth Surface Processes and Landforms*, **21**, 781–795.

Oguchi, T. 1988. Landform development during the last glacial and the post-glacial ages in the Matsumoto basin and its surrounding mountains, central Japan. *Quaternary Research*, **27**, 101–124 (in Japanese with English abstract).

Oguchi, T. 1996. Relaxation time of geomorphic response to Pleistcene-Holocene climatic change. *Transactions Japanese Geomorphological Union*, **17**, 309–322.

Onda, Y. 1989. Influence of water storage capacity in the regolith zone on hydrological characteristics and slope failure on granitic hills in Aichi, Japan. *Transactions Japanese Geomorphological Union*, **10**, 13–26. (in Japanese with English abstract).

Onda, Y. 1992. Influence of water storage capacity in the regolith zone on hydrological characteristics, slope process, and slope form. *Zeitschrift für Geomorphology Neue Folge*, **36**, 165–178.

Onda, Y. 1994. Contrasting hydrological characteristics, slope processes and topography underlain by Paleozoic sedimentary rocks and granite. *Transactions Japanese Geomorphological Union*, **15A**, 49–65.

Ono, Y. and Igarashi, Y. 1991. *Natural History of Hokkaido*, Hokkaido University Press, Sapporo, 219 (in Japanese).

Schumm, S.A. 1956. The evolution of drainage systems and slopes in badlands at Perth Amboy, New Jersey. *Geological Society of America Bulletin*, **67**, 597–646.

Sudarmadji, T., Araya, T. and Higashi, S. 1990. A hydrological study on streamflow characteristics of small forested basins in different geological conditions. *Research Bulletin Hokkaido University Forests*, **47**, 321–351.

Suzuki, T., Tokunaga, E., Noda, H. and Arakawa, H. 1985. Effects of rock strength and permeability on hill morphology. *Transactions Japanese Geomorphological Union*, **6**, 101–130.

Tanaka, S. 1957. The drainage density and rocks (granite and Paleozoic) in Setouchi sea coast region, western Japan. *Geographical Review of Japan*, **30**, 564–578 (in Japanese with English abstract).

Tanaka, Y. 1990. The relationships between the evolution of valley side slope and rock properties in Soya hill and Shiranuka hill. *Geographical Review of Japan*, **63** (Ser. A), 836–847 (in Japanese with English abstract).

Tanaka, Y. and Agata, Y. 1997. Effects of run-off characteristics on the difference in drainage density in Soya hills, northern Japan. In Yuan Daoxian (ed.), *Environmental Geology*, VSP, Utrecht, 74–85.

Tanaka, Y., Agata, Y. and Sasa, K. 1996. Relationships between rock properties of basement rocks and run-off characteristics in Soya hill, Hokkaido, Northern Japan. *Research Bulletin of Hokkaido University Forests*, **53**, 269–287 (in Japanese with English abstract).

PART 2　Sediment Transfer Dynamics

Overview: Sediment Transfer Dynamics

DAVID L. HIGGITT
Department of Geography, University of Durham, Durham, UK

GEOMORPHOLOGY AND THE SEDIMENT CASCADE

Traditionally, sediment transfer may have been viewed by geomorphologists merely as an incidental component of landscape development – the evolution of hillslope profiles and drainage networks accomplished through the cascade of sediment production, detachment, entrainment and transport. Increasingly, the sediment cascade itself has become a focus of attention, ranging from a concern with the mechanisms of sediment production through weathering to the evaluation of global patterns of sediment discharge to the world's oceans. Geomorphologists investigating aspects of sediment cascades are fundamentally concerned with sediment transfer dynamics – the processes and mechanism of sediment transfer, elucidation of the conditions that control the spatial and temporal expression of sediment transfer, the rates, pathways and residence times of sediment moving through the system and the linkages of the sediment delivery system, particularly the coupling between slopes and channels. Instinctively, geomorphological interest in sediment transfer dynamics is likely to concern evaluation of present-day rates of erosion, implications for landform development (e.g. river channel change) and interpretation from sedimentary archives such as lake and floodplain sediments. However, the implications of sediment transfer spread far beyond traditional disciplinary boundaries and offer scope for the application of gemorphological input into wider questions of resource management. In particular, the implications of sediment transfer can be divided into on-site and off-site problems. The former concern the depletion of soil resources and the latter the disruption caused as sediment is transferred to locations further down the flow path.

The seven chapters in Part 2: Sediment Transfer Dynamics address both on-site and off-site problems associated with sediment transfer. Four of the chapters are primarily concerned with the processes and patterns of soil erosion. Of these, two describe procedures to quantify erosion rates on forest floors (Dragovich and Morris) and agricultural land (Mabit *et al.*). The role of soil erosion on agricultural land in redistributing pollutants is addressed with respect to fertilizers and pesticides (Angeliaume and Wicherek) and radioactive contamination (Golosov and Ivanova). One chapter deals with mapping techniques to define landslide hazard from weathering deposits (Di Nocera and Matano). The remaining two chapters are concerned with sediment transport in fluvial systems in terms of attempting to fingerprint its source in relation to logging activity (Kurashige) and its impact on riparian wetland evolution (Piégay *et al.*). Thus, the collection represents a range across the sediment cascade from studies primarily concerned with inputs into the sediment delivery system to those dealing with the need to determine sources of fluvial sediment load.

The notion that the connectivity between fluvial sediment yield and the various sources of sediment can be expressed quantitatively is encapsulated in the sediment budget. The development of a sediment budget approach can be helpful as a means of identifying the major natural and management-related sources of sediment, the amount of sediment being generated by those sources and the volume of material in storage between origin and catchment outlet (Reid and Dunne 1996). The response of sediment transfer dynamics to changes occurring within a catchment, whether these are related to climate change, land-use changes or deliberate human modification, are of interest to both geomorphologists and to land-use managers. The impact of sediment transfer on water quality and the ecology of the aquatic environment also emphasizes the need to understand the factors that control the erosion, transport and delivery of sediment-associated contaminants. Thus, there are many aspects of land management where geomorphological expertise can be applied, though it can be argued concurrently that there remains much to be understood about the processes of sediment transfer and the ability to extrapolate results from case studies to new areas or across different scales. It can also be argued that to date geomorphology has had limited impact on reducing the problems of soil erosion globally. The themes of soil erosion and of sediment budgets are developed briefly below in the context of reviewing recent contributions to the scientific literature. The specific impacts of sediment transfer are then considered with reference to the themes developed in the chapters collected here.

SOIL EROSION – THE QUIET CRISIS

Many environmental scientists would agree that soil erosion has become a major environmental problem during the 20th century, but articulation of its extent and magnitude invite contention. Despite early attempts to draw soil erosion into the public arena through sensationalist narrative (e.g. Jacks and Whyte 1939) the problems associated with soil erosion have received less attention than higher intensity natural hazards. Indeed it is the insidious nature of progressive erosion that reduces soil quality to the point where agricultural productivity cannot be sustained.

It falls upon geomorphologists to articulate the processes, rates and dynamics of soil erosion at appropriate scales of analysis but also to contribute to the debate about solutions to the erosion problem. Recent attempts to summarize global soil erosion status (Pimental 1992; Pimental et al. 1995) have attracted castigation. Crosson (1995) launched a robust critique of the average erosion rate for the USA proposed by Pimental et al. (1995), arguing that it was grossly exaggerated. Similarly, Boardman (1998) demonstrates a tortuous ancestry of the value of $17\,t\,ha^{-1}\,a^{-1}$, proposed by Pimental et al. (1995) as the average rate of soil erosion in Europe. This figure originates from a series of small-scale plot experiments in Belgium undertaken as part of a PhD study but ends up being quoted as an average rate for a continental area which is both misleading and inaccurate. Though published data on global, regional or national soil erosion inventories may motivate government agencies to enact policy or distribute resources, overgeneralization and exaggeration inherent in large-scale audits can be counterproductive. Stocking (1995) has raised this issue in an astute challenge to geomorphologists to consider the value of their research towards the solution of erosion problems in developing countries. He argues that the majority of studies fail to generate data that are useful in planning effective conservation measures and calls upon geomorphologists to concentrate on timely and applicable information. Stocking identifies four essential ingredients for applied research:

(i)   interdisciplinarity – a willingness to place technically based results into a context for planners;
(ii)   time constraints – an ability to use rapid techniques of assessment that do not require lengthy experimentation;

(iii)   flexibility – to provide a menu of options and possibilities rather than prescriptive solutions;

(iv)   contextual analysis – where the reason for the research is stated explicitly with reference to the client's requirements and agenda.

Such an approach draws the geomorphologist away from predefined disciplinary boundaries yet requires practice to be based on a suitable range of techniques and methodology. The armoury of available techniques for soil erosion research has grown considerably in recent years and reviews are available elsewhere (Loughran 1989; Bryan 1989; Lal 1994). Two important areas of advance concern surveying technology and computer modelling.

Developments in photogrammetry, laser refraction and remote sensing offer enhanced data capture for examining changes in morphology and surface characteristics. Recent examples of land degradation studies from satellite imagery indicate that while Sujatha *et al.* (2000) are confident in reporting inventories of degraded land classifications, other studies are more cautious in linking surface reflectance from vegetation patches to indicators of degradation (Ludwig *et al.* 2000). The use of the radionuclide caesium-137 (^{137}Cs) has become widespread within the last decade, based on its ability to provide information on averaged medium-term soil redistribution from a series of soil cores (Ritchie and McHenry 1990; Walling and Quine 1991). As such it is an ideal tool for consultancy applications permitting rapid appraisal of spatial variability in soil erosion rates, though many recent papers continue to investigate its applicability to specific environments, including one of the contributions to this volume. Mabit *et al.* (Chapter 12) present results from an investigation of the vertical distribution of ^{137}Cs in organic soil in the Souse marshes of northeastern France. Finding no evidence of enhanced diffusion of the ^{137}Cs profile, its potential as a tracer on these types of organic soils is demonstrated.

Technological improvements in computer speed and memory capabilities have heralded a rapid transition from empirically based to physically based erosion models with the US Soil Conservation Service replacing the long-serving Universal Soil Loss Equation (USLE) with the Water Erosion Prediction Project (WEPP). Many of the numerous modelling options available have been reviewed by Siakeu and Oguchi (2000). Modelling approaches can be used in applied studies to zone erosion hazard or undertake 'what if' scenarios for projected land management strategies. However, the accuracy of modelling is contingent on representativeness and validation. Some critical properties influencing erosion processes such as erodibility are notoriously dynamic and require abundant data about variations in antecedent soil water, surface roughness and other hydrological properties (Torri *et al.* 1997). Models that are data-hungry for input parameters may be problematic, particularly in developing countries, where there are limited data on hydrological conditions. Model output requires validation. In a controversial attack on soil erosion research and knowledge in the USA, Trimble and Crosson (2000) criticize the over-reliance on unvalidated model estimates and demand that investment is directed towards comprehensive monitoring programmes. This suggests that soil erosion research retains a tension between field-based workers and modellers. It can only be hoped that there is further convergence.

SEDIMENT BUDGETS

Soil erosion produces the twin problem of off-site effects through sedimentation, export of nutrients and increased likelihood of flooding. In some cases, especially in temperate environments, the costs associated with sedimentation far exceed on-site effects. A major research problem concerns the uncertainties which surround attempts to link on-site rates of erosion within a drainage basin to the sediment yield at the catchment outlet. Only a small fraction of sediment eroded from slopes will reach the stream network, and opportunities for deposition on slopes, swales and floodplains increases with catchment size. Sediment delivery is the term which is used to represent the resultant of the various processes interposed between on-site erosion and the catchment sediment yield (Walling 1983).

Central to the investigation of the dynamics of sediment transfer between slopes and channel networks is the construction of sediment budgets to account for the sources and pathways of sediment as it travels from its point of origin to the catchment outlet (Reid and Dunne 1996). As physical or chemical properties of sediments, such as mineralogy, particle size, organic content, mineral magnetic characteristics and radionuclide inventories, are often strongly related to either geological or source conditions, the properties can be used to 'fingerprint' the relative contribution from different parts of the catchment or different sources (e.g. surface or subsurface soils, contrasting land-use types). The use of multiple tracers reduces some of the uncertainty associated with individual sediment properties (Collins *et al.* 1997a, b). While detailed sampling and monitoring is required for a high level of accuracy, many consultancy applications that require information about sediment dynamics are necessarily limited in time and financial resources. Writing with this purpose in mind, Reid and Dunne (1996) have produced a manual for the rapid appraisal of sediment budgets. A similar approach is adopted by Thorne (1998) in his river channel reconnaissance scheme which aims to establish the stability of selected river reaches and can be used to identify the requirement for engineering intervention (Thorne *et al.* 1996).

A much visited theme by geomorphologists interested in sediment budgets is the evidence (or lack of) for human impact on sediment dynamics (e.g. Trimble 1983; De Boer 1997). Long-term records of sediment fluxes from natural lakes enable comparison of pre- and post-land-use change impacts but also demonstrate the importance of climate change and extreme events. In the North Island of New Zealand, Page and Trustrum (1997) describe the transition from indigenous forest to fern/scrub following Polynesian settlement and then to pasture following European settlement. Lake sedimentation rates in the recent pastoral phase are five to six times higher than the fern/scrub phase and eight to 17 times higher than the forest phase. Nevertheless, much of the sediment is delivered during storms and the land-use change is associated with an increase in sediment supply during such storm events. In the Upper Mississippi, Knox (2001) suggests that agricultural practices have greatly escalated the sensitivity of the landscape and accelerated sediment transfer despite reduced sediment loads in the last half-century. In upland environments the links between land-use change and sediment yields is less clear-cut. In northern England some specific anthropogenic impacts such as Viking age clearance are reflected in headwater alluvial fan deposits (Harvey *et al.* 1981) but further downstream valley aggradation appears to be largely related to climatic variation, leading Macklin and Lewin (1993) to remark that the Holocene alluvial record is 'climatically driven but culturally blurred'. In the uplands of southeastern Australia, Wasson *et al.* (1998) demonstrate that the impacts of clearance and colonization are manifest in increased rates of channel incision rather than slope erosion. These studies emphasize the importance of controls on slope–channel coupling in determining the erosional response of catchments. Investigations of the impact of land-use change on sediment yields is therefore advised to consider the overall sediment budget rather than only the sediment output (Walling 1999).

IMPACTS OF SEDIMENT TRANSFER

The operation of sediment transfer has a number of implications for human use and management of land and aquatic resources. Soil erosion has on-site costs in terms of damage to agricultural land and off-site costs associated with the accumulation of sediment in unwanted locations. Each of the chapters in this part of the book holds some relevance to the analysis of the impacts of sediment transfer. Applied geomorphological studies offer some prospect for controlling undesired impacts of sediment transfer.

Pollution Impacts

Sediment acts as a vector for the transfer and fate of certain chemical compounds that accumulate in parts of the ecosystem. The fact that many chemical constituents are adsorbed onto colloidal matter in soils and that size-selective transport favours fine-grained particles, leads to an enrichment of the adsorbed constituent in exported sediment. This is termed the enrichment ratio (Novotny and Chesters 1989) and tends to increase as delivery ratio decreases as a function of the greater travel distance for fine particles. Sediment-associated transport of fertilizers, pesticide residues and other chemical soil treatments has implications for the maintenance of water quality standards downstream. Angeliaume and Wicherek (Chapter 8) consider the role of soil erosion on agricultural land for the transfer of potential pollutants such as nitrogen, phosphorus and the pesticides atrazine and lindane. The experiment is conducted on an instrumented agricultural slope where much previous work has given an insight into soil erosion dynamics on the loess-derived soils of the Paris Basin (Wicherek 1991). The results demonstrate that there are strong seasonal patterns in pollutant transfer dominated by spring and summer, which do not necessarily mirror periods of maximum erosion risk. Similarly, different crop covers present markedly different pollution risks with maize production associated with high concentrations of atrazine in run-off.

Though the redistribution of radionuclides has been adapted by geomorphologists as a tracer of soil erosion, the role of sediment transfer in dispersing or accumulating radioactive pollution is important following accidental releases. The Chernobyl nuclear power plant explosion in April 1986 delivered high levels of radioactivity across many regions of the northern hemisphere and a number of studies have examined the redistribution of Chernobyl fallout within headwater catchments (Higgitt et al. 1992) and through river systems (Rowan and Walling 1992). On the Russian Plain, close to the site of the accident, agricultural soils received a substantial input of ^{137}Cs that can be measured using in situ gamma detection. Following previous chapters that have used the redistribution of caesium as a means of estimating sediment transfer from agricultural slopes to balka (dry valley) systems (Golosov et al. 1999), Chapter 10 by Golosov and Ivanova considers the fate of radioactive contaminants resulting from erosional transport from slopes to valley bottoms.

Infrastructure Impacts

Not only does the transfer of sediment from field surfaces of hillslopes lead to potential pollution problems downstream but the quantity of sediment can cause difficulties for the operation of water management schemes. The most obvious example is the problem associated with sedimentation and loss of storage capacity in reservoirs. Reservoir regulation to minimize sedimentation impacts requires a detailed understanding of the spatial and temporal patterns of sediment delivery coupled to an engineering design that permits the discharge of some sediment-laden water through sluices. However, management to avoid sedimentation is constrained by operational demands (water supply, power generation and flood protection) and sedimentation control measures need to be applied in a wider remit of water resource management. Applied geomorphologists have developed a wide range of techniques for interpreting lake and reservoir sedimentation from sediment cores (Foster *et al.* 1990). Brandt and Swenning (1999) used a sediment budget approach to evaluate the dispersal of sediment from the flushing of a Costa Rican reservoir. In the Upper Yangtze River, China, where the world's largest dam is under construction, Lu and Higgitt (1998) have employed time series methods to evaluate subcatchment trajectories in sediment supply that might affect the operation of the reservoir scheme.

Tackling the problem at source through catchment-wide land-use control or through soil conservation measures is possible but difficult to achieve in practice and the use of sediment budgets provides a means of targeting specific sources of sediment. In this light, Kurashige (Chapter 11) analyses suspended sediments from streams in forested areas to evaluate the key sources of sediment arising from logging practices. A number of recent studies have identified the importance of forest roads and tracks as both sources and vectors of sediment transport (Ziegler and Giambelluca 1997). The build-up of sediment in river channels has a range of ecological impacts.

Ecological Impacts

Sediment dynamics play an important role in the generation and maintenance of habitat. Piégay *et al.* (Chapter 14) examine the characteristics of alluvial plugs on a series of channel cut-offs on the Ain River draining the French Alps. Plug development is dependent on the age of the abandoned channel, its location relative to the main channel and the instability of the main channel. The study demonstrates that alluvial plugs function as buffer zones that inhibit sedimentation in the cut-off channel and hence slow down the diminution of the aquatic zone of the abandoned cut-off channels. River managers concerned with the conservation of wetland areas on river corridors can benefit from an improved geomorphological understanding of floodplain and wetland sedimentation (Brookes 1996).

The growing recognition of the link between fluvial geomorphology and river habitat quality has encouraged cooperation between geomorphologists and ecologists in recent years (Boon *et al.* 1992), perhaps best illustrated through the developing art of river restoration (Kondolf and Michelli 1995; Petts and Calow 1996). Over the last 20 years there has been an increasing interest in more sympathetic engineering in rivers that minimizes straightening and resectioning and emphasizes the use of natural forms, particularly the meandering nature of rivers. The growing profile of geomorphological analysis in an engineering context (Thorne *et al.* 1997) is mirrored by the emerging concepts of environmental (or 'soft') engineering. As most lowland rivers have been modified by centuries of human involvement in water resources there is a demand for engineering which can restore degraded river habitats. Ultimately, river restoration can be defined as the 'complete structural and functional return to a pre-disturbance state' (Cairns 1991). In practice, there are many obstacles to full restoration and the terms rehabilitation or enhancement are used to indicate improvements to some structural or functional attributes. Philosophically there is a debate about whether 'naturalness' is an attribute that can be recreated. Nevertheless, appropriate sediment dynamics are essential for successful restoration schemes because of the impact of fine-grained sedimentation on fish spawning habitat (Kondolf 2000) and this calls upon appreciation of the geomorphological setting of the restoration project. As adjustments to restoration take time, some long-term monitoring survey data are necessary and Trimble (1997) has demonstrated how some streambank fish shelters have reduced bank erosion on Wisconsin streams over a period of two decades.

Instability Impacts

Evaluation of sediment budgets involves identification of key sources areas and estimation of the amount of sediment being derived from each area. In terrain susceptible to mass movement, sediment budget estimation may be difficult because individual landslide events may yield large amounts of sediment but occur infrequently in time. The sediment budget requires some estimation of the probability of input from particular types of slope. As a consequence, the development of methods to estimate the probability of slope failure or time elapsed since the last major landslide can be used as a guide to assessing landscape instability. Di Nocera and Matano (Chapter 13) describe a mapping exercise for evaluating the weathering grade of crystalline rocks in Southern Italy. Based on an established assumption that the thickness and characteristics of the regolith are the fundamental controls on slope instability (Ollier 1984; Thomas 1994), the classification of weathering grade has utility for engineering geomorphology such as the design and location of road cuttings and the routing of road networks. Mapping weathering and soil erosion features from aerial surveys can also be applied in mineral prospecting. Recent developments in airborne geophysical surveys of gamma ray emissions enable estimation of K, Th and U in near-surface sediments. Spatial variations are principally related to lithology but show topographically related patterns that are indicative of soil thickness and the accumulation of material in valley floors (Dickson et al. 1996; Pickup and Marks 2000).

Degradation Impacts

Finally, sediment transfer has on-site impacts in terms of the degradation of the land surface, though it should be re-emphasized that the term degradation encompasses many processes that lead to a decline in soil quality, of which erosion is but one. Nor should geomorphologists lose sight of the argument that the occurrence of accelerated soil erosion is fundamentally a political, social and economic issue (Blaikie 1985) in terms of the contexts which promote unsustainable use of land. It is clear from the critique of Stocking (1995), outlined earlier, that geomorphologists have much to prove in their contribution to solving the issues of land degradation, particularly in providing flexible technical solutions that can be implemented as part of a more holistic approach to land management (Shaxson et al. 1989). Nevertheless, recent research continues to identify controls on soils and nutrient losses.

One aspect of land degradation which is prevalent in many seasonally wet–dry regions of the world is the impact of wildfire on surface instability. Dragovich and Morris (Chapter 9) examine post-fire sediment movement in an area of eucalyptus forest in southeastern Australia. Plots on slopes burnt by high intensity fire had significantly greater amounts of run-off and sediment than moderately or unburnt slopes. The increase in run-off is likely to result from the hydrophobicity of burnt litter which reduces infiltration. Experimental studies of fire impact have demonstrated that losses of soil, organic matter and nutrients are greatest on plots affected by the most intense fire (Thomas et al. 1999; Emmerich 1999; Gimeno-Garcia et al. 2000), but the timing of the fire regime (e.g. Townsend and Douglas 2000) and its heterogeneity are important considerations. Lavee et al. (1995) reported great variability in run-off from a series of burnt plots which was attributed to the production of mosaic-like surfaces containing patches of enhanced surface roughness (where run-off was negligible) and smooth patches which generated considerable run-off. Similarly the development of soil hydrophobicity after fire has a number of effects on erosion susceptibility but its overall impact for erosion hazard is difficult to predict (Shakesby et al. 2000).

SUMMARY

Geomorphologists investigate sediment transfer dynamics at a multitude of scales and at various stages of the sediment cascade. As sound geomorphological research continues to explore the spatial and temporal dynamics of sediment delivery systems and elucidate the processes that govern sediment production, detachment and transport, the applied geomorphologist must seek to provide timely and applicable information based on the most suitable techniques and methods. There are some promising signs that rapid appraisal techniques for assessing environmental engineering or land-use management options are beginning to incorporate geomorphological principles more fully. The impact of soil erosion on land degradation is often unclear because of the limited information on site-specific erosion rates and on the link between erosion and nutrient loss. There remains a need to demonstrate the long-term impact of soil erosion on agricultural productivity.

REFERENCES

Blaikie, P. 1985. *The Political Economy of Soil Erosion in Developing Countries*. Longman, London.

Boardman, J. 1998. An average soil erosion rate for Europe: Myth or reality? *Journal of Soil and Water Conservation*, **53**, 46–50.

Boon, P.J., Petts, G.E. and Calow, P. (eds) 1992. *River Conservation and Management*. Wiley, Chichester.

Brandt, S.A. and Swenning, J. 1999. Sedimentological and geomorphological effects of reservoir flushing: The Cachi Reservoir, Costa Rica, 1996. *Geografiska Annaler*, **81A**, 391–407.

Brookes, A. 1996. Floodplain restoration and rehabilitation. In Anderson, M.G., Walling, D.E. and Bates, P.D. (eds), *Floodplain Processes*. Wiley, Chichester, 553–579.

Bryan, R.B. (ed.) 1989. *Soil erosion experiments and methods*. Catena Supplement, Catena-Verlag, Reiskirchen.

Cairns, J. 1991. The status of the theoretical and applied science of restoration ecology. *The Environment Professional*, **13**, 186–194.

Collins, A.L., Walling, D.E. and Leeks, G.J.L. 1997a. Source type ascription for fluvial suspended sediment based on a quantitative composite fingerprinting technique. *Catena*, **29**, 1–27.

Collins, A.L., Walling, D.E. and Leeks, G.J.L. 1997b. Use of the geochemical record preserved in floodplain deposits to reconstruct recent changes in river basin sediment sources. *Geomorphology*, **19**, 151–168.

Crosson, P. 1995. Soil erosion estimates and costs. *Science*, **269**, 461–464.

De Boer, D.H. 1997. Changing contributions of suspended sediment sources in small basins resulting from European settlement on the Canadian Prairies. *Earth Surface Processes and Landforms*, **22**, 623–640.

Dickson, B.L., Fraser. S.J. and Kinsey-Henderson, A. 1996. Interpreting aerial gamma ray surveys utilising geomorphological and weathering models. *Journal of Geochemical Exploration*, **57**, 75–88.

Emmerich, W.E. 1999. Nutrient dynamics of rangeland burns in south-eastern Arizona. *Journal of Range Management*, **52**, 606–614.

Foster, I.D.L., Dearing, J.A., Grew, R. and Orend, K. 1990. The lake sedimentary database: an appraisal of lake and reservoir-based studies of sediment yield. In: Walling, D.E., Yair, A. and Berkowicz, S. (eds), *Erosion, Transport and Deposition Processes* (Proceedings of the Jerusalem Workshop). IAHS Publication No. 189, International Association of Hydrological Sciences, Wallingford, 19–43.

Gimeno-Garcia, E., Andreu, V. and Rubio, J.L. 2000. Changes in organic matter, nitrogen, phosphorus and cations in soil as a result of fire and water erosion in a Mediterranean landscape. *European Journal of Soil Science*, **51**, 201–210.

Golosov, V.N., Walling, D.E., Panin, A.V., Stukin, E.D., Kvasnikova, E.V. and Ivanova, N.N. 1999. The spatial variability of Chernobyl-derived Cs-137 inventories in a small agricultural drainage basin in central Russia. *Applied Radiation and Isotopes*, **51**, 341–352.

Harvey, A.M., Oldfield, F., Baron, A.F. and Pearson, G.W. 1981. Dating of post-glacial landforms in the central Howgills. *Earth Surface Processes and Landforms*, **6**, 401–412.

Higgitt, D.L., Froehlich, W. and Walling, D.E. 1992. Applications and limitations of Chernobyl radiocaesium measurements in a Carpathian erosion investigation, Poland. *Land Degradation and Rehabilitation*, **3**, 15–26.

Jacks, G.V. and Whyte, R.O. 1939. *The Rape of the Earth: a World Survey of Soil Erosion*. Faber, London.

Knox, J.C. 2001. Agricultural influence on landscape sensitivity in the Upper Mississippi River Valley. *Catena*, **42**, 193–224.

Kondolf, G.M. 2000. Some suggested guidelines for geomorphic aspects of anadromous salmonid habitat restoration proposals. *Restoration Ecology*, **8**, 48–56.

Kondolf, G.M. and Michelli, E.R. 1995. Evaluating stream restoration projects. *Environmental Management*, **19**, 1–15.

Lal, R. (ed.) 1994. *Soil Erosion Research Methods* (2nd edition). SWCS, Iowa.

Lavee, H., Kutiel, P., Segev, M. and Benyamini, Y. 1995. Effect of surface roughness on run-off and erosion in a Mediterranean ecosystem – the role of fire. *Geomorphology*, **11**, 227–234.

Loughran, R.J. 1989. The measurement of soil erosion. *Progress in Physical Geography*, **13**, 216–233.

Lu, X.X. and Higgitt, D.L. 1998. Recent changes of sediment yield in the Upper Yangtze, China. *Environmental Management*, **22**, 697–709.

Ludwig, J.A., Bastin, G.N., Eager, R.W., Karfs, R., Ketner, P. and Pearce, G. 2000. Monitoring Australian rangeland sites using landscape function indicators and ground- and remote-sensing techniques. *Environmental Monitoring and Assessment*, **64**, 167–178.

Macklin, M.G. and Lewin, J. 1993. Holocene river alluviation in Britain. *Zeitschrift für Geomorphologie Supplementband*, **88**, 109–122.

Novotny, V. and Chesters, G. 1989. Delivery of sediment and pollutants from nonpoint sources: a water quality perspective. *Journal of Soil and Water Conservation*, **44**, 569–576.

Ollier, C.D. 1984 *Weathering*. Longman, London.

Page, M.J. and Trustrum, N.A. 1997. A late Holocene lake sediment record of the erosion response to land use change in a steepland catchment, New Zealand. *Zeitschrift für Geomorphologie*, **41**, 369–392.

Petts, G.E. and Calow, P. (eds) 1996. *River Restoration*. Blackwell, Oxford.

Pickup, G. and Marks, A. 2000. Identifying large scale erosion and deposition features from airborne gamma radiometrics and digital elevation models in a weathered landscape. *Earth Surface Processes and Landforms*, **25**, 535–557.

Pimental, D. (ed.) 1992. *World Soil Erosion and Conservation*. Prentice Hall, New York.

Pimental, D., Harvey, C., Resosudamo, P. Sinclair, K., Kurz, D., McNair, M., Crist, S., Shriptz, L., Fitton, L., Saffouri, R. and Blair, R. 1995. Environmental and economic costs of soil erosion and conservation benefits. *Science*, **267**, 1117–1123.

Reid, L.M. and Dunne, T. 1996. *Rapid Evaluation of Sediment Budgets*. Catena Verlag, Reiskirchen.

Ritchie, J.C. and McHenry, J.R. 1990. Application of radionuclide fallout cesium-137 for measuring soil erosion and sediment accumulation rates and patterns. *Journal of Environmental Quality*, **19**, 215–233.

Rowan, J.S. and Walling, D.E. 1992. The transport and fluvial redistribution of Chernobyl-derived radiocaesium within the River Wye basin, UK. *The Science of the Total Environment*, **121**, 109–131.

Shakesby, R.A., Doerr, S.H. and Walsh, R.P.D. 2000. The erosional impact of soil hydrophobicity: current problems and future research directions. *Journal of Hydrology*, **231**, 178–191.

Shaxson, I.F., Hudson, D.W., Sanders, D.W., Roose, E.J. and Moldenhauer, W.C. 1989. *Land Husbandry: a Framework for Soil and Water Conservation*. Soil and Water Conservation Society, Ankeny, Iowa.

Siakeu, J. and Oguchi, T. 2000. Soil erosion analysis and modelling: a review. *Transactions, Japanese Geomorphological Union*, **21**, 413–414.

Stocking, M. 1995. Soil erosion in developing countries: where geomorphology fears to tread! *Catena*, **25**, 253–267.

Sujatha, G., Dwivedi, R.S., Sreenivas, K. and Venkataratnam, L. 2000. Mapping and monitoring of degraded lands in part of Jaunpur district of Uttar Pradesh using temporal spaceborne multispectral data. *International Journal of Remote Sensing*, **21**, 519–531.

Thomas, A.D., Walsh, R.P.D. and Shakesby, R.A. 1999. Nutrient losses in eroded sediment after fire in eucalyptus and pine forests in the wet Mediterranean environment of northern Portugal. *Catena*, **36**, 283–302.

Thomas, M.F. 1994. *Geomorphology in the Tropics: A Study of Weathering and Denudation in Low Latitudes*. Wiley, Chichester.

Thorne, C.R. 1998. *Stream Reconnaissance Handbook*. Wiley, Chichester.

Thorne, C.R., Allen, R.G. and Simon, A. 1996. Geomorphological river channel reconnaissance for river analysis, engineering and management. *Institute of British Geographers Transactions*, **21**, 469–484.

Thorne, C.R., Hey, R. and Newson, M.D. (eds) 1997. *Applied Fluvial Geomorphology for River Engineering and Management*. Wiley, Chichester.

Torri, D., Poeson, J. and Borselli, L. 1997. Predictability and uncertainty of the soil erodibility factor using a global dataset. *Catena*, **31**, 1–22.

Townsend, S.A. and Douglas, M.M. 2000. The effect of three fire regimes on stream water quality, water yield and export coefficients in a tropical savanna (northern Australia). *Journal of Hydrology*, **229**, 118–137.

Trimble, S.W. 1983. A sediment budget for Coon Creek basin in the Driftless Area, Wisconsin, 1853–1977. *American Journal of Science*, **283**, 454–474.

Trimble, S.W. 1997. Streambank fish-shelter structures help stabilize tributary streams in Wisconsin. *Environmental Geology*, **32**, 230–234.

Trimble, S.W. and Crosson, P. 2000. US Soil erosion rates – myth and reality. *Science*, **289**, 248–250.

Walling, D.E. 1983. The sediment delivery problem. *Journal of Hydrology*, **65**, 209–237.

Walling, D.E. 1999. Linking land use, erosion and sediment yields in river basins. *Hydrobiologia*, **410**, 223–240.

Walling, D.E. and Quine, T.A. 1991. The use of ^{137}Cs measurements to investigate soil erosion on arable fields in the UK: potential applications and limitations. *Journal of Soil Science*, **42**, 147–165.

Wasson, R.J., Mazari, R.K., Starr, B. and Clifton, G. 1998. The recent history of sedimentation on the Southern Tablelands of south-eastern Australia: sediment flux dominated by channel incision. *Geomorphology*, **24**, 291–308.

Wicherek, S. 1991. New approach to the study of soil erosion in cultivated lands. *Soil Technology*, **4**(2), 99–110.

Ziegler, A.D. and Giambelluca, T.W. 1997. Importance of rural roads as source areas for run-off in mountainous areas of northern Thailand. *Journal of Hydrology*, **196**, 204–229.

8 Pollutant Transfer by Erosion on Two Agricultural Watersheds in the Northern Parisian Basin

ALEXANDRA ANGELIAUME AND STANISLAS WICHEREK
Centre de Biogéographie-Ecologie, Paris, France

ABSTRACT

Intensive farming is more and more frequently considered as responsible for negative impacts on the environment. Ongoing experiments on two intermittently flowing elementary watersheds (20 and 180 ha) of the northern Parisian Basin investigate soil erosion and its impacts on soil and water quality. Both watersheds are equipped with automatic rainfall and water flow stage recorders and automatic water samplers. Losses of suspended solids, organic carbon, different forms of nitrogen (total N, NO_3-N, NH_4-N) and of phosphorus (adsorbed, soluble, orthophosphate), chloride and some pesticides (atrazine, lindane) are monitored. This approach, considering both quantitative and qualitative aspects, is relatively new in relation to the scale of the study, i.e. Elementary Cultivated Slope of the Basin (ECSB). Due to the rainfall intensities, spring and summer are considered as high pollution risk periods, while winter erosion appears to be negligible. Topography, pedology and agrarian structures play a major role in pollutant transfer. It appears that the smaller watershed generates run-off events that are more frequent and relatively more important than the larger one. Suspended solid concentrations that may reach as much as 26% of the flow, carry important pollutant loadings (e.g. 90% of the P losses). Pesticides associated with maize production are frequently encountered, with concentrations that may reach high levels (up to 308 μg l^{-1} of atrazine). On the other hand, pesticides used for small grain production are rarely detected (<0.05 μg l^{-1} for isoproturon). It is hoped that this work on the links between agriculture and environmental quality will lead to proposals for sustainable agricultural development.

8.1 INTRODUCTION

The temperate areas of plains and hills in the northern Parisian Basin are known for their great stability of soils and landscapes. Nevertheless, these areas are affected by important problems of erosion of cultivated soil and the transferral of pollutants. This erosion is different from that observed in Mediterranean countries (Auzet 1987) or vineyard environments (Wicherek and Boissier 1991) due to the condition of the slopes and moderate climate (Monnier *et al.* 1986; Quine and Walling 1993). Although this erosion is less severe, it creates numerous problems such as mud floods (or muddy floods) and has done so for a long time (Demangeon 1905). Mud floods observed in the northern Parisian Basin are not mass movement processes: this term is not satisfactory to describe the processes that occur in the field. These mud floods are flow-loaded sediments resulting essentially from sheet erosion. Each year, 15 communes, on

Applied Geomorphology: Theory and Practice. Edited by R.J. Allison.
© 2002 John Wiley & Sons, Ltd.

average, have been declared distressed areas because of mud floods (since 1982 when the French law about natural catastrophe gives an indemnity to repair mud flow effects). It even seems, according to certain authors (Wicherek 1990, 1993), that the frequency of these flows has been increasing for several decades because of changes to agrarian systems (Chisci and Morgan 1986). Attempts to account for this erosion by scientific means are recent, since the first publications date from the conference on the erosion of agricultural soil in temperate non-mediterranean environments, which took place in Strasbourg and Colmar in 1978 (Vogt and Vogt 1979). Today, work on this theme is of interest since the costs of production losses and especially the increased clean-up costs are considerable. In addition, the effects on the environment are just as negative, particularly on the rivers that receive these mud flows, for these carry numerous pollutants with them (Dorioz and Ferhi 1994; Robert 1996). We can therefore ask what are the environmental consequences of the transferral of pollutants. Finally, we can ask about the limits of the anti-erosive planning currently being developed in the region and elsewhere, for it raises the problem of how to adapt this planning.

8.2 METHODOLOGY

The method adopted differs from the traditional one. If the scale of the work chosen, that of the Elementary Cultivated Slope of the Basin (ECSB), is classical, the place where the measuring apparatus is situated is less so. Covering 20 and 200 hectares, the ECSB contains thalwegs with intermittent run-off: a place where flow and mud floods pass. Moreover, this original study is realized in collaboration with farmers, and it is very valuable for them to be informed about the balance of fertilizer and pollutants such as pesticides in fields.

Situated in the north of the Parisian Basin (Figure 8.1), each of the two ECSB sites is representative of the morphopedological and cultural conditions encountered (Angeliaume *et al.* 1994): the hill (the Erlon site) and the plateau (the Vierzy site). Fifty-nine per cent of the slopes are below 2% at Vierzy, and 81% of the slopes are above 2% at Erlon (Figure 8.2a). Soil surface texture is essentially loam at Vierzy (69%) and clay loam at Erlon (74%) (Figure 8.2b). The two sites have some large fields (5 to 8 ha in the hill region and 10 to 60 ha in the plateau region) (Figure 8.2c). The elementary basin of Vierzy is crossed by only one road. Vegetation cover exists exclusively in relation to large-scale cultivation (cereals, beets, sugar, potatoes, peas, corn). The soil and the subsoil are permeable, and there is no drainage network. However, under certain conditions, permeability is reduced because of soil saturation by water or crust formation. The ECSBs generate flows with variable solid suspension (SS) and other elements considered to be pollutants (nitrogen, phosphorus or phytosanitary products) (Angeliaume 1995b).

An experimental station is installed downslope of each ECSB. It permits measurements of rain intensity, discharge and the sampling of run-off (Figure 8.3). When there is run-off a V-shaped overflow shoot and three sounding-lines (one pression sounding-line and two ultrasound sounding-lines) record water level (in millimetres), which is then converted to litres per minute. Above $200 \, l \, min^{-1}$, a sampler samples 200 ml of run-off water every 3 min in spring and summer (three samples per bottle, 24 bottles: automatically for 3.5 h) and every 10 min in autumn and winter (three samples per bottle, 24 bottles: automatically for 12 h). For each sample (600 ml) solid suspension (SS), organic matter (OM), total nitrogen (N), nitrates (NO_3), ammonium (NH_4), total phosphorus (P) and chlorides are measured. During spring and summer, three litres are sampled by hand to research (by gas chromatograph) pesticides in water with SS, in filtered water and in the sediment (atrazine, lindane, isoproturon, chloridazone, etc.) (Angeliaume 1995a).

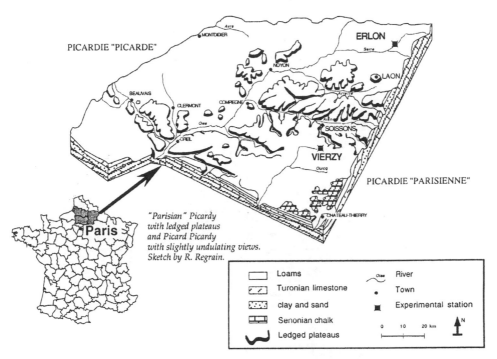

Figure 8.1 Location of the two experimental sites in the northern Parisian Basin, in the hill and plateau region

8.3 RESULTS

The frequency of flow events is variable, with a mean of three per year from 1993 to 1995 at Vierzy and 15 per year from 1993 to 1995 at Erlon. At Erlon, run-off is more frequent than at Vierzy due to morphological and pedological conditions and also agricultural practice (Tables 8.1 and 8.2; Angeliaume *et al.* 1994). In this area, accumulation of rainwater is important during winter (total rainfall of 50 mm during ten days at Vierzy; total rainfall of 20 mm during ten days at Erlon) because of soil saturation (23–27% of humidity). Rain intensity is important during spring because of crust formation (12–15 mm h^{-1} during one hour at Erlon; 50 mm h^{-1} during one hour at Vierzy). Total volume of the run-off is rarely important: only 6% of run-off at Erlon and 15% at Vierzy is more than 100 m^3 and can reach the river. Discharge can be very variable: if it is frequently above 100 l min^{-1} (77% of the run-off at Erlon and 69% at Vierzy), it can exceed 30 000 l min^{-1} during storms. During run-off, SS concentration depends on discharge and intensity of rainfall. Concerning pollutants, the risk of water pollution is linked with fixed phosphorus (>30 mg l^{-1} during storms) and pesticide concentrations. Total phosphorus is more dependent on SS than total nitrogen; nitrate and ammonium are higher during spring than during winter and summer. But orthophosphorus is constant due to exchange between soluble and bound phosphorus (Table 8.3). Concentrations of pesticides are frequently above European Union standards (0.1 μg l^{-1} in drinking water), especially for pesticides input one month before run-off (e.g. atrazine, maize herbicide input in April, 307.8 μg l^{-1} in water in May). Pesticides applied in February are rarely found (e.g. isoproturon, wheat herbicide, <0.050 μg l^{-1}). Some pollutants, such as phosphorus, are

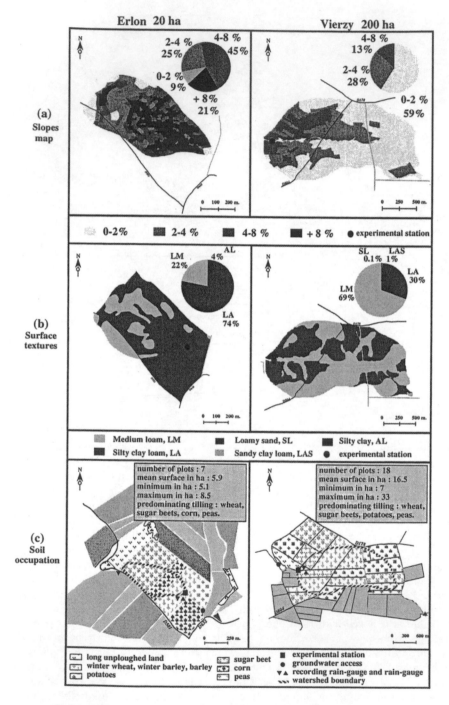

Figure 8.2   Slopes (a), surfaces textures (0–30 cm) (b) and plots and vegetation cover (c) on the ECSBs (Angeliaume 1996)

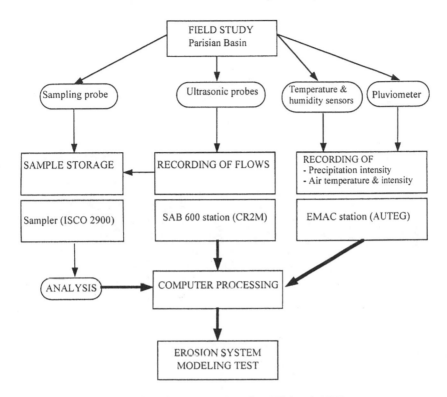

Figure 8.3 Principle of functioning of experimental station (Wicherek 1991)

closely dependent on SS concentration (e.g. fixed phosphorus 90–96%, or chloridazone, sugar beet herbicide, 88%).

8.4 DISCUSSION

8.4.1 An Attempt at Typology: Two Types of Flow

The recording of rain and stream flow on the two sites allows us to show that the signal (flow) can present two completely different faces: one is muddy and the other is essentially liquid transferral. This finding agrees with Ludwig's (1992) conclusions, which characterized two types of erosion by observing the rills and gullies formed: a winter and a spring erosion.

We can distinguish, on the one hand, the 'rise in the mud level', which is characterized by a brief flow (30 minutes to 2.5 h), a strong to very strong concentration of SS (10 to 60 g l^{-1}), a variable total rainfall (10 to 100 mm) and notable to high intensities (24 to 200 mm h^{-1} during 1 min). On the other hand, we discern the 'liquid rise' in the 'water level' that is manifested by a weak SS concentration (lower than 10 g l^{-1}), a rather long flow (several hours to one day), a variable total rainfall (several millimetres to 50 mm) but with moderate intensity (1 to 24 mm h^{-1} for 1 min).

Let us recall that two kinds of basins are being considered, one on the hill region and the other in the plateau region. The flows are much more frequent in the first – Erlon – where the

Table 8.1 Run-off, soil losses and their triggering factors (Angeliaume 1996)

Date(s) of flows	Precipitation, d day (mm)	Maximum intensity during 1 min (mm/h)	Precipitation, d-1 (mm)	Total precipitation, 10d-d (mm)	Total volume of flow (m³)	Maximum flow (l min⁻¹)	Total SS (kg)	SS maximum amount (mg l⁻¹)
Erlon 1993–1995								
10/11/93	13.5	3	0	0.5	57	1200	93	7.00
10*-11-12-13/12	15	15	1.5	23.3	407	2000	209	3.00
15-16*-17-18/12	19.2	8	2.1	57.9	142	585	90	1.4
19-20*-21/12/93	41	30	20.5	86.1	860	8500	***	***
4*-5-6/01/94	4	1	2.7	18.3	40	1250	36	0.9
10-11-12-13-14	13	3	4	19.5	408	985	469	2.6
20-21-22-23-24	5.5	15	2	15.5	430	1400	387	3.3
4/04/94	2.4	7.5	0	21.5	2.5	14	3	1.9
24/05/94	9	24	0	44	13	900	500	6.1
6*-7/8/95	75	210	0	1.5	2500	40 000	364 200	260
Erlon 1989–1992								
24/09/89	12.5	36	–	–	0.275	6	7	30
27/02/90	27.5	16.5	–	–	12.8	110	58	7
8/05/90	19	51	–	–	130	3000	9500	110
28/10/90	23	4.8	–	–	120	340	730	7
25/11/90	8.5	3.3	–	–	72	450	114	2
6/07/91	15.5	60	–	–	3	300	283	130
7/07/91	2.5	12.7	–	–	0.475	50	37	100
25/05/92	12.5	43	–	–	0.2	25	4	25
28/05/92	5.5	9.2	–	–	0.25	16	10	40
30/05/92	5	28.5	–	–	11	900	300	30
31/05/92	2	12	–	–	0.6	60	14	27
31/05/92	12.5	72	–	–	400****	5000****	40 000****	130****
2/06/92	16.5	18	–	–	160	2000	14 000	120

Date(s) of flows	Precipitation, d day (mm)	Maximum intensity during 1 min (mm/h)	Precipitation, d-1 (mm)	Total precipitation, 10d-d (mm)	Total volume of flow (m³)	Maximum flow (l min⁻¹)	Total SS (kg)	SS maximum amount (mg l⁻¹)
Vierzy 1993–1996								
14-15*/10/93	6.2	4.02	20.2	71.6	142.7	700	100	0.8
20*-21/12/93	28.7	24	20.3	89.7	1424.6	6300	1083	1.8
3/01/94	9.6	24	7.5	57.5	64.7	960	77	1.4
25/01/95	20.6	48	5.6	55.6	20.9	306	245	1.4
11/07/95	45.8	108	0	8.8	>>20 000	>>33 000**	very high**	very high**

Run-off >500 l and discharge.

* Shaded rows indicate mud floods; other rows are normal water flows. Day of maximum flow, which is also the date taken for the value of the daily rain.

** Station destroyed and village stricken by mudslices.

*** Malfunctioning of experimental station.

**** Estimation by integration of SS concentrations and recorded discharges (24 samples of SS by run-off, error <10%).

Table 8.2 Minimal rain thresholds for run-off initiation (Angeliaume 1996)

	Erlon spring/summer	Erlon winter	Vierzy spring/summer	Vierzy winter
Daily rain (mm)	16	<5	50	20
Rain accumulated during the preceding eight days (mm)	1–20	20	1–20	50
Maximum intensity during at least 10 min (mm)	60	<5	90	20

conditions for triggering the flow are much less narrow because of geomorphological and pedological conditions (Table 8.1). For the two sites, we find two types of rise in the water and SS level.

This differentiation is based essentially on the SS concentration and not on the solid flux. Indeed, a rise in the water level can present a notable SS flux (from several hundred kilograms to a tonne). An example of this occurred in December 1993 at Vierzy, where about 1100 kg (6.1 kg ha^{-1}) of earth was exported over 24 h without the concentrations having been considerable (Table 8.2).

The mud flood is not necessarily associated with rates of flow that are very high. This is the case with storms on 6 July 1991 and 24 May 1994, in which the instantaneous flow only reached 300 and 900 l min^{-1}. Intensities also play a role. Indeed, mud floods have taken place at the time of downpours of medium to very strong intensity for the region (24 to 200 mm h^{-1} for at least 1 min or 4 to 72 mm h^{-1} for at least 1 h). Likewise, observations show that factors such as soil surface state, earlier deposits of erosion in the bottoms of the thalweg or rates of covering of the soil by vegetation influence SS. This is why, in the following sections, we will characterize the transfers and define the origin of the sediments in order to understand better the mechanisms of erosion and transport.

8.4.2 Fluxes and Concentrations of the Mud Floods

The mud floods have bigger solid and soluble flux than liquid rise (except for orthophosphate). Results measured are close to field results.

For flow following moderate rain, the SS flux is modest: 0.2 to 27 kg ha^{-1} on the Erlon site and 0.4 kg ha^{-1} at Vierzy. Measurements on plots of 20 m^2 for comparable soils show higher values than on the watershed: 120 to 200 kg ha^{-1} for the months following sowing (Larue *et al.* 1995). Likewise, the estimations from the side basin with a perennial outflow of 10 km^2 in the course of a rise in the water level give a comparable or higher value: 4.2 to 88 kg ha^{-1} on the basin of Melarchez (Muxart *et al.* 1993). In comparison, losses during a storm are considerable: 20 t ha^{-1} at Erlon in August 1995 (following rain of more than 102 mm in 2 h). This figure is much higher than the highest rates of water erosion noted in the Caux area: 140 to 240 t km^{-2}, which is 1.4 to 2.4 t ha^{-1} (by an estimate of the rill and gully volume) (Ouvry 1992). It is also above the 57 t km^{-2} a^{-1} (570 kg ha^{-1} measured on the sloping basin of La Tortue, in Sarthe (Larue *et al.* 1995). The intensity of the rain constitutes the main factor explaining this strong SS load.

For a single event, the total export of phosphorus varies from 1.3 g ha^{-1} to 3 kg ha^{-1} (Table 8.3). The latter value is much higher than the annual export from the side basin (Table 8.4; Dorioz 1996). Export of phosphorus is in relation to export of SS during spring and summer (when fixed phosphorus represents 80–96% of total phosphorus). The high SS and phosphorus

Table 8.3 Balance sheet of fluxes and concentrations on two side basins (Angeliaume 1996)

Loadings

Data	Maximum flow (l min⁻¹)	Volume (m³)	SS exported (kg ha⁻¹ (kg))	Total phosphorus exported (gP ha⁻¹ (g))	Orthophosphates exported (gP ha⁻¹ (g))	Total nitrogen in filtered water exported (gN ha⁻¹ (g))	Nitrates exported (gN ha⁻¹ (g))	Ammonium exported (gN ha⁻¹ (g))
24 May 1994 – Erlon	901	12.7	28 (500)	12.8 (230)	0.46 (8.3)	9.7 (175)	7.4 (134)	2.1 (37)
25 January 1995 – Vierzy	306	20.9	1.4 (245)	1.3 (235)	0.6 (104)	6.3 (1129)	2.5 (455)	0.08 (15.6)
6–7 August 1996 – Erlon	39 042	2543.4	20 200 (364 200)	3289 (59 200)	56.7 (1200)	672.2 (12 100)	488.9 (8800)	50 (900)

Maximal concentrations

Data	Max. flow (l min⁻¹)	Max. SS (gl⁻¹)	Max. total phosphorus (mgP l⁻¹)	Max. orthophosphates (mgP l⁻¹)	Max. total nitrogen filtered water (mgN l⁻¹)	Max. nitrates (ngN l⁻¹)	Max. ammonium (mgN l⁻¹)	Max. total lindane (% soluble) (μg l⁻¹)	Max. total atrazine (% soluble) (μg l⁻¹)	Max. lindane fixed on deposited soil (μg l⁻¹)	Max. atrazine fixed on deposited soil (μg l⁻¹)
24 May 1994 – Erlon	901	60.5	22.2	0.862	20.3	13.1	3.66	1.9 (85)	307.8 (87)	307.9	690.0
25 January 1995 – Vierzy	306	1.4	1.87	0.86	1.33	1.63	0.023	–	–	–	–
6–7 August 1996 – Erlon	39 042	260	37.2	0.689	8.88	10.6	0.757	0.21 (=95)	8.9 (94)	<400	<400

Rainwater

	Total P (mgP l⁻¹)	Orthophosphates (mgP l⁻¹)	Total nitrogen filtered water (mg N l⁻¹)	Nitrates (mg N l⁻¹)	Ammonium (mg N l⁻¹)
Scale of sizes	0.02	0.003	0.7	0.01–0.3	0.006–0.02

Table 8.4 Flux and concentrations of N and P reported in the literature from agricultural basins (Dorioz 1996)

	Concentration (mg l^{-1})			Export (kg ha^{-1} a^{-1})		
	Range	Frequently cited values	Mean	Range	Frequently cited values	Mean
Total phosphorus	0.05–1.50	0.10–0.30	0.20	0.10–7.17	0.25–0.81	0.50
Soluble phosphorus	0.01–0.61	0.05–0.70	0.06	0.09–4.48	0.09–0.22	0.15
Total nitrogen	1.6–6.4	2.0–3.4	2.5	1.2–42.6	4.8–14.0	7.0

concentration during the flow of 6–7 August 1995 at Erlon is explained by the power of the flow. This sediment increased the concentration of total phosphorus. In contrast, the export of soluble phosphorus (orthophosphate), which oscillates from 0.5 to 67 g ha^{-1} (Table 8.3), is comparable to that noted by Dorioz (1996).

Finally, the total flux of nitrogen (6 to 700 g ha^{-1}) is much lower, during a specific flow, than that measured in one year. Thus, the soluble forms do not undergo the influence of the charge in SS during the flow.

The SS concentrations measured in the mud flood are extremely high (maximum 260 g l^{-1}) in comparison with those encountered in the literature. Munoz (1992) has measured the highest contents from 166 to 10.6 mg l^{-1} in a tributary basin of the Ardihres during storms (measurements between 1988 and 1989). In general, however, the concentrations measured at the outlets of the basins are lower: several milligrams to 400 mg l^{-1} during high water levels on the Orgeval (Muxart and Penven 1993) and 5 to 160 mg l^{-1} during a rise in the water level in June (Dorioz and Ferhi 1994). On fields, Larue et al. (1995) give similar values: 30 to 50 g l^{-1}, but these fall to 0.44 g l^{-1} at the outlets. Thus, during a storm, the ECSB can behave in a way resembling that of a field.

As for the SS, we can show that concentrations of N and P measured at the outlet of the basins during mud floods are very high (Tables 8.3 and 8.4): for total phosphorus the values go up to 37 mg l^{-1}. For soluble phosphorus, the concentrations measured are situated in the range, but they are, however, higher than the most frequently cited values. The total concentration of nitrogen is similar to values of other studies: they are not exceptional.

In conclusion, fluxes and concentrations are less important in winter rises than in spring or summer mud floods (except for orthophosphates).

8.4.3   Origins of the Liquid, Solid and Soluble Fluxes: Hypothesis of Transfers

Several hypotheses can be proposed for the origin of the pollutants (SS, nitrogen or phosphorus). These can derive from the slopes, from thalwegs or from existing deposits. SS and soluble pollutants do not necessarily have the same origin. The time of propagation, the duration of the flow, the conductivity and the concentration of chloride or nitrate are good indicators of the contact time between the rainwater and the soil. In particular, chloride is partly brought to the soil in mineral fertilizer and does not undergo any degradation (chloride has a conservative balance sheet and the rainwater is not highly charged). Pesticides are also good tracers for they degrade slowly. Moreover, properties of the treatment are applied by the farmer to each field (for example, only one farmer produces potatoes and uses mancozebe, a fungicide). Lindane and atrazine (respectively an insecticide and a herbicide) are associated with maize cultivation, although their usage is regulated. Isoproturon and 2.4 mcpa are cereal herbicides; metamitrone and chloridazone are sugar beet herbicides; monolinuron is a potato herbicide, etc.

The run-off derives from all fields of the side basins, even the most distant. Indeed, in the water that moves to the outlet, we can measure notable concentrations of phytosanitary products that are applied to cultivations sited furthest upstream (1 to 3 km). For example, in May 1995 at Erlon, concentrations of 0.87 μg l^{-1} of metamitrone and 6.22 μg l^{-1} of chloridazone were recorded as deriving from the plot that is most upstream (the distance travelled by the water is 900 m for a total surface of the ECSB of 18 ha). Also, in August 1995 at Erlon, concentrations of 8.9 μg l^{-1} of atrazine and 0.21 μg l^{-1} of lindane were observed. At Vierzy, in July 1995 (the distance of flow by water was 1500 m for a total surface of the ECSB of 180 ha), 0.9 μg l^{-1} of 2.4 mcpa (cereal herbicide) was measured.

The exchange between the rainwater and the soil is important since the water that crosses to the outlet is relatively rich in solid (SS) and soluble (N, P) elements. For the latter, the concentrations are clearly superior to those encountered in rainwater (Angeliaume 1995, 1996; Table 8.3). These concentrations are variable in relation to the volume of water brought into play (due to dilution and lowering of concentrations). Moreover, they are variable in relation to the time of contact between the water and the soils (there is a higher concentration with long contact). In fact, they are variable with the amount available in the soils (in the surface horizon) and with the transformations during the flow (passage from a fixed form on the sediments to a soluble form, which is the case for phosphorus, or oxidation which is the case for ammonium and nitrous nitrite).

The balance (between mineral fertilization and run-off losses) is conservative for chloride, which tends to increase during a simple flow, until the flow diminishes (Figure 8.4a), thus showing that the water which reaches the outlet has remained in contact with the soil but not for a long time. In fact, chloride concentration in water run-off is low compared to chloride concentration in rainwater (the propagation time indicating that this water derives from upstream of the basin). At the time of a complex and prolonged flow, the concentrations vary according to the nature (and therefore the origin) of the water that reaches the station (Figure 8.4b). Chloride concentrations are high when discharge is small and when contact time between the rainwater and soil is long. In contrast, they are low when rainwater arrives rapidly and in large quantities at the outlet. This suggests that chloride is diluted by less concentrated rainwater and that subsurface flow is not important but regular (small variation of the concentrations during run-off).

For nitrate, leaching occurred essentially at the time of the first outflow (strong concentrations) but the amount in the soils is probably limited, since the concentrations are clearly inferior at the time of the second flow (the time of contact with the soil also being important, for at the same time the chloride concentrations are raised (Figure 8.4b)). The chloride indicates a long contact time between the rainwater and soil but low nitrate concentration shows a using up of nitrogen. In August a large proportion of N has been concentrated by plantations.

SS can come from slopes or thalweg by concentrated erosion. At Erlon, the erosion observed mainly concerns wheel traces (which cut the thalweg perpendicularly and follow the strong slopes of the banks) and the principal thalweg (in the watershed, small scattered rills one-third or two-thirds of the way downstream; a gully can reach 6 m in width, several hundred metres in length and 15–20 cm of depth, as in August 1995). Furthermore, important deposits of sediment are noted at the intersection of the principal thalweg and the wheel traces, and at the intersection of the principal thalweg and at the limit of the plot.

At Vierzy, traces of erosion are only observed at a distance one-third of the way downstream of the principal thalweg, for 750 m, in spite of the slopes of 1 to 2% (except for the incision of some wheel traces or deposits after big storms upstream, but these are rare).

(a) Chlorides and nitrates in short runoff.

example from May 24, 1994 at Erlon.

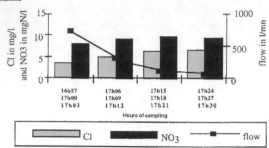

(b) Dilution of chlorides and nitrates in long runoff.

example from August 6-7, 1995 at Erlon.

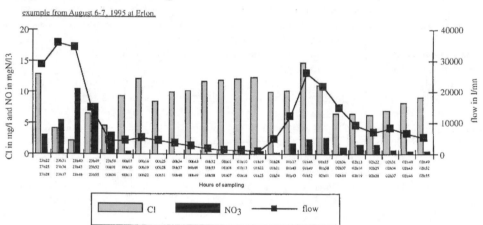

Figure 8.4 Evaluation of nitrates and chlorides in two run-off events of contrasting duration. (a) Chlorides and nitrates in short run-off. (b) Dilution of chlorides and nitrates in long run-off

If we compare the volume of the gullies and the fluxes at the station, the orders of size are comparable in winter and totally disproportionate at the time of big storms. For example, on 25 January 1995 at Vierzy, erosion was evaluated at 250 kg with sampling of flow at the station. The amount of erosion estimated by measurement in the thalweg is 250 to 500 kg (gully in the thalweg, 750 kg to 1000 kg; deposits in the thalweg around 500 kg).

In contrast, on 6 and 7 August at Erlon, the storm involved a loss from the outlet of nearly 350 t. The weight corresponding to the channels observed was much lower: 24 t (gully in thalweg, 18 000 kg; rills in wheel traces in corn field, 6000 kg; deposits in the wheel traces and at the limits of the plots, several hundred kilograms; deposits at the experimental station, 800 kg).

At the time of the winter flow, which was described as a liquid flow, the main part of the exported sediment derived from the thalwegs. In contrast, in the case of storms or stormy downpours, the origin of the SS clearly goes beyond the limits of the thalwegs.

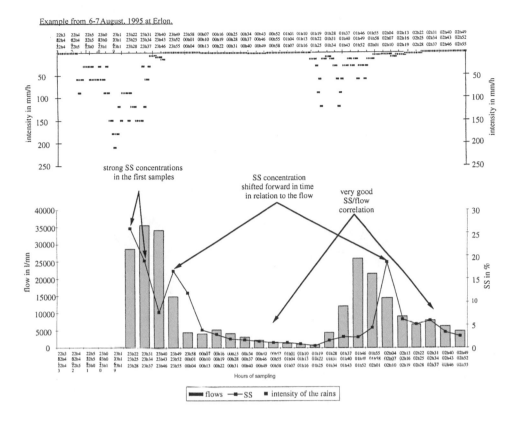

Figure 8.5   Relationship between SS, run-off and rain intensity

The correlation between flow and SS can be observed for flows above 1000 l min^{-1} (below this flow, the SS concentration remains very weak). It is a good correlation if the summer and winter flows are considered separately, probably in relation to the process of saturation of soil with rainwater during winter and crusting during spring. It is all the better since the flows are short. This correlation gives values that are much higher in SS for the same flow, as much in Erlon as in Vierzy.

At the time of the 'rising in the muddy water level', we were present, during the first period, at a redeposition of the sediment deposited in the wheel tracks and bottoms of the thalweg (very strong SS concentration in the first samples from the flow, Figure 8.5). Then the concentrations in SS follow the SS/flow correlation overall, but the correlation of SS with flows is, however, far from being excellent (Figure 8.6). This shows that the SS responds more to the intensity of the rain, and not to the flow that also responds to the intensities (Figure 8.4). Moreover, the correlation becomes excellent if we put back the SS concentrations from 18 min ($y = 0.00039x + 0.44$ and $r^2 = 0.78$). We establish, moreover, that the transfer conditions follow an evolution specific to the course of the flow (Figure 8.6). The SS concentration is weak when the flow rises. It is higher at the time of the decline: it is a negative hysteresis usually observed in rivers (Richards 1982). It is the result of run-off erosion and sediment transport. This characteristic of transportation has also been demonstrated concerning winter flows, to the extent that they are higher and long (lasting several hours).

example from August 6-7, 1995 at Erlon.
y = 0.00041x + 1.91156
r2 = 0.39

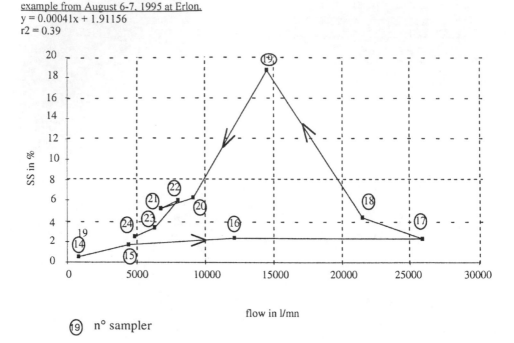

flow in l/mn

(19) n° sampler

Figure 8.6 Flows and SS: a mediocre correlation for the raised flows (Angeliaume 1996)

8.4.4 Behaviour of the Pollutants

The overall concentrations in nitrate evolve in parallel to the flows. However, nitrate can derive from soil in suspension either by sheet erosion or by incision. Nitrates can also result from throughflow (exploration of the soil reservoir by the flow on several centimetres of thickness). The soil reservoir can be exhausted in autumn at the time of successive flow as well as during the winter of 1993 (Angeliaume *et al*. 1994).

On 6–7 August 1995, the concentration of nitrate, evolving in parallel to the flows (Figure 8.4), was higher during the first flow (resumption of nitrate contained in eroded sediments and first leaching of the soil). It diminishes between the two maximum peaks, whereas the concentration for chloride remains high. This water has therefore had contact with the soil for a long time (elevated chlorides) but the stock of nitrate tends to be exhausted (a lowering of nitrate). During the second peak, the concentrations go up, but not as high as before. This water, which has still had a considerable time of contact with the soil, derives either from the same reservoir, which is lower in nitrate (interstitial water), or from a reservoir that is further away (the thickness of the explored soil is more important). However, there is probably a more complicated mixing because run-off water arrives from different plots and each plot presents a different stock of chloride or nitrate.

The ammonium, a much more unstable form, appears in high concentrations following two peaks of the flow: the first flow does not alter the mineralization or solubilization of this nitrogenized form (Figure 8.7).

example from August 6-7, 1995 in Erlon.

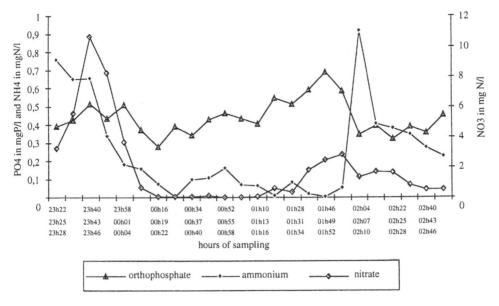

Figure 8.7 Evolution of nitrate and ammonium

Two phases or processes are apparently completed: on the one hand, the retaking of deposits containing mineralized nitrogen forms, and on the other, the exploration with taking back of interstitial water. It therefore seems that, at the time of this run-off, there are two types of water.

The total phosphorus (Figure 8.8) is constituted essentially of phosphorus fixed in sediments (Table 8.3), and is closely related to the SS concentrations. In fact, a first stock is mobilized with SS, in the first sample. At the time of this first phase, the phosphorus is diluted while the flow continues to increase. A second phase appears: the peak of specific phosphorus is produced later than that of the flow. It seems that a raised flow is necessary in order to involve transport of SS and phosphorus. At the time of the second peak of the flow, the same phenomenon is reproduced: the phosphorus concentration is shifted back in relation to the flow, but we establish this time that the concentrations are clearly superior, and a second source has probably contributed to it.

The concentration of orthophosphate has a lower order of magnitude than fixed phosphorus. Orthophosphate remains in fairly constant concentration. Although concentration increases during a second peak of the flow, its increase is explained by the higher amount of specific phosphorus, and the more it is present in a high quantity, the greater the possibility of realizing in a soluble form. There is a balance in concentrations of orthophosphate in step with those in particulate phosphorus. Transfers in soluble and fixed phosphorus seem to be coupled. The concentrations in orthophosphate, however, seems to respect a threshold, for whatever the concentration in particulate phosphorus may be, that of orthophosphates never goes beyond 0.9 μgPA.

example from August 6-7, 1995 in Erlon.

(a)

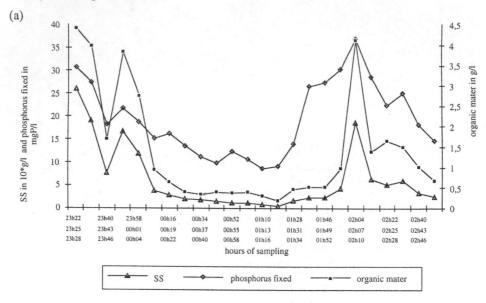

(b)

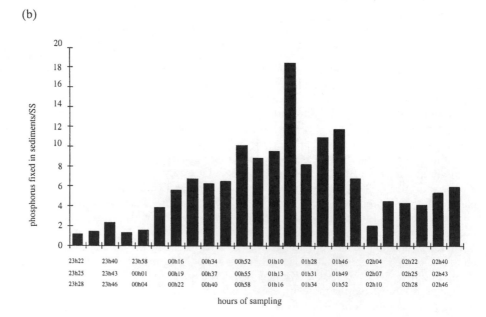

Figure 8.8   Total phosphorus, organic matter and SS concentrations

8.5   CONCLUSION: THE IMPACT OF EROSION OF CULTIVATED SOILS ON THE ENVIRONMENT

The erosion of cultivated soils constitutes an important means by which pollutants are transferred. SS, nitrogen, phosphorus and pesticides are the principal pollutants found. Two types of erosion are observed involving different transfers. The mud flood is the most harmful for the receiving environment, for it is strongly loaded with SS, as well as with nitrogen, phosphorus and certain pesticides. If nitrogen and phosphorus concentrations do not go beyond the cited thresholds, the SS and pesticide concentrations are, in contrast, exceptionally high (>260 g l^{-1} for SS, 308 μg l^{-1} in crude water and 690 μg kg^{-1} of atrazine in sediments deposited in May 1994 at Erlon).

The highest concentrations were encountered in May, at the time of short flow events, when the pollutants are not diluted. The concentrations of pesticides are much less in August (e.g. atrazine 8.9 μg l^{-1}). If we consider the rejection norms authorized for flowing water they are frequently surpassed for ammonium (3.66 mg l^{-1}) and total phosphorus (36.8 mg l^{-1}), as well as for the pesticides that constitute a source of concentrated pollution. However, the losses by run-off (in comparison to the agricultural 'input') are not enormous: for example, 0.43% of nitrogen and 3.28% of phosphorus that leave the catchment.

The rise in SS is also a rise in water level to which the entire ECSB contributes, as much for the furnishing of the SS as for nitrogen or phosphorus. The winter rise in water level involves mainly the SS from the thalweg. At the time of large flows, ones that last several hours and have a flow that exceeds 1000 l min^{-1}, the falling link conveys more SS than the rising, showing a sensitization of the soil during the flow. During the phase when run-off is generated, the water tends to be absorbed by the soil, and this infiltration limits the detachment of particles. At the time of the decline of flow, the particles of soil saturated with water are detached more easily under the pressure of the flow or of the raindrops.

Overall, these results show that for different geomorphological units, we observe flows of quite variable frequency showing similarities in the transfer of pollutants. Thus, two large types of response are characterized, one of which is muddy and the other not (or less so). The muddy outflows, the most harmful for the receiving environment, present a strong SS concentration (260 mg l^{-1} of SS), and can have either a high level of nitrogen (10.6 mg l^{-1}), pesticides, etc. (flows that, rather than being very large, are short and concentrated), or a medium level but a very large size (at the time of very big storms).

These two situations imply that greater care is needed in agricultural practice (e.g. to time fertilizer application), and to adopt anti-erosive arrangements that must respond as much to the problems of concentration as to those of volume. If we consider the volume that is subject to reaching the outlet (2500 m^3 of water and 300 t per 18 ha of earth in August 1995) we will understand the necessity for placing some sets in the area upstream of the basin by adapting their dimensions. This point is particularly important for the arrangements of types of grass-covered strips: this device, which has a 'gentle' effect, must be more widely used.

ACKNOWLEDGEMENTS

The following are thanked for their scientific and technical collaboration: MM. J.M. Dorioz and P. Blanc (INRA – Thonon-les-Bains, France), C. Bernard (MAPAQ, Québec), M.O. Boissier and M. Galochet (ENS – Saint-Cloud, France).

REFERENCES

Angeliaume A. 1995a. *Qualité des eaux de ruissellement sur petits bassins versants agricoles – rapport intermédiaire.* Centre de Biogéographie-Ecologie ENS-CNRS URA 1514 et Agence de l'Eau Seine-Normandie.

Angeliaume A. 1995b. Qualité des eaux de ruissellement sur petits bassins versants agricoles dans le nord du Bassin Parisien. *Colloque 'Qualité de l'eau',* Université de Nantes, 26–28 octobre 1995, 115–120.

Angeliaume A. 1996. *Ruissellement, érosion et qualité des eaux en terre de grande culture – Etude comparée de deux bassins versants du Laonnois et du Soissonnais.* Thèse de doctorat, Université Lille I.

Angeliaume, A., Wicherek, S. and Dacharry, M. 1994. Ruissellement, érosion et qualité des eaux en terre de grande culture. *Hydrologie continentale, ORSTOM,* **9**(2), 107–122.

Auzet, A.V. 1987. L'érosion des sols cultivés en France sous l'action du ruissellement. *Annales de Géographie,* **537,** 529–556.

Boardman, J. 1984. Soil erosion and flooding on downland areas. *Surveyer,* **64**(2), 8–11.

Chisci, G. and Morgan, R.C.P. (eds) 1986. *Soil Erosion in the European Community: Impact of Changing Agriculture.* A.A. Balkema, Rotterdam.

Demangeon, A. 1905. *La Picardie et les régions voisines Artois – Cambrésis – Beauvaisis.* Armand Colin, Paris.

Dorioz, J.M. 1996. Transfert de P dans les bassins versants. *Forum 'Sécheresse, pollution, inondation, érosion'.* Poitiers 1996.

Dorioz, J.M. and Ferhi, A. 1994. Pollution diffuse et gestion du milieu agricole: transferts comparés de phosphore et d'azote dans un petit bassin versant agricole. *Water Research,* **28**(2), 395–410.

Larue, J.P., Mahoue, J.P. and Monnier, J. 1995. Erosion des sols cultivés et dynamique fluviale dans le bassin de la Tortue (Sarthe, France). *Colloque 'Floods, slopes and rivers beds.* 22–24 March 1995, CNRS, Paris.

Monnier, G., Boiffin, J. and Papy, F. 1986. Réflexion sur l'érosion hydrique en conditions topographiques et climatiques modérées: cas des systèmes de grande culture de l'Europe de l'ouest. *Cahier Orstome, Série pédologie,* **12**(2), 123–131.

Munoz, J.F. 1992. *Méthodologie d'étude des produits phytosanitaires – Etude d'un bassin versant viticole: l'Ardières (Beaujolais).* Thèse – Lyon I, CEMAGREF.

Muxart, T. and Penven, J.M. 1993. Quantification de flux et de polluants sur le Vannetin. *Rapport GDR 1993/I 'Bassin versant ruraux',* 36–46.

Ouvry, J.F. 1992. L'évolution de la grande culture et l'érosion des terres en Pays de Caux, In *Influences des modifications des structures agraires sur l'érosion des sols,* ENS Saint-Cloud, BAGF, 2, 107–115.

Quine, T.A. and Walling, D.E. 1993. Assessing recent rates of soil loss from areas of arable cultivation in the U.K. In Wicherek, S. (ed.), *Farm Land Erosion in Temperate Plains Environment and Hills.* Elsevier, Amsterdam, 357–371.

Richards, K. 1982. *Rivers, Form and Process in Alluvial Channel.* Methuen.

Robert, M. 1996. *Le sol une interface dans l'environnement, ressource pour le développement.* INRA.

Vogt, H. and Vogt, T. (eds) 1979. *Colloque sur l'érosion des sols en milieu tempéré non-méditerranéen.* Strasbourg-Colmar, septembre 1978.

Wainwright, J. 1996. Infiltration, run-off and erosion characteristics of agricultural land in extreme storm events (SE France). *Catena,* **26,** 27–47.

Wicherek, S. 1990. Paysages agraires, couverts végétaux et processus d'érosion en milieu tempéré de plaine de l'Europe de l'Ouest. *Soil Technology,* **3,** 199–208.

Wicherek, S. 1991. New approach to the study of erosion in cultivated lands. *Catena Soil Technology,* **4**(2), 99–110.

Wicherek, S. (ed.) 1993. *Farm Land Erosion in Temperate Plains Environments and Hills.* Elsevier, Amsterdam.

Wicherek, S. and Boissier, M.O. 1991. Viticulture and soil erosion in the North of Parisian basin. Ex: the mid Aisne region. France. *Zeitschrift für Geomorphologie NF Supplementband,* **83,** 115–126.

Wicherek, S., Chene, G. and Mekharchi, M. 1993. Impact of agriculture on soil degradation: modelisation at the watershed scale for a spatial management and development. In Wicherek, S. (ed.), *Farm Land Erosion in Temperate Plains Environment and Hills*. Elsevier, Amsterdam, 137–153.

Wischmeier, W.H. 1959. A rainfall erosion index for a universal soil-loss equation. *Soil Science Society Proceedings*, **23**, 246–249.

9 Fire Intensity, Run-off and Sediment Movement in Eucalypt Forest near Sydney, Australia

D. DRAGOVICH
Division of Geography FO9, University of Sydney, Sydney, Australia
AND
R. MORRIS
Central Coast District National Parks and Wildlife Service, Gosford, NSW, Australia

ABSTRACT

Extensive bushfires occurred in eucalypt forest and woodland in eastern Australia in early 1994, during a drought period. This study investigated run-off and sediment movement on sites which experienced high, moderate and low (unburnt in 1994) fire intensities. Eight open plots and four closed plots were monitored for a six-month period. Relative to the unburnt site, run-off on the moderate and high intensity burn sites was approximately two times and four times greater respectively. Mean sediment yield from open plots on moderate and high intensity burn slopes was approximately four and 17 times, respectively, of that on the unburnt slope. Between-plot variability in sediment movement was high. Post-fire vegetation regeneration was rapid but ground cover remained low over the observation period. Fire intensity affected run-off and sediment movement by removing or reducing canopy and litter cover, and probably increasing soil hydrophobicity on intensely burnt areas. Compared with high and low fire intensities, moderately burnt slopes were associated with markedly greater amounts of bioturbation involving ants, birds and small animals. The segmented nature of the terrain contributed to similar amounts of sediment movement on upper and lower slopes.

9.1 INTRODUCTION

Bushfires are a recurring feature of the Australian environment and vegetation has largely adapted to this by developing mechanisms including rapid regeneration or fire-dependent seed dispersal (Gill 1981). Long-term human use of fire has modified Australian vegetation, with Aboriginal people deliberately setting fire to vegetation for hunting purposes and for maintaining grassland areas for game. Some fires probably also started accidentally and others resulted from lightning strikes. Following European settlement in Australia about 200 years ago, major vegetation changes resulted from tree and shrub clearance for cropping and grazing (Nadolny 1991; Kirkpatrick 1994). Forests became restricted mainly to steep country within National Parks, wilderness areas and State Forests (for timber), and as bushland around and within the major cities.

Woodlands and forests of *Eucalyptus* spp. occur near and within the Sydney metropolitan region. In the rugged Blue Mountains area to the west of the city (Figure 9.1) European

Applied Geomorphology: Theory and Practice. Edited by R.J. Allison.
© 2002 John Wiley & Sons, Ltd.

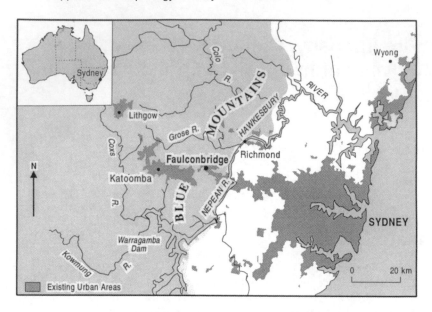

Figure 9.1   Location of Faulconbridge study site near Sydney, Australia

settlers have modified the previous (Aboriginal) fire regime which was characterized by relatively frequent, low intensity burns (Worboys and Gellie 1989). As a consequence, the area has become susceptible to recurring less frequent but higher intensity fires (Cunningham 1984). Despite the importance of the Blue Mountains for housing, national parks, and as an urban water catchment, no detailed studies of fire-related soil erosion have previously been undertaken there.

Bushfire activity was widespread in eastern Australia during 1994 when much of the region was affected by drought. In January 1994 (mid-summer), extensive fires burnt houses and other property, led to numerous household evacuations, closed some national highways for several days, and left large areas of burnt forest and heathland. More than 800 000 hectares are estimated to have been burnt out (Cheney 1995). Fires are common in the Blue Mountains, with isolated outbreaks of bushfires occurring in most years and intense fires in recent decades being reported in 1952, 1957, 1968, 1977 and 1982 (Cunningham 1984; Worboys and Gellie 1989), as well as in 1994. Worboys and Gellie (1989) documented the occurrence of more than 400 bushfires in the region during 28 years of accurate records.

Increases in soil erosion following fire have been reported previously in eastern Australia (Blong *et al.* 1982; Atkinson 1984; Prosser 1990). Rates of sediment loss in the post-fire landscape are affected by fire intensity, a variable known to influence erosion outcomes in Australia (Prosser 1990) and elsewhere (Simanton 1988; Kutiel and Inbar 1993). The aim of this study was to compare run-off and sediment movement on land which had been subjected to high, moderate and low intensity burns. Following the 1994 bushfires in the Blue Mountains, sites were selected in areas which had experienced a high intensity burn in the path of a major bushfire; a moderate intensity burn that had resulted from backburning for fire control; and an area which was not burnt in 1994 but where a low intensity fire had been deliberately lit in August 1992 as part of a bushfire hazard reduction programme.

Slope A - high intensity burn

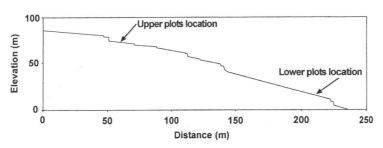

Slope B - moderate intensity burn

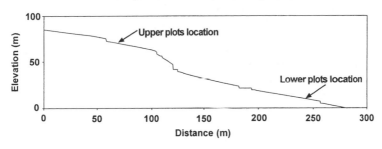

Figure 9.2 Cross-section of Slope A (high intensity burn) and Slope B (moderate intensity burn), showing positions of upper and lower slope plots. Elevation is based on local datum. Location of Slopes A and B is shown on Figure 9.3. Slope C could not be surveyed due to dense vegetation cover

9.2 STUDY AREA

Eucalypt vegetation is particularly fire-prone. In the Blue Mountains, bushfires are difficult to control in the steep and often inaccessible terrain. Both topography and soils are influenced by geology, with the mountains being composed of roughly horizontally bedded Hawkesbury sandstone (Conaghan 1980). Units of this medium- to coarse-grained quartzose sandstone contain lenticular shale lenses which occur at irregular intervals, although shale does not outcrop within the study area. The sandstone beds have varying resistance to weathering and erosion, leading to the formation of stepped hillslopes with benches and cliffs (Figure 9.2). Interfluves are flat to undulating, with steep valley-side slopes produced by stream dissection. With suitable weather and fuel conditions, fires can spread rapidly along valleys in the deeply dissected landscape. Sandstone outcrops act to divide hillslopes into several short units with abrupt within-slope changes which lead to decreased effective slope lengths and temporary storage of slope sediment (Erskine and Melville 1983). Soil development on hillslopes is therefore discontinuous and rock exposures may cover up to 25% of the surface. Where soils have formed, they are generally shallow yellow earths and earthy sands, and siliceous sands. Maximum soil depth at the study sites was 45 cm.

The study area was located near Faulconbridge (Figure 9.1), a town situated within reasonable distance of areas subjected to high, moderate and low intensity burns. The two nearest climate stations – Katoomba, approximately 23 km west, and Richmond,

Table 9.1   Temperature and rainfall data for Richmond and Katoomba

	Richmond	Katoomba
Average temperature (°C)		
July – mean minimum	2.5	2.7
– mean maximum	17.5	9.3
Jan. – mean minimum	17.1	12.9
– mean maximum	28.8	23.4
Average rainfall (mm)		
Mean monthly rainfall, driest months	48	83
(July, Aug., Sep., Oct.)		
Mean monthly rainfall, wettest months	87	156
(Dec., Jan., Feb., Mar.)		

Source: Bureau of Meteorology (1988).

approximately 19 km north-east – have widely different temperature and rainfall conditions relating principally to elevation (Table 9.1). Katoomba lies at 1030 m within the Blue Mountains and Richmond, at 20 m, is located to the east. Average annual rainfall is 1412 mm and 799 mm respectively (Bureau of Meteorology, Australia 1988). The study area at Faulconbridge lies at an elevation of between 380 and 450 m. Using unofficial figures, average annual rainfall for Faulconbridge for 57 years is 1184 mm (P. Derbyshire, personal communication 1994).

Vegetation of the study area was classified by Benson (1992) as having a Sydney sandstone ridgetop woodland structure. Main canopy species are *Eucalyptus* spp. and *Angophora costata*. The three hillslopes selected for detailed study had similar vegetation structural types before the January 1994 bushfire, as determined using aerial photography (Penrith, 1:25 000, Run 5, 4029, 14/8/91). Field survey confirmed that similar assemblages of plant species were recorded on each of the selected hillslopes, although some additional species occurred on one of the wetter lower slope positions.

The three hillslopes were all located within 2.5 km of each other, on the same geology, soil type, elevation, and vegetation structural type, with southern-facing aspects and similar gradients. Field plots were located at a considerable distance away from the road to avoid disturbance by inquisitive bushwalkers or possible vandals. The recent grading of a road surface in the area resulted in additional sediment moving down the eastern and western hillslopes; these disturbed hillslopes were avoided.

Fire intensities were described in terms of their effects on vegetation. High intensity burns left no canopy leaves or pre-fire ground cover remaining; moderately burnt areas lost most of the understorey and some sections of the canopy layer. Fire intensities were mapped as high, moderate or low (unburnt) (Figure 9.3) from an aerial photograph of the Faulconbridge area (Sydney Bushfires, January 1994, approximate scale 1:25 000, Run 20, NSW 4181 (M1977), 21/01/94). The entire study region at Faulconbridge was last burnt in 1977, with the northern section subject to intense bushfires again in 1982 and 1994 (based on historical records of bushfires and control burns kept by the Faulconbridge volunteer bushfire brigade). The fire which began on 7 January 1994 produced high intensity flames resulting in a blackened landscape, with no canopy leaves remaining. Seven buildings in the lower mountains were destroyed; in an attempt to protect surrounding properties a backburn was lit on 9 January 1994. This fire was of moderate intensity, burning mostly the understorey and some sections of the canopy layer (area with moderate intensity fire). Vegetation conditions at the commencement of this study are outlined below.

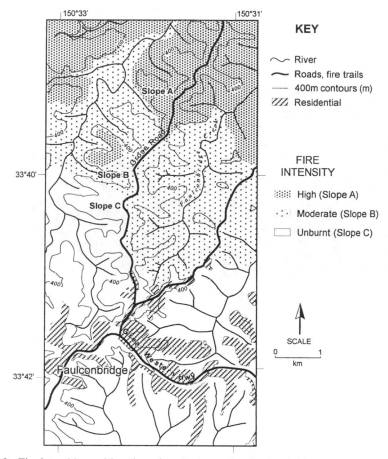

Figure 9.3 Fire intensities and location of study slopes near Faulconbridge

9.2.1 Slope A, High Intensity Burn (Bushfire on 8 January 1994)

Canopy layers had been totally removed on the upper slope section, with only a few singed leaves remaining on the lower slope. Shrubby plants were reduced to black stalks and groundcover vegetation had been lost. Charcoal fragments and fine ash were all that remained of the previously dense organic layer on the soil surface. Data collection began one month after the bushfire, by which time rainfall had already initiated the formation of litter dams and microterraces, two types of small-scale depositional features identified in local sandstone terrain (Adamson *et al.* 1983; Mitchell and Humphreys 1987). Leaves of grass and *Lomandra* spp. were approximately 1 cm high, grass trees (*Xanthorrhoea* spp.) were regrowing and epicormic buds on *Eucalyptus* spp. and *Angophora costata* were emerging (Figure 9.4).

9.2.2 Slope B, Moderate Intensity Burn (Backburn on 9 January 1994)

The upper canopy layer was mostly unaffected by the flames with green leaves still remaining. The surface organic layer consisted only of charcoal and fine ash. Shrub species had

Figure 9.4 Upper slope position on the intensely burnt Slope A. Canopy leaves were burnt but epicormic buds have commenced growth on the *Angophera costata* (centre)

blackened stalks with either singed or no leaves; groundcover species had been removed by the fire. Leaves of grass trees (*Xanthorrhoea* spp.) and epicormic buds on *Eucalyptus* spp. were already regrowing when the research began (Figure 9.5).

9.2.3   Slope C, Unburnt in 1994 (Control Burn on 15 and 16 August 1992)

Slope C was included to compare natural unburnt conditions with Slope A and Slope B. A low intensity control burn had been conducted in 1992 and, although the time interval since the last fire was not ideal, the proximity to Slopes A and B and the high frequency of fires in the Blue Mountains favoured the decision to use Slope C. In addition, vegetation condition and fuel build-up indicated the comparison would be valid (Figure 9.6).

9.3   METHODS

Soil profile descriptions were made for the upper and lower slope positions on the three study slopes; detailed descriptions were made of vegetation and its regeneration; a rain gauge was set up on Slope A; and sediment run-off plots were installed. Sediment plots were sited on upper and lower slope positions with gradients ranging from 11 to 13 degrees. The stepped topography resulting from differential weathering and erosion of horizontally bedded sandstones allowed for a cliff between 1 m and 3 m high being located directly upslope of most sediment plots.

Monitoring of run-off plot sites extended over the six-month period from 19 February to 23 August 1994. The plot design of Riley *et al.* (1981) was chosen due to its ease of

Figure 9.5 Upper slope position on the moderately burnt Slope B. Canopy leaves are present, though lower leaves have been singed. Shrub foliage was mostly burnt

Figure 9.6 Upper slope position on the unburnt Slope C. Trees, shrubs and groundcover are extensive, apart from the rock surface in the foreground

Figure 9.7   A closed plot, viewed upslope

installation, minimal expense, proven success in the local sandstone environments, and because it facilitated comparisons with the work of Blong *et al.* (1982), Atkinson (1984) and Prosser (1990), who used the same design.

Two types of plots were used. The closed plots (Figure 9.7) each measured sediment yield from an area of 8 m², and were installed on the two burnt sites (Slope A and Slope B) in both upper and lower slope positions (Figure 9.2). The open plots (Figure 9.8) measured sediment yield crossing a 2 m line oriented normal to the aspect of the hillslope. These plots were installed in upper and lower hillslope positions on both of the burnt slopes (Slope A and Slope B). On the unburnt slope (Slope C), two open plots were constructed in each of the upper and lower hillslope positions; closed plots were not feasible here due to dense vegetation and the sediment movement that would have resulted from ground surface disturbance. In all, 12 plots were installed: four closed and eight open.

Plots, based on the design of Riley *et al.* (1981), consisted of three main parts: the plot surface, the apron and the collection drum. The closed plot surface was defined by two side

Figure 9.8 An open plot, viewed upslope

walls of 4 m length, one upper wall of 2 m length and the lower apron contact. The walls were 50 mm high, made from aluminium damp course. The open plot consisted only of the apron and collection drum, without a defined upper boundary.

The plot apron was defined by a shallow triangle, with the apex facing downslope. The base length to apex height ratio was 15 to 1 for each plot. The floor of the triangle was excavated to a depth of 20 mm from the soil surface and then coated with mortar. At the apex of the apron, a 36 mm internal diameter plastic pipe was inserted to convey water and sediment to the collector drum, an approximately 20 litre plastic or metal container. A large hole was created on the drum top to allow easy measurement of run-off and collection of sediment following rainfall events. The drum was covered with plastic to prevent rainfall entry.

Sediment samples were collected from the apron by brushing into a plastic bag, and from the drum by pouring its contents through a wet sieve (63 μm). Run-off volume was measured and two 500 ml samples were collected to determine mud content (less than 63 μm fraction) and to test for water quality (pH, conductivity and turbidity). Organic percentage for the apron and drum samples was determined by loss on ignition at 550°C for six hours. The small amount of mud prevented these samples being separated into organic and inorganic fractions.

Collection of sediment and run-off from all the plots took seven hours, making collection after each rainfall event impracticable. Over the six-month study period, run-off and sediment samples were collected 11 times, following substantial rainfall. Sediment yield was combined for some smaller rainfall events occurring on consecutive days, with some collecting drums probably overflowing with heavy rainfalls. However, even when overflowing, the drums acted as sediment traps for the principally sand-sized sediment.

9.4   RESULTS

9.4.1   Soil Profile Descriptions

Soil profiles were described and sampled from the upper and lower positions on each of the three slopes; profile characteristics and surface soil properties are summarized in Table 9.2. The soil profiles on Slope A (high intensity burn) and Slope C (unburnt) were siliceous sands with uniform profile forms having little, if any, texture differentiation throughout the profiles. Slope B soils (moderate intensity burn) were yellow earths with gradational profiles which showed an increase to finer texture grades (from sand to clayey sand) with increasing depth. Soils were all thin, ranging from 12 cm to 45 cm to the sandstone parent material. Organic carbon contents ranged from 0.5% to 7.2%. On the basis of soil properties, most transported sediment is likely to be sand-sized.

9.4.2   Rainfall and Run-off

Field experiments were conducted during a relatively dry year in 1994. Total rainfall recorded at Faulconbridge for the first eight months of 1994 was 423 mm, compared with the long-term average of 914 mm for the same period. During the six-month study period, 31 days of rainfall were recorded at Faulconbridge, producing a total of 258 mm of rainfall (Table 9.3). The 1994 fires occurred following a drought and dry conditions had returned by May. Total rainfall for the four months from May to August was only 58 mm.

Rainfall amounts were measured at the field rain gauge on the same days as sediment and run-off were collected from the field plots: rainfall was left to accumulate between collection days. Rainfall measurements at the field site commenced after 2 March, with readings ranging from 12 mm to 211 mm. During the study period, substantial differences in run-off values between plot locations only occurred when rainfall registered in the field rain gauge was greater than 24 mm.

Run-off − closed plots

Run-off collected from the two closed plots on Slope A (high intensity burn) was greater than Slope B (moderate intensity burn) on all collection days. There were six occasions when one or both drums on the high intensity burn plots probably overflowed, compared with only two such occasions on the moderately burnt plots (Table 9.4). Probable overflows resulted in total run-off being underestimated for both sets of plots, but values for Slope A (high intensity burn) would be most affected as collection drums on that slope probably overflowed on four additional occasions.

Table 9.2 Profile characteristics and surface soil properties at study slopes, Blue Mountains. 'Principal profile form' is based on Northcote (1971)

Profile location	Principal profile form	Great soil group	Profile depth (cm)	Organic cover	Depth A_1 (cm)	Colour A_1	Texture A_1	pH A_1	EC A_1 ($\mu s/cm$)	Org. C A_1 (%)
Slope A upper	Uc1.21	Siliceous sands	32	Fine charcoal	20	10YR 4/1	Sand	3.35	22	2.0
Slope A lower	Uc1.21	Siliceous sands	24	Fine charcoal	1	10YR 5/1	Sand	4.07	15	3.3
Slope B upper	Gn1.14	Yellow earth	21	Scattered leaves and charcoal	2	10YR 4/2	Sand	4.48	21	3.5
Slope B lower	Gn1.14	Yellow earth	45	Scattered leaves and charcoal	2	10YR 3/1	Sand	4.35	26	7.2
Slope C upper	Uc1.21	Siliceous sands	32	2 cm thick, mostly leaves	7	10YR 5/1	Sand	3.15	36	4.4
Slope C lower	Uc1.21	Siliceous sands	12	2 cm thick, mostly leaves	9	10YR 5/1	Sand	3.88	29	5.5

Table 9.3   Rainfall at Faulconbridge, January to August 1994

	Jan.	Feb.	Mar.	Apr.	May	Jun.	Jul.	Aug.
Rainfall 1994 (mm)	28	162	141	35	8	43	3	4
Average rainfall, all years (mm)	28	95	110	91	74	69	59	53

Source: P. Derbyshire (personal communication, 1994).

Table 9.4   Mean run-off from upper and lower slope closed plots on each of Slope A (high intensity burn) and Slope B (moderate intensity burn)

Days since start of experiment	Mean run-off (l) from two plots	
	Slope A	Slope B
12	14.2*	5.5
21	16.4**	9.2
26	2.2	1.1
41	16.4*	8.3
50	1.4	0.8
57	17.0**	12.4*
90	10.0	5.5
111	17.0**	15.3**
136	8.1	2.4
146	6.9	2.2
186	16.0**	9.9
Sum***	91.6	44.9

* collecting drum at one plot probably overflowed.
** collecting drum at two plots probably overflowed.
*** sum, ignoring days 57 and 111 when probable overflows occurred on both slopes.

Run-off – open plots

Run-off collected from the two open plots on Slope A (high intensity burn), Slope B (moderate intensity burn) and Slope C (unburnt) showed that run-off was greatest on the high intensity burn plots, less on the moderately burnt slope, and least on the unburnt slope. All run-off values for the high intensity burn plots exceeded those for both moderately burnt and unburnt plots. The high intensity burn plots registered a greater number of probable overflows than the moderate intensity burn plots. There were no overflows on the unburnt area (Table 9.5).

The overflowing collection drums meant that the absolute amount of run-off from the open and closed plots could not be determined, and values listed are therefore underestimates of total run-off. After excluding recording days with probable drum overflows on the moderately burnt plots, total run-off for the high intensity fire area was 126.7 l, for the moderate intensity 44.1 l, and for the unburnt area, 25.4 l. Relative run-off figures thus followed the expected pattern.

Run-off – open and closed plots

Run-off collected from open plot designs would be expected to be greater than from closed plots. Results from different plot designs could be compared on the high intensity and

Table 9.5 Mean run-off from two open plots on each of the high intensity burn, moderate intensity burn, and unburnt slopes

Days since start	Mean run-off (l)		
	High intensity	Moderate intensity	Unburnt
12	35.5**	7.7	6.8
21	33.8**	24.5*	9.8
26	4.3	1.8	2.3
41	25.0*	11.0	7.3
50	1.9	0.4	0.6
57	33.5**	25.5*	9.0
90	20.5	9.8	3.8
111	33.6**	33.1*	17.3
136	19.0	7.3	2.3
146	20.3*	6.1	2.3
186	33.0**	26.5*	6.5
Sum***	126.7	44.1	25.4

* Collecting drum at one plot probably overflowed. Uncertain whether one plot on the high intensity burn slope had overflowed on collection day 146.
** Collecting drum at two plots probably overflowed.
*** Sum, ignoring days 21, 57, 111 and 186 when probable overflows occurred on Slope B (moderate intensity burn).

Table 9.6 Mean run-off from open and closed plots on high intensity burn, moderate intensity burn, and unburnt slopes

Plots	Mean run-off (l)		
	High intensity	Moderate intensity	Unburnt
A. All recordings			
Two open	130.2 (12)**	76.9 (4)	34.0 (0)
Two closed	122.1 (10)	72.3 (3)	n.a.
All plots	126.2 (22)	74.5 (7)	34.0 (0)
B. Excluding probable overflows on any plot * ($n=5$ collections)			
Two open	33.0	12.7	5.7
Two closed	28.5	11.8	n.a.
All plots	30.8	12.2	5.7

n.a. = not applicable. Closed plots were not installed on the unburnt slope.
* Values for all plots were excluded if any plot recorded a probable overflow; but collection day 146 was included.
** Numbers in brackets refer to number of probable overflows (plots × collection days).

moderate intensity burn slopes. In both cases, run-off from the open plots was at least 6% higher than that for closed plots (Table 9.6).

Run-off – upper and lower slope plots

Comparisons of run-off between plots in upper and lower slope positions were limited by insufficient data. Differences in run-off in relation to fire intensities were apparent (Table 9.6). However, when mean run-off for all plots was combined, the ratio of high to moderate to low intensity burn was 3.7:2.2:1. By including five collection days (when only one doubtful overflow occurred on day 146), the ratio became 5.4:2.2:1. Differences in run-off between moderate intensity burn and unburnt plots thus remained the same regardless of whether probable overflows were included or not. The consistent ranking of run-off amounts is more important than the exact relationship between absolute run-off values, as mean values for each group of plots included substantial differences in between-plot run-off collected on any given day.

9.4.3 Sediment Yield

Sediment transported to the apron and the drum of individual plots is here referred to as sediment yield. Most of the surface material collected in the drums was sand-sized sediment, having a specific gravity allowing for rapid settling and minimal loss from possible overflows.

Sediment yield – closed plots

Two closed plots on the high intensity burn area yielded 738 g and 769 g of sediment over the study period. Sediment collected on the moderate intensity burn area totalled 574 g from one plot and 1190 g from the other. A large number of ant mounds began appearing on the moderate intensity burn slope as the post-fire period lengthened, including within one of the closed plots. This resulted in loose surface material being available for removal during run-off events, increasing sediment yield.

Sediment yield – open plots

Sediment yields were greatest from plots on the high intensity burn area, where the total mean value for the two plots was 1700 g. This compared with mean totals of 449 g for plots on the moderate intensity burn slope and 83 g for the unburnt slope (Table 9.7). Cumulative

Table 9.7 Mean sediment yield from two open plots on each of the high intensity burn, moderate intensity burn, and unburnt slopes

Days since start of experiment	Mean yield* (g)		
	High intensity	Moderate intensity	Unburnt
12	190	18	4
21	584	141	12
26	16	10	1
41	44	16	7
50	6	16	3
57	122	36	8
90	40	78	20
111	164	59	13
136	21	21	8
146	22	10	1
186	495	46	8
Sum	1700	449	83

* 'Sediment yield' included mineral and organic material collected from the drum and brushed from the concrete apron. It excluded sediment delivered to the apron by obvious bioturbation (animal scratchings, edges of ant mounds) and organic fragments retained by a 4 mm mesh sieve.

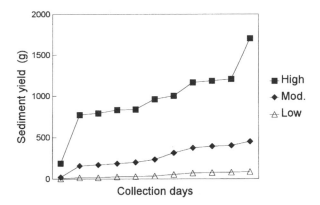

Figure 9.9 Cumulative sediment yield from open plots on slopes with high, moderate and low intensity burns. The 11 collections between days 12 and 186 of the study have been plotted at equal intervals for clarity

Table 9.8 Sediment yields from run-off values of 15 1 or more on high intensity burn, moderate intensity burn, and unburnt slopes (n = 22 collections for each slope)

Fire intensity	Sediment yield (g)				
	n^*	Mean	Max.	Min.	Range
High	13	250	860	22	838
Moderate	5	81	212	31	181
Unburnt	nil	n.a.	n.a.	n.a.	n.a.

n.a. = not applicable
* Number of collections with run-off of 15 1 or more.

sediment yield showed a continuing higher loss on high intensity burn slopes for the duration of the experiment (Figure 9.9).

Sediment yield varied between plots on slopes with the same fire intensities, between plots on slopes with different fire intensities, and between events with similar amounts of run-off. The high intensity burn plots registered more collections with at least 15 l of run-off than did other plots, and higher mean sediment values. Variability of sediment yield was also much greater on the high intensity burn plots (Table 9.8).

9.5 DISCUSSION

Run-off and sediment yield data indicated that fire intensity had a major effect on post-fire erosion in the sandstone terrain studied. Even with run-off data being underestimated by possible overflows of collecting drums, plots on slopes having been burnt by a high intensity fire recorded considerably greater amounts of run-off and sediment yield than moderately burnt or unburnt slopes. High intensity fires are likely to generate post-fire water repellence in soils with pre-fire surface organic litter, leading to reduced infiltration, increased run-off (Burch *et al.* 1989; Mitchell and Humphreys 1987) and greater potential for particle entrainment. Raindrop impact on hydrophobic soils leads to increased particle detachment which far exceeds the rate of net soil loss, indicating that rainsplash is contributing to downslope sediment movement independent of augmented run-off (Terry and Shakesby 1993). Infiltration on unburnt slopes would have been encouraged by the absence of fire-induced water repellence, a thicker organic litter layer, and a denser cover of understorey vegetation. Severe post-fire water repellence was observed in sandy soils to the south of Sydney (Zierholz *et al.* 1995), although it was noted that the extent of hydrophobicity in soils unaffected by fire could not be determined due to widespread bushfires.

Fire-induced erosional effects in other environments may differ markedly from the patterns reported for the Blue Mountains, where unburnt slopes had low run-off and sediment yields. In contrast, Kutiel (1994) reported that run-off on burnt sites may be less than for unburnt areas. Kutiel's study sites were located in Mediterranean shrublands and pine-oak forests which had undergone moderate intensity fires, and where average annual rainfall was 690 mm. Reduced run-off rates on burnt areas were explained by increased infiltration on the burnt site resulting from increased surface roughness and storage in surface depressions. In semi-arid Nevada, USA, Germanoski and Miller (1995) found that run-off and sediment loss were little altered following a fire. Pre-fire vegetation density and ground cover were low, with average annual precipitation of 327 mm. Although run-off was little affected by the fire, channel downcutting in the post-fire period was substantial, probably due to loss of grass cover on valley fills and higher than average precipitation in the two years following the fire.

In temperate climates with moderate rainfall and eucalypt forest vegetation, loss of vegetation cover and surface litter during fires reduces soil protection from high intensity rainfalls. Sandy soils are especially vulnerable. In the Sydney region, Atkinson (1984) reported a soil loss equivalent to 30 to 48 t ha^{-1} in areas of moderate to extreme fire intensity. This loss mainly occurred when 300 mm of rain fell in three days; another 85 mm fell over the observation period of ten weeks. Blong *et al.* (1982) recorded much lower losses of 2.5 to 8.0 t ha^{-1} on a moderate intensity burn area over a 12-month period, during which total rainfall was 635 mm. Differences in recorded sediment loss in the two studies can be attributed to variables including fire intensity, post-fire rainfall intensity and amount, soil texture and depth, and duration of the monitoring period.

At the scale of observations in the Blue Mountains study, sediment values represented the amount of sediment delivered to a particular point in the landscape (the apron and associated collecting drum). Between-collection variability in sediment yield for similar run-off ranges was especially high for the intensely burnt slope, suggesting that fluctuating sediment availability affected delivery to the collection point. Field observations of sediment movement within plots and on adjacent slopes provided some evidence for this, with litter dams and microterraces (Mitchell and Humphreys 1987) contributing to uneven sediment movement. Microtopographical features act as temporary storage zones for sediment in the post-fire landscape. At this small scale, organic and mineral material being transported downslope becomes trapped by logs, burnt branches and other minor surface irregularities. Litter dams are composed of mostly blackened leaf fragments and charcoal which initially accumulate upslope of vegetation remnants or other small obstacles after the fire. The litter dams in turn contribute to surface irregularities which impede the transport of quartz grains, forming microterraces. Further downslope movement and the eventual destruction of litter dams and microterraces would produce irregular patterns of sediment movement past any point on the slope. During the study, litter dams had formed, been breached and then reformed over a sequence of run-off events.

Mitchell and Humphreys (1987) described litter dams in bushfire-affected sandstone landscapes of the Sydney Basin, noting variable persistence. Litter dams formed in the first run-off event following a fire. Some dams were destroyed within 3.8 months, with dams higher than 24 mm being visible 2.2 years after the fire. At a separate site which had been burnt four years previously, no litter dams were present. Litter had broken down and bioturbation had mixed the transported quartz grains with underlying material which contained some finer sediment.

Litter dams and microterraces were common in the study area on the intensely burnt slope but less prolific on the moderately burnt slope. On these less severely burnt slopes leaf fall may have covered existing formations, and animal scratchings, including those of bandicoots and lyre birds, probably destroyed others. Scratchings mainly occurred at mid- to lower-slope positions on areas with steep gradients. Litter dam formations on the unburnt slope were covered by organic matter and could not be observed without removing this material.

Between-collection variations in sediment yield on the moderately burnt slope were apparently not dominated by litter dam and microterrace processes. However, both sediment yield and variation were influenced indirectly by bioturbation, which contributed a substantial supply of readily detachable sediment. In one of the closed plots, an increasing number of ant mounds appeared over the period of the experiment. Sediment yield from the closed plots also increased over time, an unexpected pattern if sediment yield were linked only to surface vegetation and litter cover. Although bioturbation sediment impinging on the apron was removed separately, and was the only material which could be measured as resulting directly from animal activity, disturbance of surface sediment led to this plot yielding more than twice the quantity of sediment than the other closed plot on the same slope.

The fortuitous location of plots in relation to post-fire re-establishment of small-animal populations, especially ants, appears to have had an important influence on sediment yield results on the moderate intensity burn plots. Mounding activity by ants was much lower on the high intensity burn areas where few individuals would have survived and the lack of seeds and surface vegetation following the fire would probably have delayed recolonization. Surface scratchings, probably produced by lyre birds and bandicoots, were less common than ant mounding. The lesser abundance of litter dams and microterraces on the moderately burnt slope makes it likely that variation in sediment delivery was associated with sediment provision by bioturbation. Spatially irregular recolonization of plants and animals in the post-fire landscape led to high sediment yields in one of the moderately burnt closed plots.

Small-scale differences in fire intensity may also have contributed to spatial variation in run-off and sediment yields. Using $1 m^2$ experimental plots, Kutiel et al. (1995) noted that differences may be produced from varying fire temperatures leading to a mosaic of patches with differing infiltration and run-off characteristics. In the Blue Mountains area which had been burnt by a moderate intensity fire, between-plot variability in run-off and sediment yield was high.

The influence of hillslope morphology on sediment transfer was examined by comparing upper and lower hillslope sediment yields: sediment measurements were similar. The stepped sandstone topography consisted of structural benches and cliffs resulting in temporary storage zones for run-off and sediment, with sediment leaving the catchment in any given run-off event likely to be derived from lower slope storage positions and not moved directly from upper slopes. If upper- and lower-slope sediment yield measured in this short-term study is representative of longer-term patterns, evolution of individual slope segments (benches) probably occurs at similar rates. Rates of cliff retreat would influence sediment production for each slope segment. Although sediment transfer was similar within the upper and lower slope segments in this study, between-slope variation occurred in response to different fire intensities and bioturbation activity. Increased fire frequencies may initiate a cycle of sediment entrainment causing further downslope movement of sediment and eventually more material to leave the catchment. However, fluctuating rates of sediment transfer in response to fire frequency and intensity would not necessarily lead to differential fire-induced changes in general slope morphology, if bedrock controls on slope compartments were a persistent feature. Given the between-plot variability recorded in run-off and sediment yields, predicting patterns of slope change in response to fire is difficult. It is unclear whether the short-term aggregated plot results produced a fortuitous pattern or whether similar results would be obtained from longer-term monitoring.

When compared with unburnt areas, sediment loss on the high and moderate intensity burn areas remained at substantially elevated values even at the conclusion of the study. The duration of the post-fire recovery period is not known but has been demonstrated here to exceed six months. On all of the burnt slopes, vegetation regeneration had occurred: tree regrowth was obvious and groundcover plants were established. Nevertheless rates of run-off and sediment loss remained at least twice those of the unburnt slopes. This is consistent with results for a dry sclerophyll forest area outside the Sydney region, where no decline in suspended and solute load of streams was recorded in the six months following fire (Burgess et al. 1981). In other studies, the return to pre-fire conditions has been reported to vary between two years (Belillas and Roda 1993) and three to four years (May 1990), although run-off and sediment yield in Mediterranean shrublands and pine-oak forests may decline sharply to small amounts in the year following fire (Kutiel 1994).

9.6 CONCLUSION AND IMPLICATIONS

Run-off and sediment yield increased markedly following fire in sandstone landscapes of the Sydney region. Fire-related responses are most pronounced in areas of high intensity burns, although post-fire bioturbation effects are greatest for moderate intensity burns. There are several implications of these fire-induced effects for landscape change:

(1) An erosion regime exists which is characterized by intermittent periods of accelerated sediment movement following intense and moderate burns.
(2) Differences in fire intensity and associated sediment movement are likely to lead to extreme spatial variability in rates of sediment transfer during erosional episodes.
(3) Because of the generally segmented nature of local sandstone slopes, upper and lower slopes appear to be similarly affected by fire-accelerated sediment movement when fire intensity over the entire slope is relatively uniform.
(4) Bioturbation contributes to spatial and temporal variability in sediment transfer, with most activity occurring in moderately burnt areas. However, bioturbation is not uniformly distributed even on these areas.
(5) Accelerated erosion episodes are of variable duration, with post-fire recovery not being attained after six months and in some cases of intense burns not within several years. A return to pre-fire rates may not be achieved over a long period if high and moderate intensity fires recur at frequent intervals.
(6) Accelerated rates of sediment movement occur following fire, with the amount of sediment moved being dependent on factors including probable water repellence, bioturbation, and the amount and intensity of post-fire rainfall.

ACKNOWLEDGEMENTS

The authors thank Greg Chapman, John Corbett, Chris Conoly, Phil Derbyshire and personnel from the Blue Mountains National Park for assistance with this project.

REFERENCES

Adamson, D., Selkirk, P.M. and Mitchell, P. 1983. The role of fire and lyrebirds in the sandstone landscape of the Sydney Basin. In Young, R.W. and Nanson, G.C. (eds), *Aspects of Australian Sandstone Landscapes*. Australian and New Zealand Geomorphology Group, Wollongong, 81–93.

Atkinson, G. 1984. Erosion damage following bushfires. *Journal of the Soil Conservation Service New South Wales*, **40**, 4–9.

Belillas, C.M. and Roda, F. 1993. The effects of fire on water quality, dissolved nutrient losses and the export of particulate matter from dry heathland catchments. *Journal of Hydrology*, **150**, 1–17.

Benson, D. 1992. Natural vegetation of the Penrith 1:100 000 map sheet. *Cunninghamia*, **2**, 541–596.

Blong, R.J., Riley, S.J. and Crozier, P.J. 1982. Sediment yield from run-off plots following bushfire near Narrabeen Lagoon, NSW. *Search*, **13**, 36–38.

Burch, G.I., Moore, I.D. and Burns, J. 1989. Soil hydrophobic effects on infiltration and catchment run-off. *Hydrological Processes*, **3**, 211–222.

Bureau of Meteorology, Australia. 1988. *Climatic Averages Australia*. Australian Government Publishing Service, Canberra.

Burgess, J.S., Rieger, W.A. and Olive, L.J. 1981. Sediment yield change following logging and fire effects in dry sclerophyll forest in southern New South Wales. *International Association of Science Hydrological Publication*, **132**, 375–385.

Cheney, N.P. 1995. Bushfires – an integral part of Australia's environment. *Year Book Australia No. 77*. Australian Bureau of Statistics, Canberra, 515–521.

Conaghan, P.J. 1980. The Hawkesbury sandstone: gross characteristics and depositional environment. In Herbert, C. and Helby, R. (eds), *A Guide to the Sydney Basin*. D. West, Government Printer, Sydney, 188–253.

Cunningham, C.J. 1984. Recurring natural fire hazard: a case study of the Blue Mountains, New South Wales, Australia. *Applied Geography*, **4**, 5–27.

Erskine, W. and Melville, M.D. 1983. Sedimentary properties and processes in a sandstone valley: Fernances Creek, Hunter Valley, New South Wales. In Young, R.W. and Nanson, G.C. (eds), *Aspects of Australian Sandstone Landscapes*. Australian and New Zealand Geomorphology Group, Wollongong, 94–105.

Germanoski, D. and Miller, J.R. 1995. Geomorphic response to wildfire in an arid watershed, Crow Canyon, Nevada. *Physical Geography*, **16**, 243–256.

Gill, A.M. 1981. Adaptive responses of Australian vascular plant species to fire. In Gill, A.M., Groves, R.H. and Noble, I.R. (eds), *Fire and the Australian Biota*. Australian Academy of Science, Canberra, 243–272.

Kirkpatrick, J. 1994. *A Continent Transformed: Human Impact on the Natural Vegetation of Australia*. Oxford University Press, Melbourne.

Kutiel, P. 1994. Fire and ecosystem heterogeneity: a Mediterranean case study. *Earth Surface Processes and Landforms*, **19**, 187–194.

Kutiel, P. and Inbar, M. 1993. Fire impacts on soil nutrients and soil erosion in a Mediterranean pine forest plantation. *Catena*, **20**, 129–139.

Kutiel, P., Lavee, H., Segev, M. and Benyamini, Y. 1995. The effect of fire-induced heterogeneity on rainfall-run-off-erosion relationships in an eastern Mediterranean ecosystem, Israel. *Catena*, **25**, 77–87.

May, T. 1990. Vegetation development and surface run-off after fire in a catchment of southern Spain. In Goldhammer, J.G. and Jenkins, M.J. (eds), *Fire in Ecosystem Dynamics*. SPB Academic Publications, The Hague, 117–126.

Mitchell, P.B. and Humphreys, G.S. 1987. Litter dams and microterraces formed on hillslopes subject to rainwash in the Sydney Basin, Australia. *Geoderma*, **39**, 331–357.

Nadolny, C. 1991. Tree clearing in Australia. *Search*, **22**, 43–46.

Northcote, K.H. 1971. *A Factual Key for the Recognition of Australian Soils* (3rd edition). Rellim Technical Publications, Glenside.

Prosser, I. 1990. Fire, humans and denudation at Wangrah Creek, Southern Tablelands, NSW. *Australian Geographical Studies*, **28**, 77–95.

Riley, S.J., Crozier, P. and Blong, R.J. 1981. An inexpensive and easily installed run-off plot. *Journal of the Soil Conservation Service of New South Wales*, **37**, 144–148.

Simanton, J.R. 1988. Run-off and sediment from a burned sagebrush community, USDA, Tucson, Arizona. In Wingate, G.D. and Weltz, M.A. (eds), *USDA Symposium on Effects of Fire Management of Southwestern Natural Resources*. Tucson, Arizona, 180–186.

Terry, J.P. and Shakesby, R.A. 1993. Soil hydrophobicity effects on rainsplash: simulated rainfall and photographic evidence. *Earth Surface Processes and Landforms*, **18**, 519–525.

Worboys, G.L. and Gellie, N.J.H. 1989. Setting conservation and protection objectives – conflict for management in the Blue Mountains (1983–1986), a case study. In Burrows, N., McCaw, L. and Friend, G. (eds), *Fire Management on Nature Conservation Lands*. Western Australian Department of Conservation and Land Mangement, Occasional Paper 1/89, 15–26.

Zierholz, C., Hairsine, P. and Booker, F. 1995. Run-off and soil erosion in bushland following the Sydney bushfires. *Australian Journal of Soil and Water Conservation*, **8**, 28–37.

10 Sediment-Associated Chernobyl ^{137}Cs Redistribution in the Small Basins of Central Russia

V.N. GOLOSOV AND N.N. IVANOVA
Department of Geography, Moscow State University, Vorob'evy Gory, Russia

ABSTRACT

The first results are presented of Chernobyl ^{137}Cs redistribution in small basins typical of central Russia with high levels of radionuclide. Some features of redeposition of sediment and sediment-associated ^{137}Cs are demonstrated. Approaches to the assessment of different elements of sediment budgets are discussed. Difficulties in determining the sedimentation rate of uncultivated slope are analysed. A combination of different methods is used for calculation of erosion and sedimentation rates. Erosion rates were determined using empirical models and the soil profiles method. The ^{137}Cs method was used for definition of sedimentation rates in different parts of the basin. A preliminary calculation of sediment budget demonstrates that about 90% of eroded sediment-associated Chernobyl ^{137}Cs is redeposited within the basin system. This behaviour has not changed during over the last 40–50 years. The problems and approaches for study of Chernobyl ^{137}Cs redistribution are further discussed.

10.1 INTRODUCTION

Vast areas of the Russian Plain were polluted by ^{137}Cs following the Chernobyl Nuclear Power Station accident in 1986. According to the map of ^{137}Cs contamination (Izrael *et al.* 1994) the highest levels of contamination were observed locally around the power station and in few areas within Tula, Kaluga, Orel, Bryansk and Ryazan regions. All these regions are areas with intensive agricultural activity. Because ^{137}Cs deposition to the biosphere was mainly associated with precipitation, even over relatively small areas, the total atmospheric fallout pattern was not uniform. However, detailed airborne gamma-spectrometer surveys were used to characterize the spatial variability. This was possible because ^{137}Cs is generally sorbed to soil particles within the top 2–5 cm, although there are exceptions (Timofeev-Resovskii *et al.* 1966). The depth distribution of ^{137}Cs in woodland or forest soils is very similar to that in uncultivated soils. In cultivated soils ^{137}Cs is mixed to the plough depth, so the surface concentration of ^{137}Cs in cultivated soils is usually significantly lower than that in uncultivated soils.

It is generally accepted that most radiocaesium redistribution takes place in association with the movement of soil and sediment particles (Walling *et al.* 1989; Walling and Quine 1993). Erosion processes are very active in central Russia. Two types of erosion are observed here: erosion related to snowmelt during the spring, and rain erosion from May to October. According to experimental data the annual erosion rates during the snowmelt period are

Applied Geomorphology: Theory and Practice. Edited by R.J. Allison.
© 2002 John Wiley & Sons, Ltd.

Table 10.1 Comparison of calculated soil erosion rates with catchment sediment fields obtained from pond sedimentation surveys

Pond number	Catchment area (km^2)	Catchment characteristics Slope length (m)	Mean gradient (%)	Sediment yields (t ha^{-1} a^{-1}) Field survey	Calculated	Relative accuracy (%)
1	0.27	500–600	3.4	4.9	5.8	+19
2	0.068	200–300	5.8	22.3	25	+11
3	0.148	400–500	3	5.7	6.6	+16
4	0.063	400–550	6–7	3.2	6.1	+91
5	0.055	300–400	4–5	2	4.9	+145
6	0.1	350–450	6–7	6.8	9.7	+43
7	0.085	700–750	8–9	10.6	11.8	+11
8	0.208	200–700	5–6	4.6	6.3	+37
9	0.065	500–600	6–7	4.3	8.7	+102
10	0.12	300–600	6–7	4.2	8.8	+109

2–5 t ha^{-1} in Tula region (Braude 1976) and 2–3 t ha^{-1} in southern central Russia around Kursk (Grin 1970). No direct erosion measurements exist. Instead, sediment yields were estimated from pond sedimentation surveys indicating a range of 3–22 t ha^{-1} if it is assumed that the snowmelt contribution is 2–3 t ha^{-1} (Table 10.1). This work was done in a different part of the Russian Plain. The model of the State Hydrological Institute was used for calculation of soil erosion rates for the snowmelt period. This empirical model includes water storage in snow, slope gradient, type of slope run-off and soil grain size. The intensity of rain erosion was calculated using the universal soil loss equation (USLE) modified by Larionov (1993). The principal addition to the USLE is a new approach for calculation of the erosion potential of the relief (LS). Larionov suggested coefficients for different slope configuration and recalculated the influence of slope length and gradient.

The main goal of this chapter is to elaborate the approaches for study of Chernobyl ^{137}Cs redistribution for small basins of Central Russia. The Plava River basin was chosen as a representative system heavily impacted by fallout.

10.2 THE STUDY AREA

The Plava River basin is located in Tula region 250 km to the south of Moscow (Figure 10.1). The oldest sedimentary rocks are Cretaceous limestone. Podzol Chernozem soils are widespread (about 50%) and are more clayey than standard chernozem soils which occupy about 30% of the area. Both these soils develop on carbonate loess loams. Grey forest soils are located on the northern slopes of interfluves (13%).

Mean annual precipitation for Plavsk is 550 mm. About half of the precipitation falls during the winter period (November–March) as snow. Mean annual effective depth in the snow before melting is 60 mm. The erosion index of rains, which was calculated according to the USLE, is 8.9 (the highest for all Sredne-Russkaya Vozvyshennost).

The intensive cultivation in the southern half of Tula region started towards the end of the 17th century. A slow rise in the arable land area in the second half of the 19th century happened after cultivation of steep slopes of dry and river valleys. As a result a large amount of sediment entered the river bottoms initiating widespread river aggradation. This led to

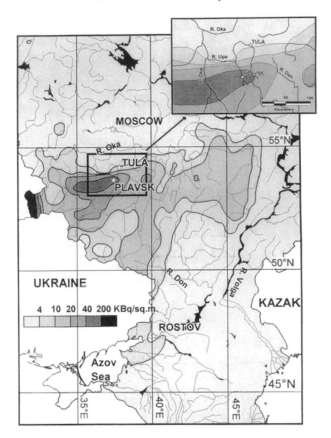

Figure 10.1   Location of the Plava River basin on Russian Plain with levels of ^{137}Cs contamination

between 30 and 40% of tributaries becoming dry valleys (or balka) during the period 1850–1900 (Figure 10.2). Consequently balka bottoms prevent sediment from the slope entering the river channels. The soil erosion rates were calculated using the modified version of USLE (Larionov 1993) and the model of the State Hydrological Institute for each small river basin in Plava River basin for recent crop rotation (Table 10.2).

Most of the Plava River basin was polluted after the Chernobyl accident with an intensity of more than 40 kBq m^{-2} with a maximum 200 kBq m^{-2} in the Lokna River basin (Figure 10.3).

Two small balka basins were chosen within the Lokna basins for field investigation of ^{137}Cs redistribution: Chasovenkov Verh balka basin and the Lapki river basin. The Chasovenkov Verh balka basin became a dry creek at the end of the 19th century. The basin area is 42.1 km^2 and comprises a series of dry river beds and stabilized gullies. The main valley flow and most of the former tributaries are relatively flat and used for pasture. It is mainly the valley slope that controls sedimentation rates within valley bottoms. Because of the very complex structure of the basin it is possible to find different types of cultivated slopes and their relationship with sedimentation zones (lower edges of cultivated slopes, uncultivated banks of valleys and valleys bottoms). Cultivated slopes can be divided into different groups: straight, concave, convex and complex (with hollow in the centre part of the slope basin). The distance

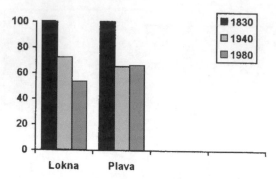

Figure 10.2   Percentage reduction of channel length since 1830 for the Plava River and the Lokna River basins

Table 10.2   Mean annual soil erosion rates for arable lands of tributaries of the Plava River

	Tributaries of the Plava River						Entire Plava basin
	Holohol'nya	Malyn'	**Lokna**	Sorochka	Plavica	Upper Plava	
Basin area (km²)	405	143	**182**	117	217	294	**1870**
Forest area (%)	3.4	6.2	**3.1**	1.8	11.5	4.0	**4.6**
Mean soil erosion rate (t ha⁻¹ a⁻¹)	7.2	7.3	**6.9**	4.0	6.4	2.4	**6.6**

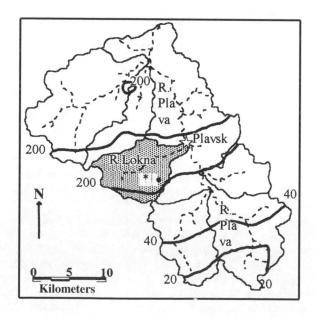

Figure 10.3   Location of the Lokna River basin in the Plava River basin and ^{137}Cs contamination (in KBq m^{-2}) of the Plava River basin. Also shown are locations of Chasovenkov Verh balka basin (•) and Lapki balka basin (*)

between the lower edge of cultivated slope and the valley bottom together with cultivated slope configuration determine the part of sediment which is redeposited on the way from cultivated slope to valley bottom. The Lapki balka basin is 2.18 km^2 and has a simpler structure with a short main valley and four tributaries, three of them located in the upper part of the basin and one located on the left bank of the main valley near the basin mouth.

10.3 THEORETICAL CONSIDERATIONS OF DEVELOPING RADIOCAESIUM BUDGET

It is necessary to consider a few points before studying Chernobyl ^{137}Cs redistribution in areas with a high level of contamination:

(a) local scale variability of ^{137}Cs fallout inputs;
(b) establishment of reference inventory;
(c) time-scale effect.

Specific studies of the variability for ^{137}Cs that originated as fallout from the atmospheric testing of nuclear weapons in non-eroding sites report that the coefficient of variation of spatial variability changed from 15% (Higgitt 1990) to 54% (Fredericks et al. 1988). The spatial variability of Chernobyl fallout in areas of high contamination requires sampling on micro- and macroscales. Reasons for local variability of ^{137}Cs inventory on the microscale include non-uniform deposition of baseline fallout to the area, uneven uptake of ^{137}Cs at the soil surface, and microscale redistribution of topsoil and bioturbation (Higgit 1995). These factors apply to both Chernobyl and bomb-derived fallout. Other factors applying to Chernobyl fallout are the relatively short period of mixing during cultivation and climatic conditions during the very short period when most ^{137}Cs fell. These reasons influence variability of the ^{137}Cs inventory on the macroscale too. Our first results (Golosov et al. 1999) demonstrate that variability of the ^{137}Cs inventory in the Plava basin is usually less than 20% at scales of hundred of metres, but it changes dramatically for basins larger than 5 km^2. Because of the high spatial variability of Chernobyl fallout, reference points should be taken in different parts of the study basin and from different landform positions such as interfluve area and creek bottoms. Because of the difficulty in finding uncultivated areas within flat interfluve areas on the Russian Plain, samples can be taken from non-eroding cultivated sites. A sufficient number of samples (20–25) should be taken to allow accurate determination of the input value. The reference points for sampling on uncultivated slopes of creek and on creek bottoms can also be located on areas without sedimentation. The sampling strategy related to areas with evidence of sedimentation is different. Here, at least, one incremental sample should be taken from each morphological unit, such as different floodplain levels or different uncultivated slopes with signs of local sedimentation.

Ten years after the Chernobyl accident Litvin et al. (1996) estimated the erosion influence on ^{137}Cs redistribution for areas with annual erosion rates of more than 20 t ha^{-1} (Table 10.3).

10.4 SAMPLING STRATEGY

According to an airborne gamma-spectrometry survey, which was done for the entire Plava basin Chernobyl-derived ^{137}Cs, the initial inventory change is five to six times from the south

Table 10.3 Percentage change of relative concentration of ^{137}Cs on plough slopes due to erosion since 1986

Erosion rate	Estimated period (years)							
$(t\ ha^{-1}\ a^{-1})$	**0**	**5**	**10**	**15**	**20**	**30**	**40**	**50**
0	100	89.09	79.37	70.71	63	50	39.69	31.5
1	100	88.92	79.08	70.32	62.53	49.44	39.1	30.92
2.5	100	88.68	78.63	69.73	61.83	48.62	38.23	30.07
5	100	88.26	77.91	68.76	60.69	47.28	36.84	28.70
10	100	87.44	76.47	66.86	58.47	44.71	34.19	26.14
15	100	86.63	75.05	65.02	56.32	42.27	31.72	23.81
20	100	85.82	73.66	63.21	54.25	39.96	29.43	21.68
30	100	84.23	70.94	59.75	50.33	35.7	25.53	17.97
40	100	82.65	68.32	56.47	46.67	31.88	21.78	14.88
50	100	81.1	65.78	53.35	43.27	28.46	18.72	12.32
60	100	79.58	63.33	50.39	40.1	25.4	16.08	10.18
70	100	78.08	60.96	47.59	37.16	22.65	13.81	8.42
80	100	76.59	58.67	44.94	34.42	20.19	11.85	6.95
90	100	75.14	56.46	42.42	31.87	17.99	10.16	5.74
100	100	73.70	54.32	40.03	29.51	16.03	8.71	4.73

of the basin to the Lokna River valley, where maximum initial inventory was defined. The main valley of the Chasovenkov Verh basin (right tributary of the Lokna River) is located along a south–north direction. So the initial ^{137}Cs inventory increases from the sources of the basin to the basin mouth. In this situation, it was only possible to use Chernobyl fallout as an important chronological marker for the study of sedimentation rates within different morphological units of the basin. The depth distribution of Chernobyl-derived ^{137}Cs for local soils is similar to the typical depth distribution of bomb-derived ^{137}Cs (Golosov *et al.* 1998). So the position of the level with maximum activity can be used to estimate the depth of the 1986 surface (Richie *et al.* 1975). Also the depth at which significant bomb-derived ^{137}Cs concentrations are first recorded can be used to estimate the position of the surface in 1954 (Popp *et al.* 1988; Golosov *et al.* 1992). Unfortunately the depth of the 1963 surface could not be identified because of bioturbation within the sedimentation layer that has taken place for about 35 years.

Two main morphological units were chosen for sedimentation study:

(a) bottoms of the main valleys and tributaries;
(b) uncultivated valley sides.

Sixteen pits were dug in the bottom of the main valley and tributary bottoms of the Chasovenkov Verh basin. The pit depths were up to the old alluvial sediments (former river bottom) or rocks. Detailed descriptions of the pits were made before sample collection in order to determine the total thickness of sediment, which was redeposited for the period of intensive cultivation of the basin slope. These sediments are characterized by a relatively similar grain size to the upper layer of cultivated soils. The samples for ^{137}Cs analysis were collected using a scraper plate at different thickness increments from an area 10 cm × 10 cm usually up to 80–100 cm depth.

Two typical uncultivated slopes were chosen for sampling within both the Chasovenkov Verh and the Lapki basin. The samples were taken using a 9.5 cm diameter core tube to provide cores from 5 cm up to 40 cm depth from the upper, middle and lower parts of each sampling slope.

The main goal of this sampling programme was to distinguish between three sedimentation episodes:

1. from the beginning of cultivation to 1954: first appearance of ^{137}Cs;
2. from 1954 to 1986: Chernobyl accident;
3. from 1986 to the moment of sampling.

Samples collected for laboratory analysis were dried, disaggregated and passed through a 2 mm sieve. Laboratory gamma spectrometry was undertaken using an Ortec HPGe coaxial detector.

10.5 APPROACHES FOR STUDY OF ^{137}CS REDISTRIBUTION

There are different approaches to ^{137}Cs redistribution studies. The main goal of study in the Chasovenkov basin was to establish the preliminary sediment budget in order to explore temporal variations in sediment redistribution. Sediment budget was established for the period since the Chernobyl accident. The sediment budget includes:

$$W_s + W_g = A_s + A_f + A_b + W_r \qquad (10.1)$$

where W_s is mass of eroded soil from the watershed (t), W_g is mass sediment from gullies (t), A_s is mass accumulation on slope (t), A_f is mass accumulation in balka tributaries (t), A_b is mass accumulation in bottom of main balka (t) and W_r is mass of sediment delivered in the river channel (t).

A few assumptions were used for calculating the sediment budget of the Chasovenkov Verh basin for the Chernobyl period.

1. Mean annual coefficient of soil protection of crops, which was determined according to the modified version of USLE, is used for calculating soil erosion rate.
2. Mean value of erosion index of rains for a 25 year period is used for calculation. This index was calculated using data of local meteorological stations and the formula for erosion index of rains from USLE.
3. Data about deposition rates on three uncultivated valley sides were used for calculating the volume of sedimentation at the uncultivated balka sides. These three slopes represent the typical types of valley sides with sediment deposition within the Chasovenkov balka basin.
4. Sedimentation in small agricultural ponds, which are located in some balka tributaries, was defined as 100%. According to investigations of pond sedimentation, which were made in different parts of the steppe and forest-steppe zones of the Russian Plain, about 95–100% sediment is retained in agricultural ponds (Prytkova 1979).
5. Physical properties of sediment on the slope and in the balka bottoms are similar. Special experiments were made in different field experimental stations in the central part of Russia (Cheremisinov 1968; Surmach 1976). They compared the physical properties of soil and sediment, which were taken directly from flow on special flumes located along the way from the slope to the balka bottom. They demonstrated that only during selective

erosion can the grain sizes of soil and sediment be different. However, the physical properties of sediment are absolutely identical to the physical properties of soil at the moment of strong sheet or rill erosion. The reason for this effect is the loess structure of soils.

The calculation of sediment budget for the Lapki balka basin was not the goal of this stage of investigation. The sampling programme aimed only to study the redistribution of ^{137}Cs in different parts of uncultivated valley sides.

10.6   RESULTS

Soil erosion rates for the Chasovenkov balka basin were established by calculation using the modified version of USLE (Larionov 1993) which was verified with experimental observation data of soil erosion rates (Table 10.1) and using the soil horizons method of Olson and Beavers (1987). The modelled annual soil erosion rate in the Chasovenkov River basin was 5 t ha^{-1}. The annual soil erosion rates estimated by using Olson and Beavers' (1987) technique were 3.6–4.9 t ha^{-1}. These independent methods therefore offer the opportunity to compare results with fallout redistribution data. The mass of sediment eroded from the slopes since 1986 is 61 000 t. Gully erosion has been restricted to the basin since the Chernobyl accident with the exception of a few gullies within the balka bottom, but their input to the sediment budget is lower than the precision of sediment budget calculations.

As previously mentioned, the study of sediment and sediment-associated ^{137}Cs redistribution was mainly based on surveys. The first geomorphological unit considered was uncultivated valley sides. As a rule it is not easy to choose the parts of uncultivated slopes with sedimentation. The reliable marks of sedimentation on these slopes are predominant weeds in grasses and the slope microrelief. However, it is difficult to use these features to assess overall deposition mass, and so Chernobyl ^{137}Cs fallout provides a good marker as shown by ^{137}Cs concentrations on slopes without sedimentation (Figure 10.4A) and with sedimentation (Figure 10.4B, C). The reference or undisturbed ^{137}Cs depth distribution curve is characterized by maximum ^{137}Cs concentration in the upper 2–4 cm (Golosov et al. 1998). So the comparison of profile depth distributions makes it possible to determine sedimentation rates. Usually the highest rates of sedimentation were observed in the upper parts of uncultivated slopes (Figure 10.4C), but where erosion intensity is extremely high the highest sedimentation layer occurs at the bottom part of the uncultivated slope (Figure 10.4B). The extent of deposition on slopes depends on cultivated slope configuration and density of hedgerows. Where hedgerows are dense or the cultivated slope has a convex form, deposition is typically less than 5–10% of the total. On the other hand the depositional area of such slopes can reach 60–80% if hedgerows are absent and slopes have a straight or concave form. The sum volume of accumulation on uncultivated parts of slopes of the Chasovenkov Verh basin for the Chernobyl period is 6000 t. We did not take into consideration accumulation within cultivated parts of slopes.

The redistribution of sediment and sediment-associated radionuclides in dry valleys changes along their length. The study of the Chasovenkov Verh basin allowed sedimentation rates to be characterized along the balka valley (Golosov 1996). It was possible to find sections within the main channel with different relationships between sediment output and accumulation (Figure 10.5). The major mass of the sediment mobilized from the field accumulates in the upper reach. Downstream, in the transit–accumulation section the sediment output increases up to 20% (Figure 10.5), but the sedimentation rate decreases. The valley

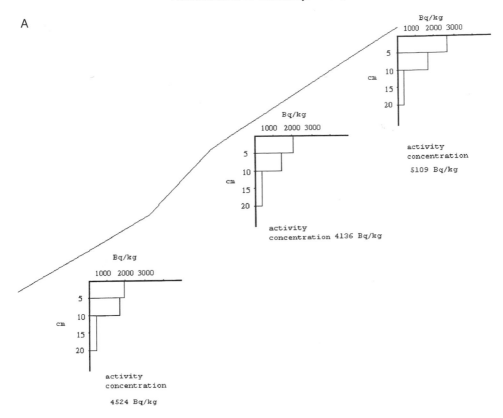

Figure 10.4 Change of ^{137}Cs concentration in the upper soil horizons on different uncultivated slopes: (A) upper reach of the Chasovenkov Verh basin; (B) lower reach of the Chasovenkov basin; (C) upper reach of the Lapki basin

(balka) bottom in this part is characterized by a flat cross-reach (Figure 10.6A), maximum bed gradient and mean values of sedimentation rates (Figure 10.7A).

Further downstream in the accumulation section, the valley bottom is characterized by a change in cross-sectional form (Figure 10.6B) and an increased sedimentation rate (Figure 10.7B). The major part of the sediment mobilized from the slopes in the lower transit zone is deposited before it reaches the incised channel which usually occurs here (Figure 10.6C) or accumulates in the incised channel (Figure 10.7C). The absence of fresh sediments on the floodplain of this area is observed (Figure 10.7D). The incised channel is the product of discharge during snowmelt or the following period. Calculations of volume of sediment in the bottom of the main balka were made for three sections (Table 10.4):

(a)   transit–accumulation, with mean gradient 0.0056;
(b)   accumulation, with mean gradient 0.0035;
(c)   transit, with mean gradient 0.0041.

The differences between these sections connected with valley gradient and type of cross-sectional profile, so they are mainly connected with morphological features of the valley.

B

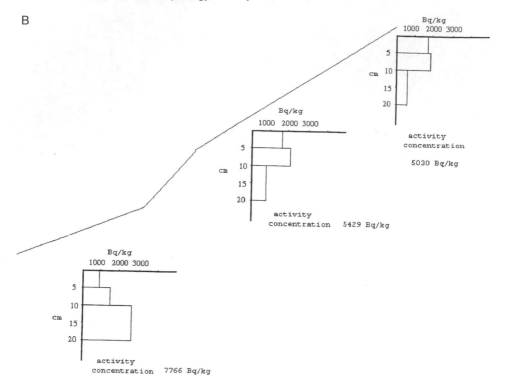

Figure 10.4 (*cont.*)

Two balka tributaries were sampled for ^{137}Cs analysis. The sedimentation rates for both types were defined by comparing ^{137}Cs depth distribution in different pits with depth distribution at the reference site. The mean layer of deposition for the period after Chernobyl was 18 cm. Then the total area of tributary bottoms was calculated. It was assumed that sedimentation rates are similar for all balka tributaries. After simple calculations the total sedimentation in balka tributaries was defined. It is equal to 19 800 t. The comparison of this mass with the mass of sediment eroded from cultivated slopes within tributary catchments showed that about 70% of sediment that entered from the cultivated slope in these balka valleys was delivered to the main balka bottom.

Totalling these results indicates that the mass of sediment deposited within the basin for the entire period after the Chernobyl accident is 6000 + 19 800 + 31 000 = 56 800 t. So about 93% of sediment eroded from the slope was redeposited within the basin and only 7% of sediment was delivered to the Lokna River channel for the period from 1986 (Figure 10.8).

It is more difficult to calculate sediment budgets for the entire period of cultivation and for the bomb-testing period. The precision of these sediment budgets was necessarily low because it was difficult to find information about crop rotation over the cultivation period. Similarly there was no information about field configuration changes over time. It was also impossible to obtain information about gully growth, because the available maps were inadequate for the period 1912–1960. However, comparison of gully volume and the volume of their fans in the main channel of the balka shows that about 90% of eroded sediment was redeposited in their fans. These minor volumes were not included in the sediment budget. But it was possible to

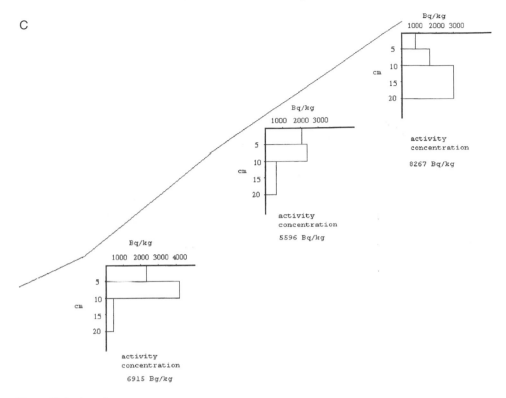

C

Figure 10.4   (*cont.*)

determine the two main components of sediment budget: mass of eroded soil and mass of sediment which was redeposited in balka bottoms. The relationship of these components determined the delivery ratio for different sections of the main balka bottom (Figure 10.5). In the results, the sediment delivery ratios for the entire basin for the Cs period and for the period from the beginning of intensive cultivation were calculated (Figure 10.5). These results demonstrate that sediment redistribution did not change during the last 40 years. But earlier, when constant flow existed in the main bottom of the Chasovenkov Verh basin, relationships between accumulation and output were typical of small river basins on the Russian Plain (Dedkov and Mozzherin 1984) or small hilly basins of the UK (Owens *et al.* 1997).

The sediment budget estimated for the Chasovenkov Verh basin is comparable to that estimated for basins where most of the sediment is redeposited within the basin valley (Trimble 1983; Golosov 1996). Such sediment redistribution in areas with high levels of radionuclide contamination promotes extremely high ^{137}Cs inventories in dry valley bottoms.

10.7   CONCLUSION

The preceding arguments indicate that the Plava River basin balka bottoms serve primarily as a ^{137}Cs sink and will continue to do so if climatic conditions and land use do not change significantly. In summer 1996 ^{137}Cs inventories in the accumulation section of the Chasovenkov Verh basin bottom increased up to 600–700 kBq m^{-2}. These values are 1.5–2

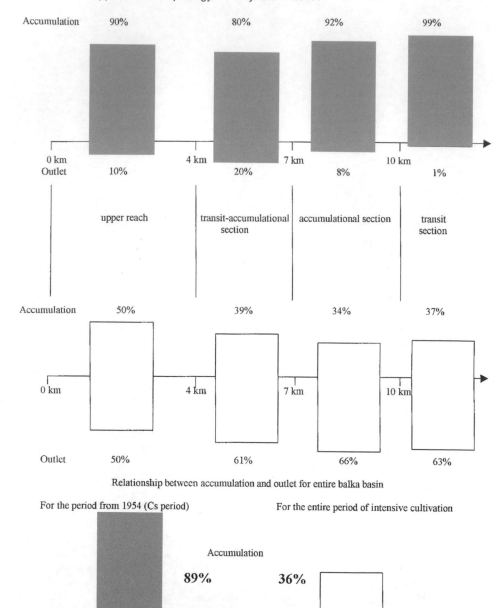

Figure 10.5 The relationship between accumulation and output for different sections of the Chasovenkov Verh balka basin. Distances given are from the dry valley source. ■ Relationship for the ^{137}Cs period; □ relationship for the entire period of intensive agriculture

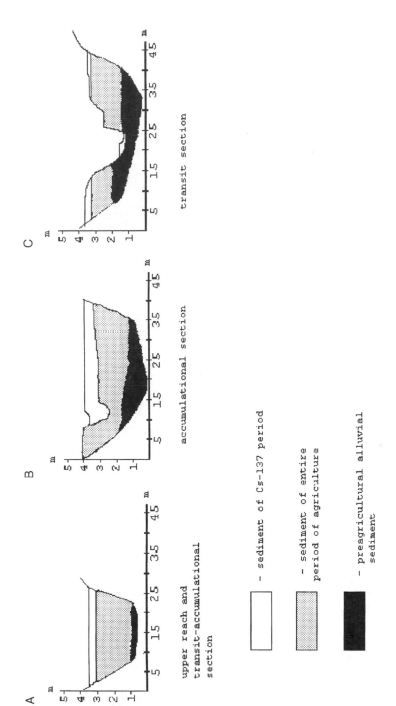

Figure 10.6 Typical cross-sectional profiles of the Chasovenkov Verh balka basin for different reaches along its course

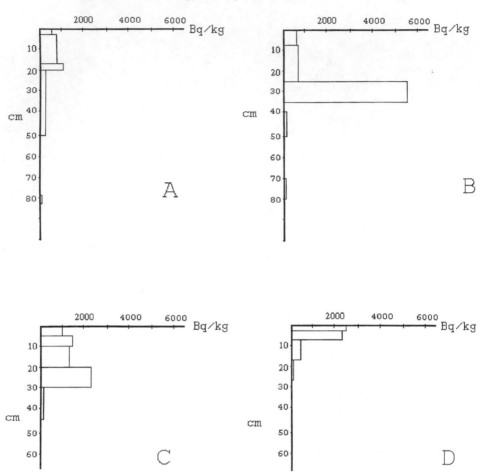

Figure 10.7 Depth distribution of ^{137}Cs at different sections along the main channel (Chasovenkov Verh basin): (A) balka bottom in the transit–accumulation section; (B) balka bottom in the accumulation section; (C) bottom of secondary incised channel in transit section; (D) high floodplain in transit section

Table 10.4 Sediment accumulation within different sections of the Chasovenkov Verh balka bottom for the period after the Chernobyl accident

Reaches of bottom	Length (m)	Mean deposition depth (m)	Mean width of bottom (m)	Bulk density (g cm^{-3})	Sediment mass (t)
Transit–accumulation	5000	0.09	14.4	1.2	7780
Accumulation	2300	0.184	23.4	1.2	11 880
Transit	3020	0.125	25	1.2	11 325
Total	10 320				30 985

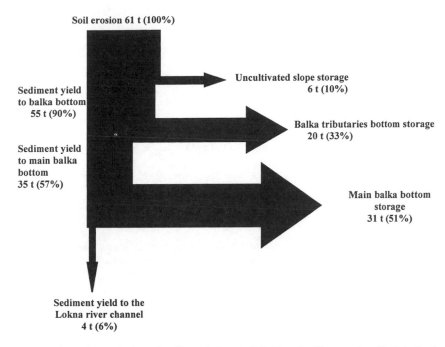

Figure 10.8 The sediment budget (for Chernobyl period) linking the Chasovenkov Verh balka basin to the Lokna River channel. Mass of erosion and sedimentation in thousands of tonnes. Percentages are given in parentheses

times higher than initial radionuclide contamination in 1986. Due to the contamination threat these sites cannot be used as pasture for cattle, and similarly water bodies cannot be used for food production. Fluvial redistribution of Chernobyl ^{137}Cs can promote secondary contamination within areas presently free from direct fallout. On the other hand, if the climatic situation or land use changes, active incision will start in balka bottoms and the whole volume of polluted sediment will enter rivers. The quality of water may become extremely dangerous. So the programme of special conservation works should be elaborated for different prognoses.

The results presented are the first attempt to study Chernobyl ^{137}Cs redistribution in central Russia. The conclusions are tentative but do show that there is potential to use Chernobyl-derived ^{137}Cs to trace environmental processes even in areas with high levels of contamination. Some of the problems which exist now, for example the difficulty in studying the relationship between erosion and change of ^{137}Cs, can be overcome with time. Now is the best time to use Chernobyl ^{137}Cs data in sedimentation research, because it is still possible to detect ^{137}Cs in the sediment.

ACKNOWLEDGEMENTS

This research has been supported by INTAS-RFBR, grant 0734, and partly by the International Atomic Energy Agency, contract 9044. For their assistance with various aspects of this work, we thank E.V. Kvasnikova and A.V. Panin.

REFERENCES

Braude, I.D. 1976. *Racional'noe ispol'zovanie erodirovannyh seryh lesnyh poch Nechernozemnoi zony RSFSR*, Lesnaya Promyshlennost', Moskva.

Cheremisinov, G.A. 1968. *Eroded soils and their productive use.* Kolos, Moskva.

Dedkov, A.P. and Mozzherin, V.I. 1984. *Erosia i Stok Nanosov na Zemle (Erosion and Sediment Yield on the Earth).* Izdatelstvo Kazanskogo Universiteta, Kazan'.

Fredericks, D.J., Norris, V. and Perrens, S.J. 1988. Estimating erosion using caesium-137: 1. Measuring caesium-137 activity in a soil. *IAHS Publication No. 174*, 225–231.

Golosov, V.N. 1996. Redistribution of sediment within small catchments of the temperate zone. *IAHS Publication No. 236.* 339–346.

Golosov, V.N., Ostrova, I.V., Silant'ev, A.N. and Shkuratova, I.G. 1992. Radionuclide method of assessment of basin sedimentation rate. *Geomorphologiya*, **1**, 30–36.

Golosov, V.N., Kvasnikova, E.V., Panin, A.V. and Ivanova, N.N. 1998. Radionuclide migration in the Chernobyl contamination zone. *Proceeding of SPERA Conference*, Christchurch, New Zealand, 15–19 February 1998, 25–34.

Golosov, V.N., Walling D.E., Panin A.V., Stukin E.D., Kvasnikova E.V. and Ivanova N.N. 1999. The spatial variability of Chernobyl-derived Cs-137 inventories in small agricultural drainage basin in Central Russia 11 years after the Chernobyl incident. *Applied Radiation and isotopes*, **51**, 341–352.

Grin, A.M. 1970. *Opyt stacionarnogo izucheniya processov stoka i smyva.* In Recent exzogenic processes relief formation. Nauka, Moskva, 89–95.

Higgitt, D.L. 1990. *The use of caesium measurements in erosion investigations.* PhD thesis, University of Exeter, Exeter, UK.

Higgitt, D.L. 1995. The development and application of caesium-137 measurements in erosion investigation. In Foster, I., Gurnell, A. and Webb, B. (eds), *Sediment and Water Quality in River Catchments.* Wiley, Chichester, 287–305.

Izrael, Yu.A., Kvasnikova, E.V., Nazarov, I.M. and Fridman, Sh.D. 1994. Global and regional radio-nuclide Cs-137 contamination of European territory of former USSR. *Meteorology and Hydrology*, **5**, 5–9.

Larionov, G.A. 1993. *Vodnaya i vetrovata eroziya (Soil and wind erosion).* State University Publishing House, Moscow.

Litvin, L.F., Golosov V.N., Dobrovol'skaya N.G., Ivanova, N.N., Kiryukhina, Z.P. and Krasnov, S.F. 1996. Redistribution of ^{137}Cs by the processes of water erosion of soils. *Water Resources*, **23**(3), 314–319.

Olson, K.R. and Beavers, A.H. 1987. A method to estimate soil loss from erosion. *Soil Science Society of American Journal*, **51**, 441–445.

Owens, Ph.N., Walling, D.E., He, Q., Shanahan, J. and Foster, I.D.L. 1997. The use of caesium-137 measurement to establish a sediment budget for the Start catchment, Devon, UK. *Hydrological Sciences Journal*, **42**(3), 405–423.

Popp, C.L., Hawley, J.W., Love, D.W. and Dehn, M. 1988. Use of radometric (Cs-137, Pb-210), geomorphic and stratigraphic techniques to date recent oxbow sediments in the Rio Puerco drainage Grants uranium region, New Mexico. *Environmental Geology and Water Science*, **11**, 253–269.

Prytkova, M.Ya., 1979. *Small Reservoirs of USSR Forest-steppe and Steppe Zones.* Nauka, Leningrad.

Richie, J.C., Hawks, P.H. and McHenry, J.R. 1975. Deposition rates in valleys determined using fallout Cs-137. *Geol. Soc. Am. Bull.*, **86**, 1128–1130.

Surmach, G.P. 1976. *Water Soil Erosion and Struggle with it.* Gidrometeoizdat, Leningrad.

Timofeev-Resovskii, N.V., Titlyakova, A.A., Timofeeva, N.A. and Timofeeva-Resovskaya, E.A. 1966. *Radioactivnost' pochv i metody ee opredeleniya (Soil Radioactivity and Methods to Determine It).* Nauka, Moscow.

Trimble, S.W. 1983. A sediment budget for Coon Creek, the Driftless Area, Wisconsin, 1853–1977. *American Journal of Science*, **283**, 454–474.

Walling, D.E. and Quine, T.A. 1993. Using Chernobyl-derived fallout radionuclides to investigate the role downstream conveyance losses in the suspended sediment budget on the River Severn, UK. *Physical Geography*, **14**, 239–253.

Walling, D.E., Rowan, J.S. and Bradley, S.B. 1989. Sediment associated transport and redistribution of Chernobyl fallout radionuclides. *IAHS Publication No. 184*, 37–45.

11 Sources of River-Suspended Sediment after Selective Logging in a Headwater Basin

YOSHIMASA KURASHIGE
Graduate School of Environmental Earth Science, Hokkaido University, Sapporo, Japan
Present address: School of Environmental Science, The University of Shiga Prefecture, Hikane, Japan

ABSTRACT

The source of river-suspended sediment after selective logging in a small mountainous forested basin was estimated from its grain-size distribution. In the basin, an unpaved road was constructed, and the sediment cut from the hillslope by bulldozer during the logging was dumped on several zero-order valleys. During storm events, surface flow on the road (SSR) with high suspended sediment content flows into the brook immediately upstream of the observation site. In contrast, in snowmelt season, SSR did not occur because the basin was under several metres of snowcover.

During storm events in summer 1992, just after the logging, the peak of suspended sediment concentration (SSC) appeared close to the peak discharge, and the river-suspended sediment was mainly supplied from both dumped sediment fines (DSF) and SSR. In contrast, in the snowmelt season of 1993, the peak SSC appeared earlier than the peak discharge. The grain-size distribution suggested that the suspended sediment was probably supplied from the river-bed sediment, but a clear determination could not be obtained from the grain-size distribution alone. However, the process-based stirring-up model, which simulates the SSC supplied from the river-bed sediment, could explain the time variation of SSC in this season. The origin of suspended sediment in this season was thus judged to be river-bed sediment. During a storm event in summer 1993, a clear peak of SSC appeared earlier than the peak discharge. The grain-size distributions of suspended sediment suggested that the origins were all of river-bed sediment, DSF and SSR, whereas the process-based model showed that the sediment supply from the river bed strongly affected the appearance of peak SSC in this case. The difference in the appearance of peak SSC in two summer cases indicates that the accumulation of river-bed fines from DSF and/or SSR was active in summer 1993.

11.1 INTRODUCTION

Some researchers have reported that environmental impact by logging affects suspended sediment supply into a river. A very high suspended sediment concentration has been observed after logging (e.g. Reid and Dunne 1984; Johnson 1988; Kurashige 1993a), as a result of run-off from hillslope surfaces on which logging was conducted, and from roads constructed for logging (e.g. Reid et al. 1981; Reid and Dunne 1984).

Applied Geomorphology: Theory and Practice. Edited by R.J. Allison.
© 2002 John Wiley & Sons, Ltd.

The source of suspended sediment is likely to change with time after logging in a basin. Even if suspended sediment is supplied both from the hillslope and the road in a logged basin, the contribution rate of each source is likely to change with time. To reveal such a process of suspended sediment supply, the source of suspended sediment should first be identified. Radioisotopes, such as caesium-137, and/or magnetic susceptibility of sediment have usually been used as tracers of suspended sediment (e.g. Walling *et al.* 1979; Walling and Woodward 1992), but these tracers are effective only to distinguish between the sediment washed out from a hillslope surface and that from a subsurface layer (e.g. Walling *et al.* 1979; Dearing *et al.* 1986; Loughran *et al.* 1987). These tracers are, accordingly, not able to determine whether the origin is, for example, a hillslope or a road surface.

However, Kurashige (1993a) has recognized that the grain-size distribution of suspended sediment in the surface flow on an unpaved road is much finer than that of the fine grains in a hillslope regolith, for a small mountainous basin in Hokkaido, Japan. He also revealed that the fine grains in the river-bed sediment are coarser than the fine grains in the hillslope regolith. Kurashige (1993b) has also shown that it is possible to distinguish seasonal differences in sediment source area, by comparing the grain-size distribution of river-suspended sediment with the aforementioned grain-size distributions. A recently developed method by Kurashige and Fusejima (1997), which has demonstrated a non-parametric statistical test for grain-size distributions, identified the percentages of suspended sediment flux supplied from two different sources. In this study, the source of river-suspended sediment was determined from its grain-size distribution by the Kurashige and Fusejima method for a series of flood events which occurred after selective logging in a small mountainous basin, and changes of the source with time after logging were also examined.

11.2 STUDY AREA

The Hiyamizusawa Brook basin, with an area of $0.93 \, km^2$, is located in a mountainous region southwest of Sapporo City, Hokkaido, Japan. The geology of the basin is predominantly andesitic agglomerate and quartz porphyry. The slope is under a fir, birch and maple forest. The period from late October to late March is the snowy season, and snow lies on the basin from December to April to a depth of several metres. The basin has two major flood seasons in a year: the snowmelt season from early April to mid-May, and the summer rainy season from mid-August to mid-October. The total precipitation in the summer rainy season is *c.* 500 mm. The basin is relatively dry from late May to mid-August, and the total precipitation is *c.* 100 mm in this period. The total annual precipitation is *c.* 1400 mm.

An unpaved road was constructed in 1989 and crosses the brook downstream of Site B, where the river water is diverted by culvert to the opposite side of the road (Figure 11.1). At this location, during storm events, turbid water with a high content of suspended sediment tends to flow into the brook from the surface of the unpaved road, according to the condition of ruts on the road. Selective logging was carried out in the southern part of the basin from May to June 1992 during which time a bulldozer cut part of the slope in order to construct a path to remove the logs. During this operation, the sediment cut from the slope was dumped on the slope, in particular in zero-order valleys in the basin (Figure 11.1). This sediment is referred to as dumped sediment. The surfaces of the path and the dumped sediment were bare in summer 1992, whereas about 10% of those surfaces had been covered with vegetation, mainly butterbur, by summer 1993. The brook was gravel-bedded before the construction of the unpaved road, whereas about 10% of the river bed had been covered with sandy material after construction (Kurashige 1994a).

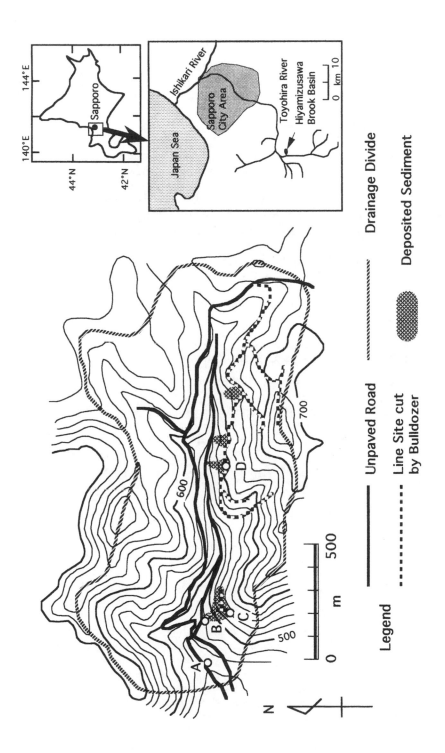

Figure 11.1 Map of the study basin

11.3 FIELD STUDY

11.3.1 Method

Water discharge was measured at Site A, and one-litre river water samples were taken manually at Sites A and B during floods on 13 August and 25 September 1992 (Figure 11.1). In contrast, from 16 to 17 April, 23 to 24 April and on 28 August 1993, the water was sampled only at Site A, by an automatic water sampler (ISCO 7200). The turbid water on the unpaved road was manually sampled once on 25 September 1992.

A half-litre of the sampled water was vacuum-filtered through a pre-weighed cellulose nitrate filter paper (0.45 μm pore size), and its dry weight was measured in a laboratory to calculate the suspended sediment concentration (SSC). The suspended sediment flux (SSF) was calculated from the discharge and SSC data at Site A. The remaining half-litre was also vacuum-filtered to provide a suspended sediment sample for grain-size analysis. The suspended sediment trapped on the filter paper was recovered using a supersonic cleaner, dispersed in Calgon, and its grain size was measured using a Shimazu centrifugal particle size analyser SA-CP3L.

Dumped sediment was sampled at Sites B, C and D in summer 1992 by Kurashige and Fusejima (1997). The sandy river-bed sediment was also sampled 300 m upstream of Site B in July 1993 by a handy scoop. The grain-size distributions of these sediments were obtained using the sieving method for the fraction coarser than 4.5ϕ (44 μm). The fraction finer than 4.5ϕ was analysed using the centrifugal analyser.

11.3.2 SSC, SSF and Grain-size Distributions in Summer 1992

Figure 11.2 shows the time variations of the river discharge and SSC. In the summer rainy season in 1992 (Figure 11.2A, B), the peak value of SSC appeared close to the time of peak discharge on 13 August, and a high SSC appeared at peak discharge on 25 September.

The grain-size distributions obtained in this season were explained by Kurashige and Fusejima (1997). They considered only the distributions of grains smaller than 4.0ϕ (62.5 μm) for the dumped sediment (the dumped sediment fines, DSF), to compare with those of suspended sediment, because the suspended sediment is all smaller than 4.0ϕ. The grain-size distributions of DSF sampled at Sites B, C and D were all similar to each other. In the suspended sediment from the river, the percentages of grains finer than 8.0ϕ (3.9 μm) at Site A were larger than those at Site B. In the suspended sediment associated with the road surface flow (SSR), the percentage was much higher. These findings indicate that the inflow of SSR increased the percentage of very fine grains of river-suspended sediment, and consequently the percentages at Site A were larger than those at Site B. In contrast, the percentages of very fine grains were small in the DSF. SSR was thus determined to be the major source of very fine grains. The grain-size distribution of suspended sediment at Site B more closely resembled that of DSF than SSR. The suspended sediment at Site B was thus judged to be supplied mainly from DSF in summer 1992.

The SSF at Sites A and B is also reported by Kurashige and Fusejima (1997). They calculated the SSF at Site B (SSF$_B$) by assuming that the river discharge at Site B was almost equal to that at Site A, because the SSC of SSR was very high (c. 6000 mg l^{-1}) and the discharge of SSR was very small (c. 0.0005 m^3 s^{-1}). The SSF at Site A (SSF$_A$) was larger than SSF$_B$ at each sampling time. SSF from the road (SSF$_{road}$) was also calculated, assuming that the SSF from the road is equal to the difference of SSF between Sites B and A. At each sampling time, SSF$_{road}$ occupied about 20 to 40% of SSF$_A$.

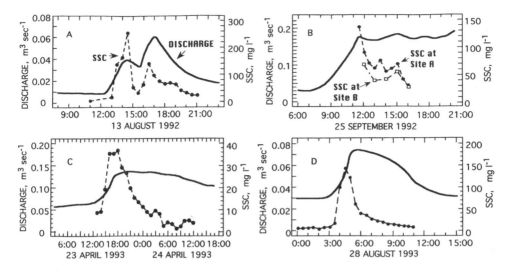

Figure 11.2 Time variations of river-suspended sediment concentration and river discharge: (A) 13 August 1992; (B) 25 September 1992; (C) 23 April 1993; (D) 28 August 1993

11.3.3 SSC, SSF and Grain-size Distributions in 1993

In the snowmelt season in 1993, SSC increased to $37\,\text{mg l}^{-1}$ at 16:00 on 23 April while the discharge was on the rising limb (Figure 11.2C). The SSC remained on this level until 18:00, and decreased to $c.$ $10\,\text{mg l}^{-1}$ until 0:00 on 24 April while the discharge was still on high stage. The peak discharge appeared at $c.$ 20:00 on 23 April in this case. In the summer rainy season in 1993, the peak of SSC appeared at 04:30 on 28 August, whereas the peak discharge appeared at 05:30 (Figure 11.2D). The peak SSC on 28 August 1993 ($150\,\text{mg l}^{-1}$) was much higher than that on 23 April 1993 ($39\,\text{mg l}^{-1}$), whereas the peak discharge on 28 August ($0.075\,\text{m}^3\,\text{s}^{-1}$) was lower than that on 23 April ($0.14\,\text{m}^3\,\text{s}^{-1}$). Accordingly, the SSF at the time of peak SSC on 28 August ($8.1\,\text{g s}^{-1}$) was about 60% larger than that on 23 April ($5.1\,\text{g s}^{-1}$). The SSF at the time of peak discharge on 23 April and that on 28 August were $2.9\,\text{g s}^{-1}$ and $3.9\,\text{g s}^{-1}$, respectively.

Figure 11.3 shows the representative grain-size distributions of river-suspended sediment sampled in each event and the grain-size distribution of fine grains smaller than 4.0ϕ in the river-bed sediment. In the river-suspended sediment, the percentage of very fine grains (smaller than 8.0ϕ) was $c.$ 50% at 16:00 on 23 April and at 04:30 on 28 August 1993, and was $c.$ 65% at 06:00 on 28 August 1993. In contrast, the river-bed sediment fines included only about 10% of grains smaller than 8.0ϕ.

11.4 FINGERPRINTING THE SOURCE OF SUSPENDED SEDIMENT

11.4.1 Method

The ratios between SSF from DSF and that from SSR in SSF_A (i.e. the ratio DSF:SSR) were estimated from grain-size distributions of river-suspended sediment using non-parametric statistical tests (Kurashige and Fusejima 1997). When the major sources of river-suspended

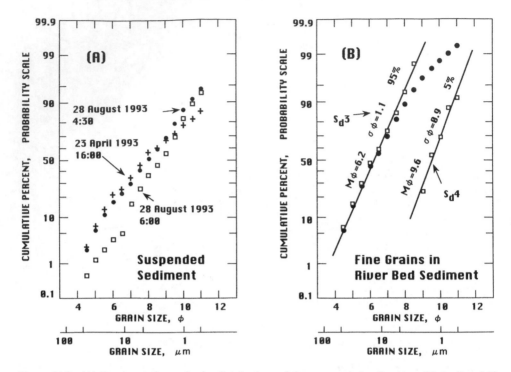

Figure 11.3 (A) Representative grain-size distributions of river-suspended sediment on 23 April and 28 August 1993; (B) grain-size distribution and lognormal subpopulations of river-bed sediment fines

sediment were considered to be DSF and SSR, the grain-size distribution of DSF and that of SSR were compounded, and this compounded distribution was further compared with the grain-size distribution of river-suspended sediment. On the other hand, the grain-size distribution of DSF can be separated into two lognormal subdistributions (S_d1 and S_d2) by the method of Inokuchi and Mezaki (1978), which was briefly introduced by Kurashige (1994b, 1996). S_d1 had $M\phi = 6.2$ and $\sigma\phi = 0.8$, and S_d2 had $M\phi = 8.9$ and $\sigma\phi = 1.5$. The original ratio of $S_d1:S_d2$ in DSF was 85:15. Then, the ratio $S_d1:S_d2$ was also changed to consider the sorting of grains supplied from DSF. The compounded distribution was further compared with the grain-size distribution of river-suspended sediment by the Kormogorov–Smirnov two-sample goodness-of-fit test with the pseudo-sample number $n_p = 100$, and the case with the smallest χ^2-value concurrent with the significant level $p > 0.5$ was accepted as similar to the grain-size distribution of river-suspended sediment. The resolution of the determined ratio was $c. \pm 10\%$. More details on this method are described in Kurashige and Fusejima (1997).

The grain-size distribution of the river-bed sediment fines was also separated into two lognormal subdistributions (S_d3 and S_d4, Figure 11.3B). However, since the river-bed sediment was sampled in July 1993, the ratio $S_d3:S_d4$ before this time was likely to be different from this ratio (i.e. $S_d3:S_d4 = 95:5$). Thus, when the river-suspended sediment was supposed to be supplied mainly from river-bed sediment, the compounded distributions with various $S_d3:S_d4$ ratios were compared with the grain-size distributions of river-suspended sediment.

Table 11.1 The ratio DSF:SSR and $S_d1:S_d2$ estimated from the grain-size distribution of river-suspended sediment in summer 1992

13 August 1992			25 September 1992		
Time	DSF:SSR	$S_d1:S_d2$	Time	DSF:SSR	$S_d1:S_d2$
13:30	70:30	70:30	11:30	80:20	70:30
14:00*	60:40	40:60	12:00	80:20	60:40
14:30	80:20	50:50	12:30	70:30	60:40
15:00	60:40	80:20	13:00*	60:40	70:30
16:00	NF		13:30	NF	
16:30	NF		14:00*	80:20	50:50
17:00	60:40	30:70	14:40	NF	
			15:00*	80:20	70:30

NF = not found
* Data after Kurashige and Fusejima (1997)

11.4.2 The Origin of Suspended Sediment in Summer 1992

Table 11.1 shows the resultant ratios of DSF:SSR and $S_d1:S_d2$ obtained at around the peak or high SSC in summer 1992. On 13 August and 25 September 1992, DSF:SSR ranged between 60:40 and 80:20, and $S_d1:S_d2$ ranged from 30:70 to 80:20. In these cases, a clear peak or high SSC appeared close to the time of each peak river discharge (Figure 11.2A, B). This type of relationship has been observed previously by Kurashige (1994b), who reported that the peak of SSC appeared close to the peak of discharge when the suspended sediment was washed out from the entire basin area. In the Hiyamizusawa Brook basin, the dumped sediment is sporadically distributed, thus the suspended sediment supplied from DSF can result in a peak SSC close to the peak discharge. The unpaved road is adjacent to the length of the brook, thus the peak discharge of road-surface flow occurs close to the peak river discharge, and thus the maximum concentration of SSR can also be expected to occur close to peak river discharge. Accordingly, both the DSF and the SSR produced the peak SSC close to the peak river discharge.

At many sampling times, the percentage of S_d2 washed out from the DSF was estimated to be greater than 15%, which is the original percentage in DSF (Kurashige and Fusejima 1997). Since the resolution of the method is c. ±10% (Kurashige and Fusejima 1997), the percentages of S_d2 ranging from 0 to 30% cannot be judged to be significantly different from the original percentage. However, in six cases in Table 11.2, the percentage of S_d2 exceeded 30%, indicating that the grains of S_d2 were easily washed out from DSF, and consequently the percentage of S_d2 in the mobilized sediment became larger than that in the grain-size distribution of DSF (Kurashige and Fusejima 1997).

11.4.3 The Origin of Suspended Sediment in the 1993 Snowmelt Season

Table 11.2 shows the ratios DSF:SSR and $S_d1:S_d2$ estimated from grain-size distributions of river-suspended sediment sampled on 23 April 1993. In this case, only the grain-size distribution of samples obtained at 16:00 and 19:00 could be determined, probably because of the low SSC on this day. The DSF:SSR was estimated to be 60:40 and 70:30, and $S_d1:S_d2$ to be 50:50 and 60:40 for 16:00 and 19:00, respectively.

However, on this day, the basin had been covered with snow, and accordingly both the

Table 11.2 The ratio DSF:SSR and $S_d1:S_d2$ under the assumption that the origin was both DSF and SSR, and the ratio $S_d3:S_d4$ under the assumption that the origin was the river-bed sediment for RSS obtained in 1993

	23 April 1993				28 August 1993		
	Slope and road origin		River-bed origin		Slope and road origin		River-bed origin
Time	DSF:SSR	$S_d1:S_d2$	$S_d3:S_d4$	Time	DSF:SSR	$S_d1:S_d2$	$S_d3:S_d4$
16:00	60:40	60:40	50:50	04:00	100:0	60:40	75:25
				04:30	80:20	40:60	50:50
19:00	70:30	60:40	60:40	05:00	90:10	30:70	45:55
				06:00	60:40	10:90	NF

NF = not found

dumped sediment and the unpaved road were under snowcover. As a result, road surface flow was not observed on this day. These facts indicate that DSF and SSR could not both behave as active sources of river-suspended sediment in this case.

In this event, the high SSC appeared at the rising limb of the hydrograph, and SSC tended to decrease while the river discharge was on high stage (see Figure 11.2C). Kurashige (1985, 1993b, 1994b, 1996) reported that the peak of SSC can appear earlier than the peak discharge when the suspended sediment is supplied from bed sediment. This indicates that the river-suspended sediment in this event was most likely supplied from river bed. To check this hypothesis, the two subdistributions separated from the grain-size distribution of river-bed sediment fines (i.e. S_d3 and S_d4, see Figure 11.3B) were compounded to find a distribution similar to the grain-size distribution of river-suspended sediment (Table 11.2). The compounded distributions with $S_d3:S_d4$ of 50:50 and 60:40 were judged to be similar to the grain-size distributions of river-suspended sediment at 16:00 and 19:00, respectively.

Kurashige (1996, 1998) considered the mechanism of suspended-sediment supply from a sandy river bed, and explained the time variation of SSC using a process-based physical model. In this model, all grains finer than a critical size which can be stirred up by bed shear stress were assumed to be suspended in the river water. This model requires that sufficient grains of subpopulations S_d3 and S_d4 be contained in the river-bed sediment to explain the grain-size distribution of river-suspended sediment obtained on 23 April 1993.

The ratio $S_d3:S_d4$ was 95:5 in the bed sediment fines sampled in July 1993 (Figure 11.3B). This indicated that the amount of grains of S_d4 in the bed sediment was small. The sandy river-bed sediment was, however, sampled in July 1993 at only one site, and accordingly it does not represent the grain-size distribution of river-bed sediment fines before the snowmelt season. The large percentages of S_d4 in the river-suspended sediment suggested that very fine grains had accumulated in the bed sediment before the snowmelt season.

Kurashige (1998) simulated the time variation of SSC in the snowmelt season of 1993 by the process-based physical model (Kurashige 1996). In this model, the maximum grain size d_{max} which will be supplied from river-bed sediment was calculated, and the cumulative percent of d_{max} was found from the grain-size distribution of river-bed sediment. Next, the mass of grains finer than d_{max} stirred up from the tractive layer was calculated to obtain SSC.

Kurashige (1998) simulated not only the time variation of SSC during the snowmelt season in 1993 but also that in 1990, when the origin of SSC was also determined to be river-bed sediment fines. For the 1990 case, he used the amount of river-bed sediment fines in autumn 1989 surveyed by Kurashige (1996) as an initial condition of the simulation. The initial condition was set for the start of the snowmelt season, and the SSC was calculated from time to time. The simulation explained well the trend of SSC change measured in this season. In contrast, in the 1993 case, he could not obtain the amount of river-bed sediment fines present before the snowmelt season. The trial-and-error method was thus used to find the initial amount of river-bed sediment fines (i.e. the initial condition of the simulation) which explains all the peak values of SSC and their timing measured in this season, and as a result the initial amount was set to be 25% of the initial condition of the 1989 case. Consequently, Kurashige (1998) determined that the origin of suspended sediment in the snowmelt season in 1993 was the river-bed sediment fines.

11.4.4 The Origin of Suspended Sediment in Summer 1993

The ratio DSF:SSR was estimated to range between 100:0 and 60:40, and the ratio $S_d1:S_d2$ between 60:40 and 10:90 for river-suspended sediment on 28 August 1993 (Table 11.2). However, on this day, the peak of SSC appeared earlier than the peak discharge (Figure 11.2D), suggesting that the river-suspended sediment may have been supplied from the river-bed sediment. The compounded distributions at a certain ratio of $S_d3:S_d4$ which is similar to the grain-size distribution of river-suspended sediment was thus found at around the time of peak SSC (Table 11.2). The ratio $S_d3:S_d4$ ranged between 45:55 and 75:25.

In addition, the process-based physical model results closely matched the peak of SSC on this day (Figure 11.4). In this calculation, the initial conditions on 26 August were made identical to those of the snowmelt season of 1993. The simulation results indicate that the river-suspended sediment around peak SSC was supplied mainly from river-bed sediment.

In the process-based model, d_{max} is calculated and all the grains smaller than d_{max} in the tractive layer are stirred up. The d_{max} calculated from 04:00 to 06:00 on 28 August 1993 ranged between 4.0ϕ and 4.4ϕ. The grain-size distribution of the river-bed fines can be represented by two subdistributions S_d3 and S_d4, and the calculated d_{max} coincides with the diameter of the second to fifth percentile of S_d3 (Figure 11.3B). This shows that almost all grains of subdistributions S_d3 and S_d4 could be stirred up during this period. Accordingly, the ratio $S_d3:S_d4$ in the suspended sediment cannot change much unless the ratio $S_d3:S_d4$ in the river-bed fines itself changes. The rapid change of the ratio in the river-bed fines is unlikely to occur unless a new sediment with a different ratio is supplied. The simulation results with unique initial conditions indicate that a new supply of river-bed fines is unlikely to have occurred. This suggests that we cannot determine whether almost all the suspended sediment was supplied from the river-bed fines during this period.

In contrast, at 06:00 on 28 August (i.e. close to the peak discharge of this day), the compounded distributions with various $S_d3:S_d4$ ratios were not similar to the grain-size distribution of river-suspended sediment, whereas the compounded distribution with DSF:SSR = 60:40 and $S_d1:S_d2 = 10:90$ was judged to be similar to the grain-size distribution of river-suspended sediment. The origin of river-suspended sediment at this time was thus judged to be both DSF and SSR, suggesting that the suspended sediment during this event was also supplied from DSF and SSR.

The response of the hydrograph to rainfall is quick in this basin (Ishii 1990). In addition, the non-uniform distribution of rainfall intensity in the basin leads to a time lag of only about ten minutes for increases in the river discharge (Kurashige 1993a). These findings show that DSF and/or SSR supplied by the highest intensity or those supplied by the maldistributed intensity of rainfall cannot produce peak SSC about one hour earlier than the peak discharge. The origin of suspended sediment at peak SSC in this event cannot be determined only to be DSF or SSR.

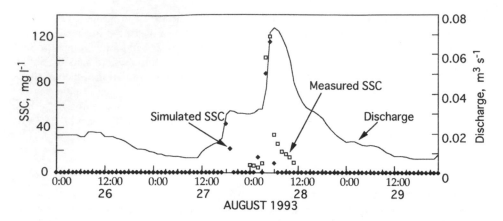

Figure 11.4 Simulated SSC for the event in summer 1993 based on the process-based stirring-up model. River discharge and measured SSC are also shown to compare with the simulated results

Therefore, the suspendaged sediment in this event was determined to be supplied not only from river-bed fines but also from DSF and SSR. In addition, the occurrence of peak SSC earlier than the peak discharge was judged to be strongly affected by the stirring up of river-bed fines.

11.5 DISCUSSION AND CONCLUDING REMARKS

In summer 1992 (i.e. just after the logging), the peak of SSC appeared close to the peak of river discharge, and the river-suspended sediment was supplied from both DSF and SSR. In contrast, in the early snowmelt season and in summer 1993, the peak of SSC appeared earlier than the peak discharge, and the suspended sediment supply from the river bed strongly affects the appearance of the peak SSC. Work by Kurashige (1993b, 1994b) conducted in this basin from summer 1989 to the snowmelt season of 1990 has shown that the peak of SSC does not appear earlier than the peak discharge in summer, whereas it does in the snowmelt season. This shows that the river-bed sediment fines were exhausted in summer 1989, whereas they accumulated before the snowmelt season of 1990 (Kurashige 1994b). From this analogy, the river-bed sediment fines were considered to have scarcely accumulated in summer 1992, whereas the fines had been accumulated before both the snowmelt season and the summer of 1993.

Davies and Nelson (1993) buried gravel-filled traps in the beds of streams draining steep logged and unlogged basins, and revealed that the amount of trapped sediment fines for all logged stream sites were significantly greater than those for unlogged sites. Moreover, the amount of trapped sediment fines at the sites in the basins logged three to four years before the measurement were higher than those logged one to two years or five to six years before. These results indicate that the accumulation rate of river-bed fines from the logged bare surface was largest at three to four years after the logging. Analogously, in the Hiyamizusawa Brook basin, the accumulation rate of river-bed fines from DSF and/or SSR in summer 1993 was likely to be much larger than that in summer 1992. Consequently, in summer 1993, the river-bed fines could be a major source of river-suspended sediment.

The difference in the time variations of SSC in the two summer cases indicates that the relationship between SSC and river discharge in summer 1993 was very different from that in summer 1992. Such a difference in the relationship was also reported by Johnson (1988). He obtained a significant correlation between SSC and discharge one year after the logging, whereas no apparent relationship could be found two years after the logging. Since the process of suspended sediment supply strongly affects the appearance of the hysteresis between SSC and discharge (Kurashige 1994b) and the hysteresis affects the trend of the sediment rating curve (e.g. Walling and Webb 1987), the difference in the relationship in Johnson's case indicates that the process of suspended sediment supply in the second year was likely to be different from that in the first year.

The relationship between grain size of suspended sediment and river discharge and that between grain size and SSC were tested by Walling and Moorehead (1987) and Clark (1995), respectively. Walling and Moorehead (1987) examined the relationships at eight measuring sites in the Exe basin, UK. At four sites, the percentage of clay tended to decrease and the percentage of sand tended to increase with the increase of discharge, but such tendencies were not clear at the other four sites. They considered that the percentage of sand is increased by basic hydraulic controls such as increase of turbulence or shear stress, whereas the latter cases are caused by the erosional behaviour of cohesive fine-grained material or the change of dynamic contributing areas. Clark (1995) carried out tests in a small basin in the UK, and found that the percentages of very fine silt and fine silt tended to decrease with increased SSC to more than $600 \, mg \, l^{-1}$, whereas the percentages of medium silt, coarse silt and very fine sand tended to increase with increased SSC. He considered that these tendencies are due to an expansion of the area of bank covered by the floodwater on the rising limb of the hydrograph, or by an increasing contributing area of run-off on the valley-side slopes. However, in these two studies, the explanation for the grain-size change could not be proven.

In this study, the ratio DSF:SSR ranged between 60:40 and 80:20 when the origin of suspended sediment was judged to be DSF and SSR. No clear relationship between the ratio and discharge or SSC could be found (Table 11.1 and Figure 11.2). However, the percentage of S_d2 in DSF was more than 50% at around the peak discharge (14:00, 14:30 and 17:00 on 13 August, see Table 11.1). This shows that very fine grains in DSF easily flowed out during high intensity rainfall. Also on 25 September when the discharge had been maintained at high stage, the percentage of S_d2 exceeded 30% at 12:00, 12:30 and 14:00, indicating that very fine grains easily flowed out from DSF. In contrast, in summer 1993, the percentage of S_d2 was determined to be 90% at 06:00 on 28 August, suggesting that the amount of available grains of S_d1 had been decreased in this case, but no clear judgement is possible with only this unique data. In any case, the change in grain size could be explained by the size sorting of grains when the suspended sediment was supplied from the origin.

On 13 August 1992, the SSF at peak discharge was $10.5 \, g \, s^{-1}$ at 14:30, and $6.0 \, g \, s^{-1}$ at 17:00. On 25 September 1992, SSC tended to decrease while the discharge was at high stage (Figure 11.2B). These findings show that the amount of available sediment in DSF and SSR tended to decrease during each rainfall event, as found for the sediment supply in a UK basin by Walling and Webb (1982). At 11:30 on 25 September, SSF was $25.2 \, g \, s^{-1}$, and this value was larger than the SSF at the first peak discharge on 13 August. In contrast, the SSC at 11:30 on 25 September was about half of the SSC at 14:30 on 13 August, whereas the peak discharge on 25 September was much larger than that on 13 August (Figure 11.2A, B). This shows that the availability of fines in DSF and/or SSR for rainfall intensity had already decreased on 25 September. However, the decrease in availability was not great, and accordingly the SSF on 25 September was larger than that on 13 August.

This study has made clear that the contribution rate of a source of river-suspended sediment and the ratio between subdistributions of grain size supplied from sediment origin change with time after logging. In addition, the accumulation of fine grains in river-bed sediment is also affected by logging. This suggests that the nature of suspended sediment supply is very complicated when a basin is impacted by logging. Further study to reveal these complicated processes should be pursued in order to better understand the impact of logging in a forested basin.

ACKNOWLEDGEMENTS

The author is much indebted to the very kind review reading of the early version of this chapter by two anonymous referees. Part of this study was made possible through Special Coordination Funds for Promoting Science and Technology (Joint Research Utilizing Scientific and Technological Potential in the Region) of the Science and Technology Agency of the Japanese Government.

REFERENCES

Clark, C. 1995. Sediment sources and their environmental controls: In Foster, I., Gurnell, A. and Webb, B. (eds), *Sediment and Water Quality in River Catchments*. Wiley, Chichester, 121–141.

Davies, P.E. and Nelson, M. 1993. The effect of steep slope logging on fine sediment infiltration into the beds of ephemeral and perennial streams of the Dazzler Range, Tasmania, Australia. *Journal of Hydrology*, **150**, 481–504.

Dearing, J.A., Morton, R.I., Price, T.W. and Foster, I.D.L. 1986. Tracing movements of topsoil by magnetic measurements: two case studies. *Phys. Earth Planet. Int.*, **42**, 93–104.

Ishii, Y. 1990. Run-off characteristics of subsurface flow from forested hillslopes in headwaters. *Geophysical Bulletin of Hokkaido University*, **53**, 1–24 (in Japanese with English abstract and illustrations).

Johnson, R.C. 1988. Changes in the sediment output of two upland drainage basins during forestry land use changes. *IAHS Publication*, **174**, 463–471.

Kurashige, Y. 1985. Model for pulling up fine particles from armour-coated gravel bed in the early snowmelt season. *Transactions of Japanese Geomorphological Union*, **6**, 287–302.

Kurashige, Y. 1993a. Mechanism on delayed appearance of peak suspended sediment concentration in a small river. *Transactions of Japanese Geomorphological Union*, **14**, 385–405.

Kurashige, Y. 1993b. Mechanism of suspended sediment supply to headwater rivers and its seasonal variation in West Central Hokkaido, Japan. *Japanese Journal of Limnology*, **54**, 305–315.

Kurashige, Y. 1994a. Minute structure of river-bed sediment surface and its temporal variation in a headwater river. *Annals of Hokkaido Geographical Society*, **68**, 15–20 (in Japanese).

Kurashige, Y. 1994b. Mechanisms of suspended sediment supply to headwater rivers. *Transactions of Japanese Geomorphological Union*, **15A**, 109–129.

Kurashige, Y. 1996. Process-based model of grain lifting from river bed to estimate suspended sediment concentration in a small headwater basin. *Earth Surface Processes and Landforms*, **21**, 1163–1173.

Kurashige, Y. 1998. Process-based estimation of suspended-sediment concentration during the thaw season in a small headwater basin. *IAHS Publication*, **249**, 415–422.

Kurashige, Y. and Fusejima, Y. 1997. Source identification of suspended sediment from grain size distribution. I: Application of nonparametric statistical tests. *Catena*, **31**, 39–52.

Loughran, R.J., Campbell, B.L. and Walling, D.E. 1987. Soil erosion and sedimentation indicated by caesium 137: Jackmoor Brook catchment, Devon, England. *Catena*, **14**, 201–212.

Reid, L.M. and Dunne, T. 1984. Sediment production from forest road surfaces. *Water Resources Research*, **20**, 1753–1761.

Reid, L.M., Dunne, T. and Cederholm, C.J. 1981. Application of sediment budget studies to the evaluation of logging road impact. *New Zealand Journal of Hydrology*, **20**, 49–62.

Walling, D.E. and Moorehead, P.W. 1987. Spatial and temporal variation of the particle-size characteristics of fluvial suspended sediment. *Geografiska Annaler*, **69A**, 47–59.

Walling, D.E. and Webb, B.W. 1982. Sediment availability and the prediction of storm-period sediment yields. *IAHS Publication*, **137**, 327–337.

Walling, D.E. and Webb, B.W. 1987. Suspended load in gravel-bed rivers: UK experience. In Thorne, C.R., Bathurst J.C. and Hey, R.D. (eds), *Sediment Transport in Gravel-Bed Rivers*. Wiley, Chichester, 691–732.

Walling, D.E. and Woodward, J.C. 1992. Use of radiometric fingerprints to derive information on suspended sediment sources. *IAHS Publication*, **210**, 153–165.

Walling, D.E., Peart, M.R., Oldfield, F. and Thompson, R. 1979. Suspended sediment sources identified by magnetic measurements. *Nature*, **281**, 110–113.

12 Vertical Redistribution of Radiocaesium (^{137}Cs) in an Undisturbed Organic Soil of Northeastern France

LIONEL MABIT
Département des sols et de génie agroalimentaire, Université Laval, Sainte-Foy, Canada

CLAUDE BERNARD
Institut de recherche et de développement en agroenvironnement, Sainte-Foy, Canada

STANISLAS WICHEREK
Centre de Biogéographie-Ecologie, Paris, France

AND

MARC R. LAVERDIÈRE
Département des sols et de génie agroalimentaire, Université Laval, Sainte-Foy, Canada

ABSTRACT

Caesium-137 (^{137}Cs) is a particularly well suited indicator for environmental studies involving assessments of soil erosion by water. Soil movement and sediment budgets have been estimated in many parts of the world over the last 15 to 20 years from ^{137}Cs spatial redistribution data, but mainly on mineral soils. Very few studies have reported the applicability of the ^{137}Cs technique to organic soils. It was thus necessary to run preliminary tests before initiating any large-scale study for organic soils.

To evaluate the potential of this technique to quantify redistribution of organic soils, we compared total inventories of ^{137}Cs in three profiles of an organic soil with that in a nearby uncultivated mineral soil with a typical ^{137}Cs depth distribution. They confirm the potential use of ^{137}Cs for the investigation of erosion/deposition processes on organic soils and of the impacts of agrosystems on humid areas.

12.1 INTRODUCTION

The preservation of high quality soils has not received the attention it deserves, despite the quantitative and qualitative limitation of the resource worldwide. Intensification and specialization during the last few decades has made agriculture more efficient, but at the cost of a decrease in soil and water quality.

The best soils of the world are now threatened not only by urbanization but also by a decrease of natural fertility through erosion, flooding and mud slides. The north European plain is no exception, despite the rather mild climatic conditions of the area (Mabit 1999). Changes in agricultural practices have considerably modified the behaviour of run-off waters. With a reduced protective plant cover, direct run-off has increased, carrying sediments, organic matter, nutrients and pesticides to pollute surface and underground waters.

Applied Geomorphology: Theory and Practice. Edited by R.J. Allison.

Caesium-137 (^{137}Cs) was introduced into the environment mainly through atmospheric nuclear tests, in the 1950s and 1960s. In Europe, the Chernobyl accident in 1986 added to the inventory (Anspaugh et al. 1988). Except for rare cases of intensive biodisturbance, the mobility of radiocaesium in soils is very restricted and ^{137}Cs is mainly redistributed in the landscape with sediments which are eroded and redeposited. Consequently, it has been widely used as a soil movement tracer over the last 15–20 years (Ritchie and McHenry 1990; Bernard et al. 1998a).

Humid areas play an important ecological role by sustaining biodiversity. Those located in agricultural landscapes are under the direct influence of cultivated areas. They filter pollutants, such as sediments and nutrients, carried by waters. Pollutant accumulation in these zones can then be used as a measure of the impact of upstream agricultural areas. ^{137}Cs is an interesting indicator of these processes.

However, most of the previous studies involving the use of this isotope were on mineral soils, in which ^{137}Cs is strongly retained in the upper layer of the soil and vertical migration is very limited (Ritchie et al. 1972; Bachhuber et al. 1982; Bunzl et al. 1989). The capacity of organic soils to retain the fallout radiocaesium is less well known. It was therefore necessary to investigate the behaviour of fallout ^{137}Cs in organic soils prior to any further widespread investigations.

12.2 MATERIAL AND METHODS

12.2.1 The Study Area

The Souche marshes are located in the Parisian Basin, some 100 km northeast of Paris, in the Aisne Department (Figure 12.1). They are located on Tertiary chalk plateaux of the circular sedimentary megastructure of the Parisian Basin. Quaternary loess was deposited on this Lutetian structure. These aeolian deposits are the parent material of a brown soil which developed between 5000 and 10 000 years ago.

The marshes cover some 3000 ha with peat depths reaching several metres. Peat extraction from the marshes was important between 1830 and 1914. From the early 1930s, it gradually decreased and then stopped completely after World War II (Sajaloli 1993). Recently, the ecological importance of the marshes has been recognized, as they are one of the 37 sites of the Life/Natura 2000 programme in France.

The marshes are surrounded by large and intensively cropped areas where stock breeding has almost disappeared. The main rotations include wheat, sugar beet and potatoes. The average annual precipitation in the area is 700 mm. The slopes of the cultivated mineral soils are moderate and the surface textures are dominated by silt loams that are prone to crusting and water erosion. Soil erosion estimates based on ^{137}Cs measurements in a 1.8 km^2 basin located 30 km from the Souche area, but with similar agri-environmental conditions, suggest that several tonnes per hectare of sediments and associated pollutants coming from surrounding cultivated areas could reach the marshes annually (Bernard et al. 1998b).

12.2.2 Characteristics of the Peat in the Souche Marshes

In summer 1995, physico-chemical characterization was done on a peat sample. The measured parameters were: bulk density, water content (w/w dry basis), organic matter (loss on ignition), total N, C/N ratio, pH (H_2O), cation exchange capacity (CEC), Mehlich-3

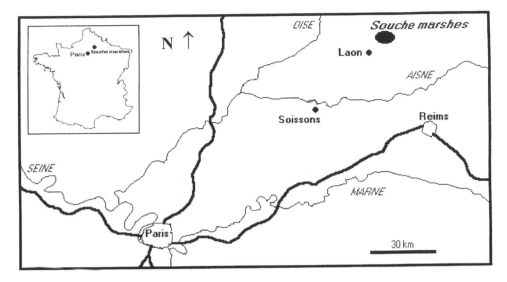

Figure 12.1   Location of the experimental site in France

extractable P, K, Ca and Mg (Carter 1993). Humic and fulvic acids were extracted following the procedure of N'Dayegamiye and Côté (1989). Absorbances were measured at 465 and 665 nm and expressed as the E4/E6 ratio. A low ratio is indicative of mineralized and condensed organic matter, while higher ratios indicate a low degree of condensation and the presence of important proportions of aliphatic structures (N'Dayegamiye and Côté 1989).

The results are reported in Table 12.1. The high Ca values reflect the nature of the substratum and contribute to the near-neutral pH values. The organic matter content ranges from 62 to 87%. The relative content of humic and fulvic acids, expressed as the E4/E6 ratio, as well as the C/N ratio and the high content of nutrients (P, K, Mg) all suggest that the organic matter has reached a high degree of decomposition. This, in turn, would explain the high CEC measured. The increases, with depth, of organic matter content and of the E4/E6 and C/N ratios, together with decreases in CEC, bulk density and nutrient content, indicate that mineralization has proceeded from the top downwards. A lower water content near the surface, and consequently better aeration, explains this.

12.2.3   Soil Sampling and Gamma Analysis

Three peat profiles, located within 5 m of each other, were sampled in an area that is uncultivated, topographically flat and hydrologically isolated from the surrounding agricultural land. Samples were taken with a cylinder at 10 cm depth increments over 0–40 cm depth, and with an auger, by 20 cm increments, between 40 and 100 cm. A permanent and non-eroded meadow on a nearby mineral soil was also sampled at 10 cm increments, to a depth of 40 cm.

^{137}Cs was determined on all the samples from the four soil profiles (three organic and one mineral) using the procedure of de Jong et al. (1982). The areal activity (in Bq m^{-2}) was obtained from the product of specific activity (in Bq kg^{-1}), the bulk density (in kg m^{-3}) and the depth (in m).

Table 12.1 Physico-chemical parameters of the peat in the Souche marshes

Depth (m)	Bulk density (kg m^{-3})	Water content (% w/w)	Ignition loss (%)	E4/E6	Total N (%)	C/N	pH	CEC (cmol kg^{-1})	P (mg kg^{-1})	K (mg kg^{-1})	Ca (mg kg^{-1})	Mg (mg kg^{-1})
0–0.1	235	315	65.2	5.51	2.35	13.9	6.68	168	61	387	14 234	341
0.1–0.2	242	312	62.6	5.68	2.40	13.0	6.27	117	33	75	8337	309
0.2–0.3	132	600	87.3	5.60	2.55	17.0	5.91	138	23	47	3707	243
0.3–0.4	120	725	89.2	5.95	2.56	17.4	5.81	132	17	31	2929	239
0.4–0.6	115	741	85.1	5.64	2.71	15.6	6.16	194	8	29	1291	137
0.6–0.8	147	620	76.0	5.85	2.53	15.0	6.02	217	7	49	1773	106
0.8–1.0	115	741	87.3	6.58	2.78	15.6	6.27	94	4	24	928	80

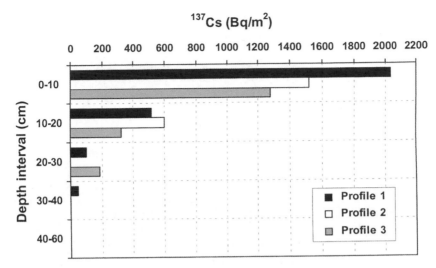

Figure 12.2 ^{137}Cs inventory and vertical distribution in the three organic soil profiles

12.3   RESULTS AND DISCUSSION

Figure 12.2 shows the vertical distribution of ^{137}Cs in the peat profiles. All three profiles exhibit a similar pattern, with an exponential decrease of the ^{137}Cs content with depth. Between 71% (profile 3) and 75% (profile 1) of the total inventory was found in the first 10 cm. No radiocaesium was detected below 0.2 m, for profile 2, and below 0.3 m, for profile 3. For profile 1, 2% of the total inventory was found in the 0.3–0.4 m increment.

The total ^{137}Cs inventories amounted to 2695, 2115 and 1785 Bq m^{-2} for profiles 1, 2 and 3 respectively, for a mean value of 2198 ± 461 Bq m^{-2} (mean ± standard deviation). The high inventory in profile 1 can be explained by the important concentration measured in the 0–0.1 m increment, which amounted to 86.5 Bq kg^{-1} and translated into a 2035 Bq m^{-2} inventory. In the other two profiles, the concentration for the same depth increment was significantly lower, with values of 59 and 69 Bq kg^{-1}.

The profile in the mineral soil shows a vertical distribution of radiocaesium similar to that of the peat profiles (Figure 12.3). The total inventory is 2150 Bq m^{-2} and 84, 14 and 2% of it are found in the 0–0.1, 0.1–0.2 and 0.2–0.3 m increments respectively.

It therefore seems that the organic soil encountered in the Souche marshes can efficiently retain ^{137}Cs fallouts with very limited vertical migration. The total inventory, averaged over the three peat profiles, and the vertical distribution are similar to those measured in the mineral soil under a nearby permanent meadow. Other researchers have reported ^{137}Cs inventories for uneroded mineral soils that are consistent with those of the organic soils of the Souche marshes. For example, Bernard et al. (1998b) measured an average inventory of 2000 ± 200 Bq m^{-2} for 13 grassland sites located some 30 km southwest of the Souche area, and Vanden Berghe and Gulinck (1987) reported a 2550 Bq m^{-2} value (approximately 2020 Bq m^{-2} when decayed to 30 June 1996) for control stations in Belgium, before the Chernobyl accident. The ^{137}Cs inventories we measured therefore suggest that the Chernobyl fallout was minimal in the Souche area. The only actual measurement of the Chernobyl fallout near the study area is 260 Bq m^{-2} or 200 Bq m^{-2} on 30 June 1996 (Cambray et al. 1987) for the city of Soissons which is 30 km away (Figure 12.1).

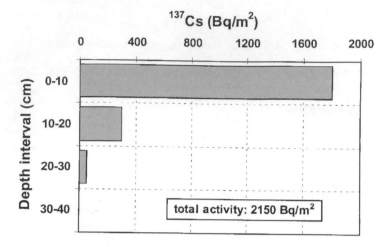

Figure 12.3 ^{137}Cs inventory and vertical distribution in an uneroded mineral soil profile

The retention capacity of the organic soils of the Souche marshes decreases the risk of groundwater contamination by radiocaesium, although the water table is frequently encountered at 0.3–0.5 m depth. The high humification degree of the organic matter, revealed by its physico-chemical characteristics, and the high CEC values of the soil explain this retention efficiency. In less humified ombrotrophic peat, Oldfield *et al.* (1979) noted some evidence of downward diffusion of ^{137}Cs.

12.4 CONCLUSION

It appears from the measurements reported in this study that no major vertical migration of ^{137}Cs took place in the organic soils of the Souche marshes, and that these soils were efficient at retaining the radiocaesium from the historic fallout. These undisturbed organic soils thus seem to behave in an identical manner to undisturbed mineral soils. This agrees with the fact that caesium is strongly adsorbed by soils and that its vertical mobility is generally limited.

These results also suggest that erosion studies using ^{137}Cs are valid on highly organic soils, using uncultivated sites to establish the ^{137}Cs base levels. The technique could then be used to assess the impacts of the surrounding cultivated fields on the inputs of sediment in humid areas and to investigate long-term soil movement in organic soils.

ACKNOWLEDGEMENTS

The authors wish to thank Dr Adrien N'Dayegamiye and Pierre Audesse, of the Institut de recherche et de développement en agroenvironnement (Québec City, Canada), for useful advice and comments and for the soil analyses respectively.

REFERENCES

Anspaugh, L.R., Catlin R.J. and Goldman, M. 1988. The global impact of the Chernobyl reactor accident. *Science*, **242**, 1513–1519.

Bachhuber, H., Bunzl, K. and Schimmack, W. 1982. The migration of ^{137}Cs and ^{90}Sr in multilayered soils: results from batch, column, and fallout investigations. *Nuclear Technology*, **59**, 291–301.

Bernard, C., Mabit, L., Laverdière, M.R. and Wicherek, S. 1998a. Césium-137 et érosion des sols. *Cahiers Agricultures*, **7**, 179–186.

Bernard, C., Mabit, L., Wicherek, S. and Laverdière, M.R. 1998b. Long-term soil redistribution in a small French watershed as estimated from ^{137}Cs data. *Journal of Environmental Quality*, **27**, 1178–1183.

Bunzl, K., Schimmack, W., Kreutzer, K. and Schierl, R. 1989. The migration of fallout ^{134}Cs, ^{137}Cs and ^{106}Ru from Chernobyl and of ^{137}Cs from weapons testing in a forest soil. *Journal of Plant Nutrition and Soil Science*, **152**, 39–44.

Cambray, R.S., Cawse, P.A., Garland, J.A., Gibson, J.A.B., Johnson, P., Lewis, G.N.J., Newton, D., Salmon, L. and Wade, B.O. 1987. Observations on radioactivity from the Chernobyl accident. *Nuclear Energy*, **26**, 77–101.

Carter, M.R. (ed.) 1993. *Soil Sampling and Methods of Analysis*.

de Jong, E., Villar, H. and Bettany, J.R. 1982. Preliminary investigations on the use of ^{137}Cs to estimate erosion in Saskatchewan. *Canadian Journal of Soil Science*, **62**, 673–683.

Mabit, L. 1999. *Estimation de l'érosion hydrique des sols par la méthode du ^{137}Cs. Application aux bassins versants de Vierzy (France) et Lennoxville (Québec)*. Doctoral thesis, Université de Paris I Panthéon-Sorbonne.

N'Dayegamiye, A. and Côté, D. 1989. Effect of long-term pig slurry and solid cattle manure application on soil chemical and biological properties. *Canadian Journal of Soil Science*, **69**, 39–47.

Oldfield, P.G., Appleby, G., Cambray, R.S., Eakins, J.D., Barber, K.E., Battarbee, R.W., Pearson, G.R. and Williams, J.M. 1979. ^{210}Pb, ^{137}Cs and ^{239}Pu profiles in ombrotrophic peat. *Oikos*, **33**, 40–45.

Ritchie, J.C. and McHenry, J.R. 1990. Application of radioactive fallout cesium-137 for measuring soil erosion and sediment accumulation rates and patterns: A review. *Journal of Environmental Quality*, **19**, 215–233.

Ritchie, J.C., McHenry, J.R. and Gill, A.C. 1972. The distribution of ^{137}Cs in the litter and upper 10 cm of soil under different cover types in northern Mississippi. *Health Physics*, **22**, 197–198.

Sajaloli, B. 1993. *Les zones humides laonnoises. Fonctionnement, usages, gestion*. Doctoral thesis, Université de Paris I.

Vanden Berghe, I. and Gulinck, H. 1987. Fallout ^{137}Cs as a tracer for soil mobility in the landscape framework of the Belgian loamy region. *Pedologie*, **37**, 5–20.

13 Weathering Surveys in Geomorphology: Mapping Examples from Sila Massif (Calabria, Italy)

SILVIO DI NOCERA AND FABIO MATANO
Dipartimento di Scienze della Terra, Università Federico II, Napoli, Italy

ABSTRACT

In weathered crystalline rocks the evolution of slopes, and particularly of instability phenomena, is primarily controlled by the characteristics and thickness of the weathering profile. Field analysis, classification and mapping of the weathering grade are therefore basic tools in general geomorphological and engineering morphological studies. The importance of a comprehensive overview of the weathering patterns also concerns territorial planning studies and natural resource management, which need a cartographical representation of the weathering profile. This requirement prompted the definition of a methodology for the survey and mapping of the weathering grade of crystalline rocks, applicable to various scales. This chapter describes a methodology for the classification and survey of the weathering grade in gneiss. It consists of a detailed survey in the field, which is based on a careful analysis of the outcropping rocks and soils through observations of strength, discoloration and texture of the regolith and the results of Schmidt hammer tests. Six weathering classes have been adopted: fresh (class I), slightly weathered (class II), moderately weathered (class III), highly weathered (class IV) and completely weathered (class V) rock, and residual and colluvial soils (class VI). The weathering grade survey on cuttings has resulted in the collection of useful data regarding thickness and features of the weathering profile horizons.

This approach has been used for deeply weathered crystalline rocks of the Sila Massif (northern Calabria, southern Italy), which are strongly affected by landsliding. After a preliminary study of the weathering conditions along the western slope (gneiss) and on the plateau (granite) of the massif, detailed weathering surveys of selected sectors of the slope have been performed and various weathering maps with scales of 1:5000 to 1:200 have been plotted. The surveys provide evidence of complex and deep weathering profiles. This approach can improve our understanding of engineering geomorphological problems, such as rockwall stability and landsliding.

13.1 INTRODUCTION

Slopes are the result of complex relationships between denudational and weathering processes. In regions where deeply weathered crystalline rocks crop out (as in Hong Kong, England and Australia), slope evolution and particularly instability phenomena are mainly controlled by the characteristics and thickness of the weathering profile (Ollier 1984).

Classification and mapping of the weathering grade of crystalline rocks are very useful tools both in general morphological and in engineering geomorphological studies. Many

Applied Geomorphology. Theory and Practice. Edited by R.J. Allison.
© 2002 John Wiley & Sons, Ltd.

weathering classification schemes for engineering, geological and geomorphological purposes are available in the scientific literature (Dearman 1976; Gamon 1983; GSEGWP 1995). Such schemes use classification parameters and criteria which differ according to the aims for which they have been set (Lee and de Freitas 1989). They can be based on the assessment of engineering properties by simple index tests and/or on the visual evaluation of the geological conditions. Examples of studies concerning mainly gneiss include: Knill and Jones (1965) for dam foundation investigations in Sudan; Hall (1985, 1986, 1987) for analysis of the geomorphological significance of the weathering patterns at different scales in Scotland; Dobereiner et al. (1993) for the description of the weathering profile in relation to its mineralogical and textural changes as well as its main geotechnical parameters in Massif Central (France); and Thomas (1966) for the study of the geomorphological implications of deep weathering patterns in crystalline rocks in Nigeria.

In Calabria, southern Italy, granitic and metamorphic rocks form the bulk of the main mountains and massifs. Deep weathering profiles have developed on these rocks and the territory contains various and complex landslide phenomena (Ippolito 1962; Nossin 1972; Guzzetta 1974; Ietto 1975; Verstappen 1977).

In recent years, multidisciplinary research (Cascini et al. 1992a, 1992b, 1994) has been carried out on the weathering and landsliding of gneissic rocks in a sample area located on the western slopes of the Sila Massif (northern Calabria, Italy). The studied area is characterized mainly by the outcropping of gneiss of the Palaeozoic, which has undergone a complex metamorphic history. Veins of aplite and pegmatite are interbedded with gneiss; migmatite gneiss and mylonite gneiss have also been found in the field. The compressive and extensional tectonic events have induced an intense state of fracturing in the rock mass.

Surveys of the weathering grade (Cascini et al. 1992b; Gullà and Matano 1994, 1997) and geological–structural (Matano and Tansi 1994; le Pera et al. 2001), morphologic (Matano 1991; Critelli et al. 1991a), petrographic (Critelli et al. 1991b), geotechnical (Cascini and Gullà 1993) and hydrological (Cascini and Versace 1988; Cascini et al. 1995) investigations have allowed us to describe a complex weathering profile extending to a depth of 80 m and a tendency to impulsive landslides involving c. 29% of the studied area.

Surveys on slope cuttings integrated with analysis of the stratigraphs of continuous core boreholes allowed us to estimate the total thickness of the gneiss involved in the weathering process (corresponding to c. 80 m), and to disclose the vertical sequence of the various weathering horizons, which is variable from site to site (Cascini et al. 1992b).

The evolution of the slopes is controlled by neotectonic uplifting, by the consequent deepening of the drainage network and by the development of the weathering processes; landslide phenomena play a major role in this evolution. We have postulated a morphological evolution model (Cascini et al. 1992b) which synthesizes the main processes that have contributed to the definition of the present setting of the territory: (a) development of the weathering profiles on the high portions of the slopes and of downcutting along the stream channels (Wurmian Interglacial stages); (b) severe physical–mechanical degradation of the gneiss and filling of the valley bottoms by debris which fossilized the river beds (Wurmian Glacial stages); (c) erosion of the debris covers and renewed deepening of the river beds. Where the discharge of watercourses has been sufficiently high, a complete erosion of the detritic accumulations occurred. On the contrary, along the streams where the discharge has been limited, only a partial erosion of the detritic sheets took place (Holocene–Actual).

A close relationship between slope failures, weathering profiles and water table oscillations has been proved. The piezometric and geotechnical data fit into the proposed geological model of slope evolution (Cascini et al. 1994), which shows two types of landslides. Along the slopes where weathered rock outcrops, rock slides and rock slumps are activated in fractured and weathered gneiss; such phenomena are activated by downcutting processes determined by stream waters at the slope base, which then progressively migrate towards the high portion of the slope. Earth and debris slides, soil slips and earth flows occur along the slopes mantled by saprolitic, residual and colluvial soils. The landslides take place directly in the cover soils, generally because of downcutting of the slope or because of variations of the groundwater regime. A detailed analysis of these landslides has demonstrated that they are controlled by buried structures such as hollows or structural steps.

With reference to intense rainfall, one of the key factors triggering landslides, we would point out that hydrologic research (Cascini and Versace 1988; E. Cascini et al. 1992) has allowed a definition of the relationships among rainfall, critical piezometric levels and the triggering phases of landslide movements. On the other hand, more complex models such as the hydrogeological models (Patton 1984) have allowed us to recognize the relationships between meteoric events, piezometric levels of the acquifer and landslide reactivation in gneissic covers (Cascini et al. 1995).

In this chapter, on the basis of the data collected during the above-mentioned research, we intend to provide evidence of the importance of using the weathering grade for mapping during studies on slope failures related to weathering of crystalline rocks. New data are presented with reference to the regional pattern of the weathering profile in Sila Massif and we also illustrate some examples of detailed weathering grade maps at various scales related to the lower sector of the Sila western slope.

13.2 GEOLOGICAL AND MORPHOLOGICAL OUTLINE OF THE SILA MASSIF (NORTHERN CALABRIA)

The geology of northern Calabria is characterized by crystalline allochthonous nappes which overthrust Apennine sedimentary formations in the Miocene. They were subsequently (from the Messinian to the Quaternary) covered by evaporitic and terrigenous sediments (Amodio Morelli et al. 1976). From the Pliocene a tectonic uplifting produced normal fault systems which are still active and give the general shape to the mountainous massifs. Tectonics is the most important factor of relief construction, but it also greatly influences the distribution and magnitude of dismantling processes. In particular the crystalline rocks are affected by widespread jointing and faulting so that granites and gneiss are very prone to weathering. Current tectonic activity plays an important role in intensifying the intrinsic weakness of the parent rocks and the landslide-prone geomorphological conditions.

The geomorphology of the territory of Northern Calabria is very complicated, presenting evidence of new forms due to active processes and ancient inherited forms, which have been greatly modified. The main landforms of Calabria are morphostructures generated by spatially and temporarily discontinuous tectonic uplifts which produced mountain massifs and highlands alternating with lowlands. The relief at the boundaries of elevated blocks like the Sila Massif is due to normal faulting arranged as narrow steps or degrading horst and graben sequences. Tectonic depressions, such as the River Crati valley, are sectors where the average rate of uplift is lower (Sorriso Valvo 1990).

Remnants of morphological surfaces (palaeosurfaces) are preserved in the highlands. The Sila Massif emerged after the Upper Miocene and its evolution is at a very advanced stage; indeed its morphology is gentle and remnants of a 'peneplane' can be seen. The age of this palaeosurface is ascribed to the end of the Lower Pleistocene (Dramis et al. 1990). The surface is truncated by several erosional episodes, including glacial events for which scarce and faint evidence has been found on the highest tops of Sila (Boenzi and Palmentola 1975). The present Calabrian climate is of mediterranean type with warm summers (Csa). It is apparent that the climatic differences in the Little Ice Age and in older glacial times have influenced the rates of morphodynamic processes.

The highlands are characterized by forest cover, whereas badlands dominate those slopes affected by intensive landlsiding. A large number of non-active alluvial fans can be found at the mouths of canyons where they enter intramontane tectonic valleys. Rapid mass-movement is widespread throughout, with humid temperate climate in the highlands intensifying soil creep and chemical weathering. Mass-movement phenomena appear as the most effective present factor of slope and stream channel development; in the last few decades, there has been a high incidence of landsliding and, as a consequence, a tendency to aggradation of stream channels (Sorriso Valvo 1988). The present widespread distribution of landslides is the result of previous periods of maximum landsliding so that, while nearly 30% of the land surface is affected by landsliding, only 30% of the landslides are still active (Carrara *et al.* 1982; Sorriso Valvo 1985).

13.3   MORPHOLOGICAL ASPECTS AND WEATHERING PROFILE IN WESTERN SILA

In the western sector of the Sila Massif, between the Crati River and the Trionto River, three main morphostructural sectors have been identified (Figure 13.1): (i) the Sila highland; (ii) the western Sila slope; and (iii) the Crati River valley bottom. Figure 13.2 shows the weathering characteristics of the area, which have been described in part in CAS.MEZ. (1968).

The *Sila highland* sector lies between 1200 m and 1900 m a.s.l.; the top plateau is the relict of a wider Lower Pleistocene palaeosurface (Dramis *et al.* 1990) and is bordered by steep slopes (Figure 13.1). Some Wurmian moraine deposits and glacial cirques have been described near Botte Donato Mountain, on the plateau, where the Wurmian snow line lies at 1650 m a.s.l. (Palmentola *et al.* 1990).

Granites and subordinately gneiss, slates and sedimentary rocks crop out in the highlands; granitic saprolite and residual soils crop out widely across the palaeosurface, where they are covered with Pleistocene terraced lacustrine and alluvial deposits (Calderoni *et al.* 1989) and by Holocene alluvial deposits in the topographic lows (Figure 13.2). In some localities, such as near Lake Cecita and Piano del Barone, saprolite and residual soils, which have been produced by the *in situ* weathering of the granites, crop out and are tens of metres thick.

The granites are extremely weathered and exhibit deep weathering profiles, which are characterized by thick horizons of saprolite and residual soil with rounded corestones (spheroidal weathering).

Weathering profiles are tens to some hundreds of metres deep (Guzzetta 1974), and are typically characterized by an upper horizon of organic sandy colluvial and residual soil, which overlies completely decomposed rock of a sandy texture within which the original discontinuities are still preserved (saprolite). A horizon of spheroidally weathered rock boulders, surrounded by sandy soil, underlies the saprolite and in the lowermost part of the profile there is jointed rock with discoloration along the major joints at depth.

The *western Sila slope* sector of the study area is characterized by an uneven topography with a criss-cross pattern of second-order morphological highs and lows (Figure 13.1).

Weathered gneiss, granite, schist and slate crop out along the slope (Figure 13.2). They are mainly characterized by moderate to high weathering grades, while fresh and slightly weathered rocks crop out only along deeply incised streams. Residual soils are only some

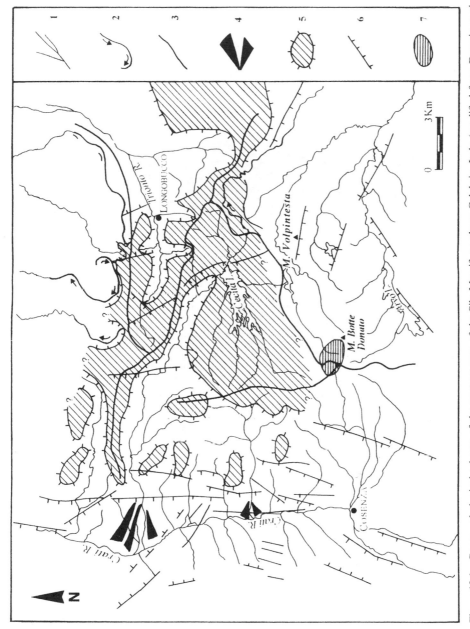

Figure 13.1 Geomorphological scheme of the western area of the Sila Massif, northern Calabria, Italy (modified from Dramis *et al.* 1990). Legend: 1, hydrographic network; 2, river capture; 3, divide; 4, fan; 5, Lower Pleistocene palaeosurface; 6, normal fault; 7, moraines

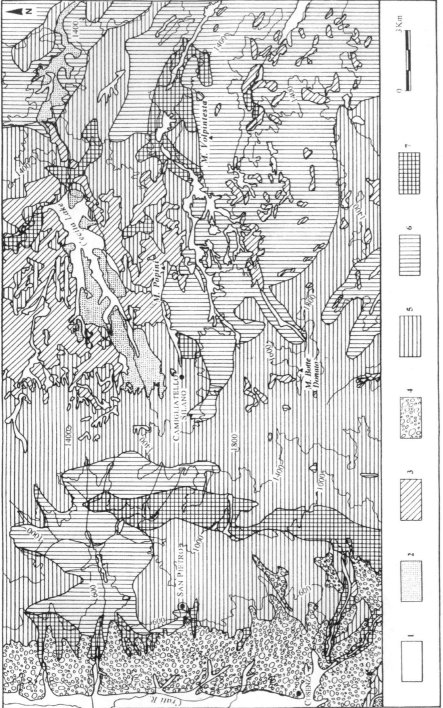

Figure 13.2 Weathering patterns and geolithological sketch map of a sector of the western area of the Sila Grande Massif, northern Calabria, Italy. Legend: 1, current and Holocene alluvial deposits; 2, Pleistocene terraced alluvial and lacustrine deposits; 3, continuous and deep covers of colluvial soils, residual soils and saprolite with local strips of alluvial deposits and few fresh or weathered crystalline rock outcrops (classes V and VI); 4, silty clays, conglomerates and arenites (Pliocene and Upper–Middle Miocene); 5, deeply weathered crystalline rocks with few partially weathered rock outcrops (classes IV and V); 6, variously weathered crystalline rocks with pockets of deep weathering and saprolite and isolated fresh rock outcrops (classes II to V); 7, generally fresh to moderately weathered crystalline rocks with rare pockets of deep weathering (classes I to III). For classes see Table 13.1

centimetres thick, while colluvial soils, slope debris and landslide debris are widespread and are metres to decametres thick because of the very active morphodynamics of the slopes.

In the foothills, at an altitude of about 750 m a.s.l., there is a mature and complex weathering profile about 80 m thick (Cascini *et al.* 1992b). On the basis of petrographical analysis, Critelli *et al.* (1991b) consider the weathering to have developed through chemical processes under humid climatic conditions.

The *Crati River valley bottom* sector has developed along an asymmetrical graben, which is bordered by several active faults (Figure 13.1). The tectonic valley is filled by silico-clastic deposits of Pliocene and Pleistocene ages and by Quaternary terraced alluvial deposits. Crystalline rocks do not crop out in this sector (Figure 13.2).

On the whole, the Sila Massif crystalline rocks are characterized by a deep and complex weathering profile. Guzzetta (1974) hypothesized an ancient tropical weathering in Calabria, so that the deep weathering profile would represent a fossil, partially eroded mantle. On the basis of petrographical analysis, Critelli *et al.* (1991b) consider the weathering profile of the Sila gneiss to have developed through chemical processes under humid climatic conditions.

Today Calabria has a mediterranean-type climate, but a reliable analysis of the observed decomposition and degradation features of the weathered rocks must conclude that they result from the accumulation of the weathering effects produced under climatic conditions since the emergence of the Sila Massif to the present time.

The regional tectonic evolution shows that the Sila Massif structural high had already emerged after the Tortonian (Lanzafame and Zuffa 1976). The development of the weathering profiles in Sila could have occurred between the Messinian and the Pleistocene under various climatic conditions. Under the current climatic conditions the erosional processes, as activated by tectonic uplifting, prevail over the weathering processes, although not enough to completely remove the exposed, fossil, deeply weathered mantle. The saprolite and regoliths crop out widely along the top palaeosurface of the Sila Massif, where they are *c.* 100 m thick, while the fossil weathering profile is highly eroded along the Sila western slope.

13.4   THE METHODOLOGY OF WEATHERING ANALYSIS AND MAPPING

A field procedure of simple and rapid classification of the crystalline rocks cropping out in western Sila has been defined for the purpose of mapping the weathering grade in order to understand the relationship between the state of weathering and the tendency of the rock mass to landsliding. Reference has been made in particular to the methodology proposed by the Geotechnical Control Office (1984, 1988) of Hong Kong for granitic and volcanic rock masses, which has been suitably modified and adapted for the lithologies of the Sila Massif.

13.4.1   Weathering Grade Classification and Field Survey

After an experimental phase, in which mainly gneissic rocks were tested, six weathering classes are identified: fresh gneiss (class I), slightly weathered gneiss (class II), moderately weathered gneiss (class III), highly weathered gneiss (class IV), completely weathered gneiss (class V), and residual and colluvial soils (class VI). This identification is based on criteria, such as visual analysis of geological characteristics and strength assessment by simple tests with geological and Schmidt hammers.

The characteristics of the different weathering classes are summarized in Table 13.1. The definitions of classes I to V are similar to those proposed by the Geotechnical Control Office

Table 13.1 Weathering classification of gneiss

Class	Rock material	Rock mass
I – Fresh gneiss	Rock unchanged from original state or only slightly stained along major joints.	Behaves as rock; 0% soil.
II – Slightly weathered gneiss	Rock discolored along discontinuities and slightly weakened; strength approaches that of fresh rock; greenish grey colour and brown colour only along discontinuities; N value more than 50.	Strength, stiffness and permeability affected; less than 10% saprolite and residual soil.
III – Moderately weathered gneiss	Rock with penetrative discoloration and considerably weakened but large pieces; cannot be broken by hands; greenish grey colour in mass and reddish brown in discontinuities; does not slake in water; N value 25–50.	Rock framework is still locked and controls strength and stiffness; matrix controls permeability; 10–30% saprolite and residual soils; moderately to highly weathered gneiss with slightly weathered to fresh corestones and with saprolite and residual soils in discontinuities.
IV – Highly weathered gneiss	Rock completely discoloured but largely weakened so that large pieces can be broken by hand; greyish to reddish brown colour; does not slake readily in water; N value 10–30.	Rock framework contributes to strength while soil controls permeability and stiffness; 30–50% saprolite and residual soils; highly weathered gneiss with rare less weathered corestones and more weathered rocks and soils in discontinuities.
V – Completely weathered gneiss	Rock wholly decomposed and disintegrated having soil consistency but original texture apparent and structural discontinuities relict; reddish to greyish brown colour; sandy gravel to gravelly sand grain size; slakes in water; considerably weakened; N value 0–15.	Weak grades control behaviour; more than 50% saprolite and residual soils with corestones of less weathered rock.
VI – Residual and colluvial soils	Soil derived by *in situ* weathering, which has lost original texture and fabric, and soil reworked and transported by colluvial processes; yellowish, reddish or greyish dark brown colour; from gravelly sand to sandy silt grain size.	Behaves as soil although relict fabric may still be significant; about 100% colluvial and residual soils with random saprolite relicts and very rare disordered corestones.

(1984, 1988) and the GSEGWP (1995). Class VI groups both the loose rocks formed by the weathering processes *in situ* (residual soils) and the soils made up of weathered material transported by slope processes (colluvium).

The evaluation of the weathering grade of a gneiss exposure and its mapping is carried out by rating the results of some observations (lithology, texture, grain, rock material discoloration, strength) and measurements by the Schmidt hammer with reference to the characteristics described in Table 13.1.

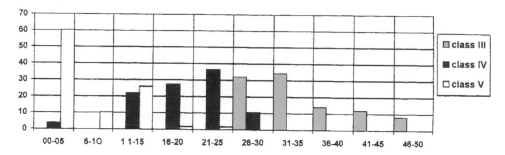

Figure 13.3 Frequency of readings (vertical axis) versus Schmidt hammer N rebound value ranges (horizontal axis) measured in gneiss of the Sila Massif

The Schmidt hammer has proved to be a very useful tool for weathering classification purposes. The measurements of the Schmidt hammer 'N' index (Geotechnical Control Office 1984), in conjuction with the other tests and visual observations, allow the gneiss weathering classes to be distinguished (Figure 13.3).

The weathering grade must be estimated both at the mass scale and at the material scale. However, in order to classify the rock in a given exposure and to map its weathering grade, it is necessary to consider the weathering grades of the outcropping rock material and also their percentage of distribution in the rock mass.

In order to obtain more data about the thickness and the geometrical characteristics of the weathering horizons, we have integrated the field surveys with a detailed analysis of cuttings.

13.4.2 Cutting Survey

In engineering geological and geomorphological studies on weathered rocks, a useful contribution to surface field study is represented by the detailed survey of the weathering grade on cuttings of the slope. The cuttings investigation comprises three elements: (a) cuttings inventory and preliminary weathering condition analysis; (b) structural analysis; (c) detailed survey of the most significant cuttings and measurements with the Schmidt hammer. These surveys provide several indications of the features and the thickness of the horizons of the weathering profile.

The inventory of cuttings found in the study area represents a preliminary but important part of the study, which permits the identification of exposures which are suitable for further research and introduces the geological and weathering conditions of the studied area.

For the detailed weathering grade survey of the most significant cutslopes a methodological approach has been proposed, based on Gullà and Matano (1994, 1997), enhanced by suggestions contained in ISRM Commission (1978), IAEG Commission (1981) and in Geotechnical Control Office (1984, 1988). Gullà and Matano (1997) have proposed a semi-quantitative procedure for the characterization of weathering grade which can be used on gneiss cuttings to identify zones of uniform engineering and morphological behaviour. This approach consists of three phases:

Phase 1: General characterization of the cutting
    1.1   Location on a topographic map and definition of the altitude
    1.2   Photographic survey
    1.3   Geological and morphological study with reference to the whole slope

Phase 2: Preliminary delimitation of the weathering zones
    2.1   Survey of lithology and structural discontinuities
    2.2   Preliminary delimitation of the zones referable to the different weathering classes on the basis of visual observations (discoloration and texture) and tests by geological hammer
Phase 3: Checkpoints, final delimitation and characterization of the different weathering grade zones
    3.1   Checkpoints arranged on the cutslope according to, if possible, a regular 2 m sided square mesh
    3.2   Confirmation or modification of the limits of the different weathering grade zones, by eventually utilizing further checkpoints
    3.3   Definition and characterization of the different weathering grade zones
    3.4   Photographic documentation of the different weathering grades with chromatic and size reference marks.

In the first phase we have to single out the geological and morphological elements which could have locally affected the weathering profile. In the second phase a preliminary delimitation of the weathering zones is made on the cutting on the basis of visual observations (discoloration, strength and texture), using a procedure very similar to the one used for the field survey.

In the last phase some observations and measurements are carried out in order to verify the correctness of the preliminary delimitation of the weathering zones. The first step consists of the identification of some checkpoints on the cutting. Then, the observations and measurements, summarized in the form showed in Figure 13.4, are carried out on these points. Every checkpoint is marked with a progressive identification number; then the preliminary weathering grade (column 1), as defined in phase 2, the lithology (column 2), the colour of the rock material (column 3) and its discoloration (column 4: A, complete; B, partial; C, only along discontinuities; D, none) are indicated.

Column 5 indicates if the point of a geological hammer: (A) easily indents in depth, (B) indents at the surface, or (C) produces a superficial scratch on the surface of the rock.

Column 6 indicates in which way a rock fragment about 15 cm long breaks away from the surface of the rock by geological hammer: (A) easily, (B) with difficulty, (C) with much difficulty, or (D) does not break away.

Column 7 describes the sound emitted from the rock when it is struck by geological hammer head, as: (A) dull, (B) intermediate, or (C) ringing.

Column 8 indicates if the breakage of a rock fragment about 15 cm long, struck by a geological hammer head, happens: (A) easily with a slight blow, (B) with a blow, (C) with a firm blow, (D) with two or more firm blows, or (E) does not happen.

Column 9 indicates if a rock fragment about 15 cm long worked by hand: (A) crumbles into constituent grains; (B) breaks up into smaller pieces, or (C) does not break at all.

Column 10 indicates if on the surface of the rock a pocket-knife point produces: (A) a deep peeling, (B) a superficial peeling, (C) a scratch, or (D) no significant trace.

Column 11 indicates if the immersion in water of some small rock samples produces: (A) complete loss of shape; (B) disaggregation into smaller fragments which maintain their shape, or (C) no modification of the original shape.

Column 12 reports the Schmidt hammer test from 1 to 10: the ten measured rebound values are indicated, while the value of the rebound index N of the Schmidt hammer test is calculated, according to the suggestion of the Geotechnical Control Office (1984), as the mean value of the five highest rebound values among the ten values measured.

Test point N°	1 Preliminary weathering grade	2 Lithology	3 Colour	4 Discolouration				5 Geological hammer point				6 Geological hammer head				7 Sound				8 Break by geological hammer					9 Break by hands				10 Pocket knife				11 Slaking			12 Schmidt Hammer rebound values										13 Final weathering grade	Note
				A	B	C	D	A	B	C	D	A	B	C	D	A	B	C	D	A	B	C	D	E	A	B	C	A	B	C	D	A	B	C	1°	2°	3°	4°	5°	6°	7°	8°	9°	10°	N		

Figure 13.4 Data collection form for observations and measurements at the checkpoints (from Gullà and Matano 1997). For the legend see text

Table 13.2 Criteria for the classification of weathering grade at checkpoints (from Gullà and Matano 1997)

Weathering class	Observation								
	4	5	6	7	8	9	10	11	12
I	C–D	C	C–D	C	D–E	C	D	C	>50
II	B–C	C	C	C	D	C	C–D	C	>45
III	A–B	B–C	B–C	B–C	C–D	B–C	C	C	26–45
IV	A	B	A–B	A–B	B–C	B	B–C	B	11–29
V	A	A–B	A	A	A–B	A	A–B	A	0–16
VI	A	A	A	A	A	A	A	A	0

See Figure 13.4 and text for abbreviations.

Column 13 assigns the final weathering grade, on the basis of the previous observations and measurements.

The overall examination of the acquired data allows confirmation or modification of the preliminary limits of the weathering zones; in cases that are more difficult to interpret supplementary checkpoints can be used.

At the end of the procedure operations the exposure is subdivided into uniform zones which are referred to different gneiss weathering grades. The final evaluation of the weathering grade is made by rating the results of the described observations, tests by the geological hammer and measurements by the Schmidt hammer, which have been performed in checkpoints (Figure 13.4).

The rating for the observations and the value range for the measurements have been defined for the various weathering classes of the gneiss and are shown in Table 13.2. The suggested methodology and the respective reference criteria ensure a sufficient homogeneity of the physical–mechanical and morphological behaviour of the cutting zones, which have been definitively assigned to the same weathering class. Besides, the framework so defined allows both the choice of the laboratory and *in situ* test techniques to be optimally addressed, and the localization of the sampling points for the petrographic and geotechnical characterization of the rock materials of the cutslope.

The preliminary weathering grade of the checkpoints, which is defined only by qualitative observations (column 1), is generally modified after check operations on the cutting in only 20% of cases (Gullà and Matano 1997). Therefore we can ascertain that the preliminary definition allows a narrow margin of error and that it can be usefully utilized to recognize the various weathering classes in large-scale investigations and field surveys.

13.4.3 Weathering Grade Mapping of Gneiss Terrain

The mapping representation of weathering profiles and grades permits a general and comprehensive overview of the effect, intensity and distribution of weathering processes. On the basis of the described methodology of weathering field analysis and classification, it is possible to survey and map the weathering condition of gneiss, or of other crystalline rocks, at different scales. Detailed field survey is based on a careful analysis of rock and soil outcrops, through visual observations (discoloration, texture, grain), geological hammer tests for qualitative strength evaluation, and Schmidt hammer measurements.

A 1:5000 scale map of gneiss weathering grade of a sample area has been prepared by Cascini *et al.* (1992b) by using the described weathering survey methodology. The area is located in the western slope sector of the Sila Massif (Figure 13.2). The map shows the surface distribution of gneiss weathering classes and gives some indications of the weathering profiles. The chosen scale has proved very useful for the acquisition of an effective synthesis framework and a helpful tool for the singling out of the research themes (i.e. landslide hazard and risk assessment, landslides and slope evolution, preliminary evaluation of rock mass physical condition and morphological behaviour) to be developed.

The necessity of carrying out other weathering grade surveys to a more detailed scale has resulted from the comparison of the distribution of the weathering classes and of the land-slides on the analysed territory, which has provided evidence of a correlation between the weathering conditions and the typologies of the landslides. A field survey on more detailed scale topographic maps (1:2000, 1:1000, 1:200) has therefore been carried out for sectors of particular interest in relation to different landslide types and combined with the characterization of selected cuttings to provide information on the weathering profiles.

13.5 EXAMPLES OF WEATHERING GRADE MAPS AT VARIOUS SCALES

By using the classification methodology described, some detailed weathering grade maps at different scales (from 1:5000 to 1:200 and on cuttings) have been realized with reference to a sample area, located at the foot of the Sila Massif western slope.

The 1:5000 scale has been used for morphological and landsliding studies on entire slopes and municipal territories. The 1:2000 to 1:1000 scales have been useful in the landsliding and morphological analysis of particular sectors of the slope or of inhabited areas, but also for infrastructural planning or slope management. The more detailed scales from 1:500 to 1:200 have been used for the study of single landslides or for particular engineering problems, such as road cuttings and quarry fronts.

An example of the 1:5000 scale map is shown in Figure 13.5, where the survey revealed that 9.9% of the outcropping gneiss belongs to class VI, 41.5% to class V, 47.7% to class IV and 0.9% to class III, whereas there was a total lack of outcropping fresh or slightly weathered gneiss (classes I and II), which have been found only in some boreholes 50 to 80 m in depth.

As for the areal distribution of the various weathering classes, the soils of class VI are present on flattened areas or along ridges (residual soils) as well as at the base of the main structural slopes, along the bottoms of watercourses and in palaeovalleys (colluvial soils). Alternations between the horizons of class IV (highly weathered) and class V (completely weathered) gneiss are very diffuse along the slopes. Class III rocks outcrop to a limited extent in zones of rapid erosion along the main streams.

In some sectors, located inside the 1:5000 map area, such as the Altavilla Hill area (Figure 13.6), a more detailed field survey based on a 1:2000 scale topographic map has been used. In this area, moderately to completely weathered gneiss, residual and colluvial soils, pegmatite veins and landslide debris are present.

The class III gneiss appears to be the least widespread among the outcropping rock grades; it is found along the Castiglionese street as small and discontinuous exposures and mainly on the left side of the Santo Ianni torrent valley and along its thalweg. Classes IV and V gneiss are discontinuously widespread along the ridges and the slopes, where they alternate in an irregular way. About half of the mapped area is covered by colluvial and residual soils; the colluvial soils prevail over the residual ones. The class VI soils are present as hollow fillings along the eastern slope of the hill and the Santo Ianni torrent valley and as little strips located at different altitudes.

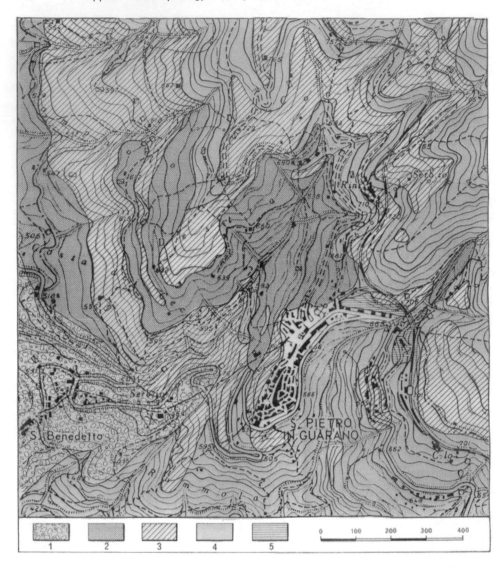

Figure 13.5 Extract of the weathering grade map of the gneiss in the San Pietro in Guarano area (reduced from 1:5000 scale, from Cascini *et al.* 1992b). Legend: 1, clays, sands and conglomerates of Pleistocene age; 2, residual and colluvial soils (class VI); 3, completely weathered gneiss or saprolite (class V); 4, highly weathered gneiss (class IV); 5, moderately weathered gneiss (class III)

On the basis of the field surveys it is possible to evaluate the thickness of the weathering horizons. The class IV–III weathering horizon, outcropping at the foot of the Altavilla Hill, is about 60 m thick; class V–IV horizon, which develops at approximately 640 m a.s.l., is at least 70 m thick. Class VI horizon is usually 10–15 m thick.

A more detailed survey – at 1:200 scale – of the gneiss weathering grades was performed in an area located on the slope which develops eastward of San Pietro in Guarano (Figure 13.7).

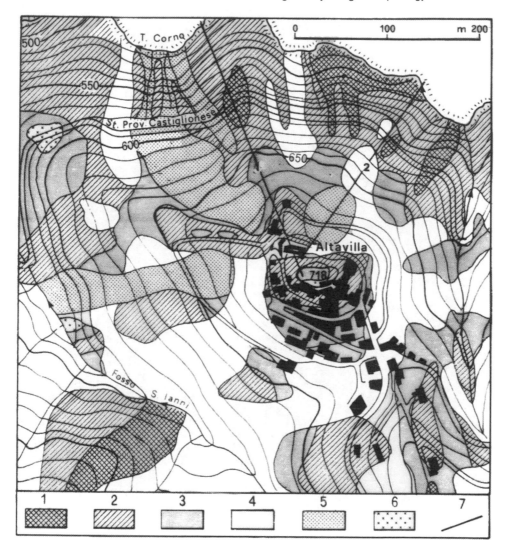

Figure 13.6 Weathering grade map of the gneiss outcropping in the Altavilla hill (reduced from scale 1:2000, from Critelli *et al.* 1991a). Legend: 1, moderately weathered gneiss (class III); 2, highly weathered gneiss (class IV); 3, completely weathered gneiss or saprolite (class V); 4, residual and colluvial soils (class VI); 5, landslide body; 6, pegmatite; 7, section track. For classes see Table 13.1

Two small active landslides, which are located above the football field, are located in this sector.

Highly and moderately weathered gneiss crop out along the little ridge which forms the westward limit of the landslide area. In the central sector, where landslides have evolved on sandy to gravelly-sandy soils, these soils are mostly either colluvium or previous landslide debris. In the higher part of this sector, isolated spurs of highly to completely weathered pegmatite and gneiss crop out; they show the weathering grade (IV–V) of the buried *in situ*

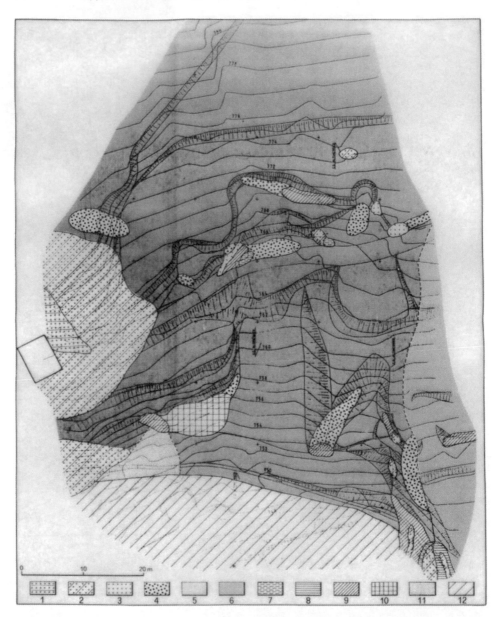

Figure 13.7 Weathering grade map of the pegmatite gneiss outcropping at the San Pietro in Guarano village football field (reduced from scale 1:200). Legend: 1, moderately weathered gneiss (class III); 2, highly weathered gneiss (class IV); 3, completely weathered gneiss or saprolite (class V); 4, gneissic pebbly landslide debris; 5, gneissic pebbly–arenaceous landslide debris; 6, gneissic colluvial soil (class VI); 7, fault gouge (cataclastic pegmatite and gneiss); 8, moderately to highly weathered pegmatite (class III–IV); 9, highly to completely weathered pegmatite (class IV–V); 10, completely weathered pegmatite (class V); 11, pegmatite saprolite and sandy colluvial soil (class V–VI); 12, debris

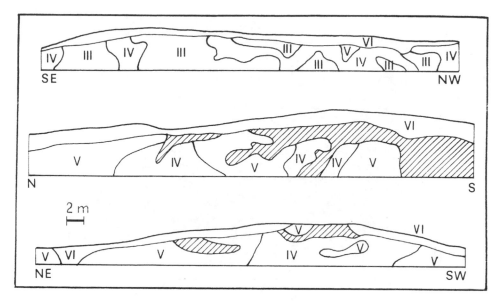

Figure 13.8 Examples of gneiss cuttings with detailed weathering grade survey

weathered rock. In the lower part of the sector a highly fractured pegmatite vein outcrops at about 755 m a.s.l., which is highly to completely weathered; at about 748 m a.s.l. highly weathered gneiss with many minor veins of less weathered pegmatite crop out.

Landslide debris is less than 10 m thick at the foot of the slope (about 750 m a.s.l.) and is formed by a gravelly to clayey matrix which includes some boulders of gneiss and pegmatite, which are variously weathered (classes III–V).

In the study area an inventory was taken and a preliminary analysis with qualitative criteria carried out of over 100 cuttings, of which gneiss consistency, discoloration and textural characteristics were recorded. Fourteen cuttings were studied with more detailed techniques and with Schmidt hammer tests. These investigations have allowed the testing and verification of the proposed procedure to better define weathering profiles. It has also allowed the collection of useful data for the characterization of weathering grades for the outcropping gneiss. In Figure 13.8 the weathering profiles of some cuttings are shown. The compressive and distensional tectonic events have induced a particularly intense state of fracturing in the gneiss. The severe fracturing, which gave rise to polyhedric elements with variable dimensions from a centimetre to several cubic metres, has favoured the development and deepening of the weathering processes and the consequent physical–chemical decay of the gneiss.

For a detailed description of the studied cuttings see Gullà and Matano (1994, 1997).

13.6 THE GNEISS WEATHERING PROFILE AT THE BASE OF THE WESTERN SLOPE OF THE SILA MASSIF

The detailed surveys carried out in the sample study area have disclosed a complex and variable weathering profile at the base of the western slope of the Sila Grande. This may be related to several factors, i.e. the compositional heterogeneity of the gneiss, the diffuse

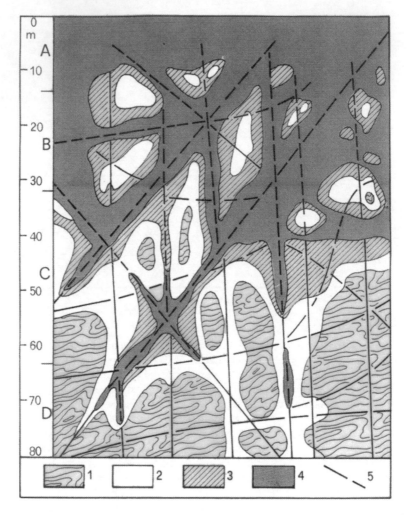

Figure 13.9 Schematic outline of the weathering profile of the gneiss outcropping along the western slope of the Sila Massif. Legend: 1, fresh or discolored gneiss (classes I and II); 2, weathered gneiss (classes III and IV); 3, saprolite gneiss (class V); 4, residual and colluvial soil (class VI); 5, joints; A, soil horizon; B, transition horizon; C, weathered gneiss horizon; D, bedrock

pegmatite veins, the state of fracturing, the erosional and climatic events. The profile schematically assumes the trend indicated in Figure 13.9.

The weathering grade of the gneiss increases in relation to depth and it is strongly influenced by the structural elements present in the rock mass. In fact, along a given vertical line one can find out-of-sequence weathering horizons giving rise to a partial, or even complete inversion of the 'normal' weathering profile. Moreover, the lithological horizons do not have a wide lateral continuity; in particular, in correspondence with the main structural discontinuities, bands of soil can be observed.

The weathering profile presents a complex vertical trend which is variable from site to site. The complexity of the weathering profile makes it difficult to locate well-defined horizons of a given weathering class. The various lithological horizons that may be schematically singled out present extremely variable and articulated thicknesses and geometric relationships.

The horizons are intersected by bands of saprolitic and residual soils and by bands of cataclastic material (fault breccia) no more than 1 m wide, through which there is active water circulation. At times centimetric levels of fault gouge, forming impermeable limits, are evident along tectonic discontinuities.

13.7 CONCLUSIONS

This chapter described a survey and mapping methodology of the gneiss weathering grade. The wide lithological variation of the Sila Massif weathered rocks, and particularly of granites and gneiss, has required the use of a weathering classification, as already in use in other countries where deep weathered crystalline rocks crop out. The methodology is based on qualitative and semi-quantitative observations of a geological kind and is useful for both detailed and large-scale studies in the field. We have carried out weathering field surveys, formulated some hypotheses about the weathering profiles and plotted weathering maps at various scales.

The study experience has shown the good potential of the weathering grade survey in the specific gneissic lithology and in the geomorphological study of landsliding. In particular, the reliability of rapid techniques, which refer to simple parameters, for large-scale studies and investigations has been verified. The detailed survey of the weathering grade on the cuttings has allowed us to gather significant information in regard to the weathering profiles of the studied gneiss.

The 1:5000 map has proved very meaningful for the arrangement of morphological – applied or theoretical – problems, such as landsliding in slope evolution.

More detailed weathering grade surveys (1:2000–1:1000 scales) have proved to be useful tools in the collection of significant information on slopes, such as type, nature, thickness and geometry of the weathering profile, and on landslide typologies.

The methodology of weathering grade surveys has also proved to be very useful for extremely detailed maps (1:500–1:200) and cutting studies. At these scales it allows collection of data about the main characteristics of a single landslide, and about the superficial trend of the weathering profile, especially on road cuttings, quarry fronts, excavations, trenches, etc.

This approach represents a typical example of the interface between geomorphological, engineering, geological and geotechnical aspects. As a matter of fact the procedure is particularly interesting for the understanding of engineering geomorphological problems, such as rockwall stability and landsliding, and allows the preliminary outline of the problems connected with the realization of engineering works.

REFERENCES

Amodio Morelli, L., Bonardi, G., Colonna, V., Dietrich, D., Giunta, G., Ippolito, F., Liguori, V., Lorenzoni, S., Paglionico, A., Perrone, V., Piccarreta, G., Russo, M., Scandone, P., Zanettin Lorenzoni, E. and Zuppetta, A. 1976. L'arco calabro-peloritano nell'orogene appenninico-maghrebide. *Memorie della Società Geologica Italiana*, **17**, 1–60.
Boenzi, F. and Palmentola, G. 1975. Osservazioni sulle tracce glaciali della Calabria. *Bollettino della Società Geologica Italiana*, **94**, 961–977.
Calderoni, G., Ciccacci, S., Fredi, P. and Pambianchi, G. 1989. Geomorphologic features and

radiocarbon dating of some quaternary deposits at Difesella del Trionto (Calabria, Italy). *Geografia Fisica e Dinamica Quaternaria*, **12**, 87–89.

Carrara, A., Sorriso Valvo, M. and Reali, C. 1982. Analysis of landslide form and incidence by statistical techniques, Southern Italy. *Catena*, **9**, 35–62.

Cascini, E., Cascini, L. and Gullà, G. 1992. A back-analysis based on piezometer response. In Bell, D.H. (ed.), *Proceedings of the 6th International Symposium on Landslides*. A.A. Balkema, Rotterdam, 1123–1128.

Cascini, L. and Gullà, G. 1993. Caratterizzazione fisico-meccanica di terreni prodotti dall'alterazione di rocce gneissiche. *Rivista Italiana di Geotecnica*, **27**(2), 125–147.

Cascini, L. and Versace, P. 1988. Relationship between rainfall and landslide in a gneissic cover. In Bonnard, C. (ed.), *Proceedings of the 5th International Symposium on Landslides*. A.A. Balkema, Rotterdam, 565–570.

Cascini, L., Critelli, S., Di Nocera, S. and Gullà, G. 1992a. A methodological approach to landslide hazard assessment: a case history. In Bell, D.H. (ed.), *Proceedings of the 6th International Symposium on Landslides*. A.A. Balkema, Rotterdam, 899–904.

Cascini, L., Critelli, S., Di Nocera, S., Gullà, G. and Matano, F. 1992b. Grado di alterazione e franosità negli gneiss del Massiccio silano: l'area di San Pietro in Guarano (CS). *Geologia Applicata e Idrogeologia*, **27**, 49–76.

Cascini, L., Critelli, S., Di Nocera, S., Gullà, G. and Matano, F. 1994. Weathering and landsliding in Sila Massif gneiss (Northern Calabria, Italy). In Oliveira, R., Rodrigues, L.F., Coelho, A.G. and Cunha, A.P. (eds), *Proceedings of the 7th IAEG International Congress*. A.A. Balkema, Rotterdam, 1613–1622.

Cascini, L., Gullà, G. and Sorbino, G. 1995. Modellazione delle acque sotterranee di una coltre di detrito in frana: risultati preliminari. *Rivista Italiana di Geotecnica*, **29**(3), 201–223.

CAS.MEZ. 1968. Various Geological Sheets: Foglio 229 II SE, Foglio 229 II SO, Foglio 230 III SE, Foglio 230 III SO, Foglio 236 I NE, Foglio 236 I NO, Foglio 236 IV NE and Foglio 236 IV NO. *Carta Geologica della Calabria alla scala 1:25.000*. Cassa per il Mezzogiorno, Roma.

Critelli, S., Di Nocera, S., Gullà, G. and Matano, F. 1991a. La frana di Altavilla: esempio di una tipologia di frana diffusa nelle rocce cristalline alterate della Sila Grande (Calabria settentrionale). *Geologia Applicata e Idrogeologia*, **26**, 85–110.

Critelli, S., Di Nocera, S. and Le Pera, E. 1991b. Approccio metodologico alla valutazione petrografica del grado di alterazione degli gneiss del massiccio silano (Calabria settentrionale). *Geologia Applicata ed Idrogeologia*, **26**, 41–70.

Dearman, W.R. 1976. Weathering classification in the characterisation of rock: a revision. *Bulletin of International Association of Engineering Geology*, **13**, 123–127.

Dobereiner, L., Durville, J.L. and Restituito, J. 1993. Weathering of the Massiac Gneiss (Massif Central, France). *Bulletin of International Association of Engineering Geology*, **47**, 79–96.

Dramis, F., Gentili, B. and Pambianchi, G. 1990. Geomorphological scheme of the river Trionto basin. In Sorriso Valvo, M. (ed.), *I.G.U.-COM.TA.G. and C.N.R. Symposium on Geomorphology of Active Tectonics Areas, Excursion Guide-book*. IRPI, Cosenza. *Geodata*, **39**, 63–66.

Gamon, T.I. 1983. A comparison of existing schemes for the engineering description and classification of weathered rocks in Hong Kong. *Bulletin of the International Association of Engineering Geology*, **28**, 225–232.

Geotechnical Control Office. 1984. *Geotechnical Manual for Slopes*. Geotechnical Control Office, Engineering Development Department, Hong Kong.

Geotechnical Control Office. 1988. *Guide to Rock and Soil Descriptions*. Geotechnical Control Office, Civil Engineering Department, Geoguide, **3**, Hong Kong.

GSEGWP. 1995. The description and classification of weathered rocks for engineering purposes. Geological Society Engineering Group Working Party Report. *Quarterly Journal of Engineering Geology*, **28**, 207–242.

Gullà, G. and Matano, F. 1994. Proposta di una procedura per il rilievo del grado di alterazione di gneiss su fronti rocciosi. *Geologica Romana*, **30**, 227–238.

Gullà, G. and Matano, F. 1997. Surveys of weathering profile on gneiss cutslopes in Northern Calabria, Italy. In Marinos, P.G., Koukis, G.C., Tsiambaos, G.C. and Stournaras, G.C. (eds), *Proceedings of the*

International Symposium on Engineering Geology and the Environment, Greek National Group of IAEG. A.A. Balkema, Rotterdam, 133–138.

Guzzetta, G. 1974. Ancient tropical weathering in Calabria. *Nature*, **251**(5473), 302–303.

Hall, A.M. 1985. Cenozoic weathering covers in Buchan, Scotland and their significance. *Nature*, **315**(6018), 392–395.

Hall, A.M. 1986. Deep weathering patterns in north-east Scotland and their geomorphological significance. *Zeitschrift für Geomorphologie*, **30**(4), 407–422.

Hall, A.M. 1987. Weathering and relief development in Buchan, Scotland. In Gardiner, V. (ed.), *International Geomorphology*. Wiley, Chichester, 991–1005.

IAEG Commission on Engineering Geological Mapping. 1981. Rock and soil description and classification for engineering geological mapping report. *Bulletin of International Association of Engineering Geology*, **24**, 235–274.

Ietto, A. 1975. Geologia e pianificazione urbana in Calabria. *Memorie della Società Geologica Italiana*, **14**, 421–490.

Ippolito, F. 1962. Sulla geologia della Calabria. In Ippolito, F. (ed.), *Saggi e studi di geologia*. Neri Pozza Editore, Venezia, 145–151.

ISRM Commission on Laboratory Texts. 1978. Suggested methods for determining the uniaxial compressive strength and deformability of rock materials. *International Journal of Rock Mechanics and Mining Science, Geomechanics Abstracts*, **16**, 135–140.

Knill, J.L. and Jones, K.S. 1965. The recording and interpretation of geological conditions in the foundations of the Roseires, Kariba and Latiyan dams. *Geotechnique*, **15**, 94–124.

Lanzafame, G. and Zuffa, G.G. 1976. Geologia e petrografia del foglio Bisignano (Bacino del Crati, Calabria) con carta geologica alla scala 1:50.000. *Geologica Romana*, **15**, 223 270.

Lee, S.G. and De Freitas, M.H. 1989. A revision of the description and classification of weathered granite and its application to granite in Korea. *Quarterly Journal of Engineering Geology*, **22**, 31–48.

Le Pera, E., Critelli, S. and Sorriso-Valvo, M. 2001. Weathering of gneiss in Calabria, Southern Italy. *Catena*, **42**, 1–15.

Matano, F. 1991. Prime osservazioni sulla geomorfologia dell'area di San Pietro in Guarano (CS). *Geodata*, **38**, 51–73.

Matano, F. and Tansi, C. 1994. Influenze delle strutture tettoniche sul profilo di alterazione e sulla franosità negli gneiss dell'area di S.Pietro in Guarano (Calabria). *Geologica Romana*, **30**, 361–370.

Nossin, J.J. 1972. Landsliding in the Crati basin, Calabria, Italy. *Geologie en Mijnbouw*, **51**(6), 591–607.

Ollier, C.D. 1984. *Weathering*. Geomorphology Texts, Longman, London.

Palmentola, G., Acquafredda, P. and Fiore, S. 1990. A new correlation of the glacial moraines in the Southern Apennines, Italy. *Geomorphology*, **3**, 1–8.

Patton, F.D. 1984. Climate, groundwater pressures and stability analysis of landslides. *Proceedings of the 4th International Symposium on Landslides*. 43–59.

Sorriso Valvo, M. 1985. Mass-movement and slope evolution in Calabria. *Proceedings of the 4th International Conference and Field Workshop on Landslides*. Tokyo, 23–30.

Sorriso Valvo, M. 1988. Landslide-related fans in Calabria. *Catena Supplement*, **13**, 109–121.

Sorriso Valvo, M. 1990. Geomorphology of Calabria – A scheme. In Sorriso Valvo, M. (ed.), *I.G.U.-COM.TA.G. and C.N.R. Symposium on Geomorphology of Active Tectonics Areas, Excursion Guidebook*. IRPI, Cosenza. *Geodata*, **39**, 5–15.

Thomas, M.F. 1966. Some geomorphological implications of deep weathering patterns in crystalline rocks in Nigeria. The Institute of British Geographers, *Transactions and Papers*, **40**, 173–193.

Verstappen, H.T. 1977. Sulla geomorfologia della parte sud-occidentale della provincia di Cosenza. *Bollettino della Societa' Geografica Italiana*, **10**(6), 541–561.

14 Assessment of Silting-up Dynamics of Eleven Cut-off Channel Plugs on a Free-Meandering River (Ain River, France)

H. PIÉGAY
UMR 5600 – CNRS, Lyons, France

G. BORNETTE
Laboratoire d'écologie des eaux douces et des grands fleuves, Université C. Bernard, Villeurbanne, France

AND

P. GRANTE
UMR 5600 – CNRS, Lyons, France.

ABSTRACT

Managers and decision-makers have recognized the importance of preserving and restoring wetland areas because of their high ecological value and their increasing rarity. In the case of riverine wetland areas such as abandoned channels, there is a need to improve understanding of their evolution, particularly with regard to the silting-up dynamics of these environments, as this determines their life span and their potential for preservation.

This chapter compares the dynamics of the upstream plugs of 11 cut-off channels on the Ain River, France. Indicators describing cut-off channels (i.e. age, form) and sedimentation, soil and vegetation characteristics of plugs were measured. Multivariate analysis (nPCA, co-inertia analysis) were used to examine variations in fill and silting-up patterns within individual sites and to compare sites.

The results indicate that the vegetation, soil and sedimentation characteristics of plugs are conditioned by: (i) time or the number of flooding/geomorphic events having affected a plug; (ii) abandoned channel geometry (pattern, plug length) which controls its roughness and sedimentation dynamics; (iii) channel evolution (i.e. channel aggradation or displacement) which favours rejuvenation processes (growth of psammophile plants, soil regression, increase of overbank sediment size and volume).

14.1 INTRODUCTION

Large rivers of Alpine mountains and their piedmonts are characterized by high energy, abundant gravel bedload, and braided or wandering patterns (Kellerhals and Church 1989; Church 1992; Best and Bristow 1993; Bravard and Peiry 1993). The lateral instability of these systems favours channel chute cut-off or avulsion (Lewis and Lewin 1983) and explains the occurrence of abandoned channels within the floodplain. This horizontal mobility of the main

Applied Geomorphology: Theory and Practice. Edited by R.J. Allison.
© 2002 John Wiley & Sons, Ltd.

channel also increases ecological diversity of riparian zones, leading to the coexistence of units of different age distributed in various chronosequences (Pautou 1984; Shankman 1991, 1993). Fluvial mosaics are characterized by a great number of vegetal or sedimentary patches in aquatic, semi-aquatic and terrestrial environments (Amoros and Petts 1993). The complexity of riparian landscapes is thus in large part due to floodplain history and the presence of open or aquatic areas within a predominantly wooded corridor.

Natural margins of this sort have become rare in Europe where the floodplains of large rivers are often cultivated or disconnected from the main river channel to decrease risks of flooding and erosion (Yon and Tendron 1981). Riverine wetland areas are transformed, drained or dried because of irrigation or channel incision, in turn linked to gravel-mining, damming, etc. (Babinski 1992; Bornette and Heiler 1994). The stability of these rivers has also been increased by regulation and decreased bedload supply following the end of the Little Ice Age and modifications in watershed land use. All these factors have led to the increasing rarity of abandoned channels (Bravard 1986).

These riverine patches are today recognized as having high biodiversity, and as being necessary for the equilibrium of the whole river system (Amoros and Roux 1988). Organizations such as the Ministry of Environment, water agencies and the European Union are taking measures to preserve wetland areas because of their rarity, their high ecological value (Point 1993; Barnaud 1996) and their major role in maintaining floodplain biodiversity (Furness and Breen 1980; Castella et al. 1991; Ward and Stanford 1995). Recent studies have contributed to the understanding and the conservation of abandoned channels, particularly those of great ecological value, or have laid the groundwork for restoring abandoned channels that could be of such high value (Glitz 1983; Möller and Wefers 1983; Henry and Amoros 1995a, b).

As explained by Henry and Amoros (1995b), one of the essential considerations in defining a successful riparian wetland restoration project lies in assessing the potential life span of abandoned channels, which is in large part dependent on the rate at which they are filled with sediment. In cut-off channels that continue to be inundated during floods, most restoration strategies are mainly based on ecological descriptors, but should also integrate geomorphological dynamics. Indeed, this would provide better understanding of silting-up processes and thus potential life span, but also the dynamics of fluvial forms and sediment transfers within the floodplain on a long-term scale (sedimentation rate, sediment mosaic). Although the study of cut-off channels has provided increased understanding of the Holocene history of some rivers (Starkel and Thornes 1981; Peiry 1990; Salvador 1991), few results are available on the geomorphology and dynamics of recently cut-off channels that are still influenced by the river and its hydrological fluctuations (Weihaupt 1977). Most case studies focus mainly on the channel adjustment following cut-off rather than abandoned channel infill (Lewis and Lewin 1983; Shields and Abt 1989; Hooke 1995).

This chapter focuses on the silting-up dynamics of the upstream part of cut-off channels. Our hypothesis is that the characteristics of silting-up deposits (i.e. length, sedimentation patterns and soil characteristics of plugs) are dependent upon: (i) the age of the abandoned channel which is related to the number of flood events affecting the plug after cut-off; (ii) the form and the location of the abandoned channel, notably its position in the fluvial corridor relative to the main channel; and (iii) the stability or evolution of the main channel (i.e. aggradation or incision, horizontal mobility) which may increase or decrease the frequency of overbank flow and bedload transport capacity in the floodplain (Figure 14.1a). We assume that the amount and calibre of sediment supply available are the same from one former channel to another because of the absence of major tributaries along the studied reach.

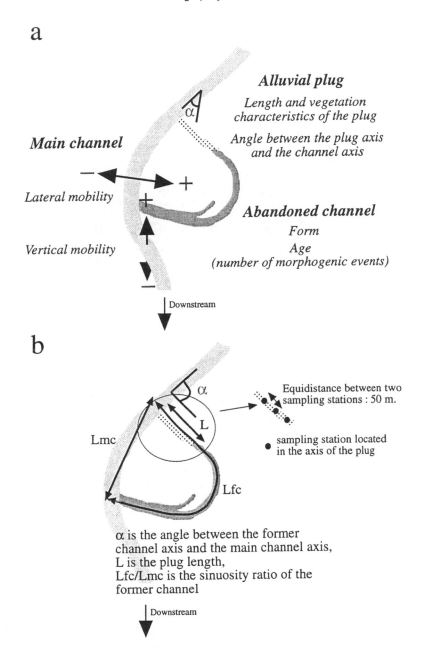

Figure 14.1 (a) General parameters documented on the abandoned channels and the related river reach; (b) morphological characteristics of cut-off channels and alluvial plugs

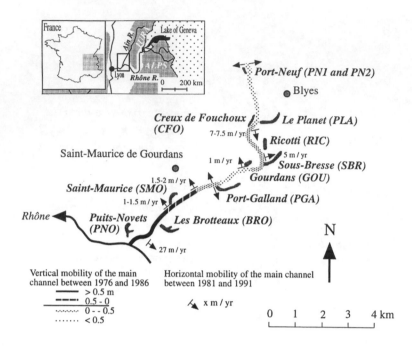

Figure 14.2 Location of the studied river reach and of the 11 abandoned channels; information on the recent instability of the river channel is superimposed on the river course

14.2 MATERIAL AND METHOD

14.2.1 Study Sites

The Ain River is one of the main tributaries of the Rhône river upstream from Lyons (France), and drains a watershed of 3672 km² located essentially in the Jura Mountains. After a course of 160 km in a mountainous V-shaped valley, the lower 40 km of the Ain River meanders in a wide glacial outwash plain (Figure 14.2). In this reach, the river is characterized by a high number of cut-off channels that vary in form, distance to the main channel, and in elevation relative to the channel. The floodplain width varies from 0.5 to 2 km.

The Ain River is a torrential river characterized by a mean annual discharge of 123 m³ s⁻¹, and very short but strong floods with winter maxima and an irregular regime. This sixth order river has a wide shallow gravel bed with a wooded corridor and a slope varying from 1.8‰ to 1.2‰ all along its lower course.

The Ain River has changed considerably since the end of the 19th century. Braiding patterns which were still active in 1945 have been replaced by a unique wandering channel. The river is now characterized by a sinuous single bed over most of its course, only wandering or meandering on short sectors. This phenomenon is explained by the decrease of peak flows and bedload mobility due to both natural (end of the Little Ice Age) and anthropogenic (upstream dams) factors. Moreover, decreased grazing on river margins since 1945 has led to the development of alluvial forests, which accelerates the narrowing and degradation of main river channels (Marston *et al.* 1995; Tsujimoto and Kitamura 1996).

The Ain being a high energy river with very unstable, poorly cohesive banks, this rapid fluvial metamorphosis explains the presence of a great number of cut-off channels that present various shapes, ages and connections with the main channel (Figure 14.2).

Terrestrial vegetation is characterized by a complex mosaic of patches of various ages, characterized by the coexistence of arboreous stages (ash units with *Fraxinus excelsior* and *Acer pseudoplatanus*), shrubby stages (dry units with *Crataegus monogyna* and *Salix eleagnos* or pioneer and post-pioneer units with *Salix* sp., *Populus nigra, Alnus glutinosa*) and semi-aquatic stages (sedge communities dominated by *Carex* sp., *Phalaris arundinacea* or *Phragmites australis*) (Pautou and Girel 1986; Bornette *et al.* 1996).

14.2.2 Data Collected

Plugs located at the upstream end of 11 cut-off channels were studied. Field measurements, land survey maps, aerial photographs and long profiles were analysed in order to evaluate links between descriptors of plug formation (length, sedimentological and soil characteristics of plugs) and structural factors (channel cut-off age, abandoned channel form, mobility of the main channel).

Field measurements

A total of 107 sampling plots were studied on 11 cut-off channel alluvial plugs. The sampling plots were located every 50 m along the plug axis, the first one being situated a few metres from the aquatic zone. The number of sampling plots varies from three to 26 depending on the length of the plug. This sampling frame induces the disparity of the number of sampling plots from one site to another, but makes it possible to evaluate longitudinal variations within the longer plugs. A minimum of three sampling plots were used, even on plugs of less than 150 m in length (in one cut-off channel: PN1).

Six sedimentological, soil and biological variables were measured in each sampling plot, and classified for analysis (Table 14.1).

Nine vegetal communities were observed; the modality nine (plantation) was considered as the most mature or terrestrial one.

Two soil descriptors of the soil surface were quantified:

(1) the presence or absence of the A1 soil horizon, a descriptor of both plug age and autogenic processes; and
(2) the thickness of the upper sediment layer, an indicator of rejuvenation intensity, age, and autogenic processes.

Sediment characteristics were assessed by three variables:

(1) the thickness of the entire fine sediment deposit;
(2) the number of sediment layers; and
(3) the granulometry of the upper layer.

The number of sediment layers and the granulometry of the upper layer are both indicators of plug age and of overbank flow energy. Samples were collected using a conventional soil drill of 5 cm in diameter. The granulometry was classified according to the sets defined by Monnier and Stengel (1982).

Table 14.1 Modality of the variables measured on the study sites

Sets	Vegetal unit VEG	Upper soil horizon A1 HOR	Thickness of the upper sediment layer (cm) DUL	Number of sediment layers NSL	Thickness of overbank deposits (cm) DOD	Granulometry of the upper sediment layer* GRA
1	*Carex* com.**	Missing	0–3	1	0–5	<24.73%
2	*Phalaris* com.	Mixed	3–5	2	5–15	<45.59%
3	*Phragmites* com.	Developed	5–10	3	15–30	<54.84%
4	Willow unit		10–15	4	30–50	<64.09
5	Poplar unit		>15	5	50–90	<74.62
6	Alder unit			6	>90	≥74.62
7	Young ash unit			7		
8	Aged ash unit			8		
9	Plantation			>8		

* Percentage of sands
** Com. = Community

Land survey maps, aerial photographs and long profile analysis

Five series of vertical photographs (from National Geographic Institute, France), taken respectively in 1945/46, 1954/56, 1965/71, 1980 and 1991 (scales: 1/15 000 to 1/30 000), were studied. The combinations of photographs from one date to another allowed dating of channel cut-offs (precision: ±10 years in most cases) and identification of the geomorphological pattern at the date of cut-off. Land survey maps drawn after 1807 and renewed between the First and Second World Wars were also used to date the oldest channel cut-offs. The limited precision of datings for these older cut-offs, identified as having occurred during the periods 1830 to 1920–1930 or 1920–1930 to 1945–1946, led us to define three sets of former channels (Table 14.2).

Patterns of cut-off channels (i.e. length of the alluvial plug or of the abandoned channel, angle between the present main channel axis and the upstream abandoned channel axis, sinuosity of the abandoned channel) were also measured from 1991 aerial photographs (Figure 14.1b). The length of the plug is defined as the length of the vegetated corridor that

Table 14.2 Modality of the variables measured on the aerial photos and the long profiles

Sets	Vertical instability between 1976 and 1986 (m) VEI	Lateral instability between 1980 and 1991 HOI	Angle between the plug and the main channel ANG	Sinuosity ratio of the abandoned channel SIN	Original geomorphological pattern STY	Age of the cut-off channel (years) AGE
1	−0.5/−1	Enlarged plug*	0–30°	<1.3	Meandering	<30
2	0/−0.5	Stable plug	30–60°	>1.3	Braided	30–50
3	0/+0.5	Eroded plug*	60–90°			>50
4	+0.5/+1					

* Due to channel displacement.

separates the main channel and its gravel bars from the aquatic part of the abandoned channel. Angle and sinuosity ratio were also simplified for statistical analysis (Table 14.2).

Contemporary river channel instability was evaluated by overlaying 1980 and 1991 channel courses as observed on aerial photographs. The maximum mobility of the main channel banks was assessed at upstream and downstream confluences for each cut-off channel. Thirty river channel cross-sections (one per kilometre, surveyed in 1976 and 1986 by CNR (1987)) were also used to evaluate recent river degradation or aggradation (Figure 14.2) following the method developed by Piégay and Peiry (1997). The plugs were classified in several sets according to (i) their proximity to an aggraded or degraded main channel, and (ii) the degree of lateral main channel migration towards or from the abandoned channel. This last variable pertains to the evolution of plug length (growth, stable or eroded) for the period of observation (Table 14.2). Instability was only evaluated over a short timespan due to the lack of reliable older data that could serve to evaluate vertical mobility. Given available data sources, variables were thus collected for both long-term and short-term observation.

14.2.3 Data analysis

As we worked on two large data sets, multivariate analyses were used to synthesize their structure by discriminating the main patterns occurring in these complex data sets. First the two data sets were considered independently by multiple correspondence analysis (MCA) in order to identify: (i) plug characteristics based on vegetation, soil and sedimentological variables (Table 14.1); and (ii) cut-off channel types based on structural variables (Table 14.2).

MCA is regularly used in ecology to analyse the structure of categorical multivariate data (Pialot et al. 1984; Tenehaus and Young 1985). Structural variables were also combined with the plug descriptors in order to delineate differences in the plug characteristics (Table 14.3). MCA of the structural variables (11 plugs × 6 variables), made it possible to explain variations in plug length. Spearman rank correlation tests and ANOVA were done to validate statistical links between variables.

Inter-battery analysis was used to determine the common structure between the field data table and the structural data table. The method is based on simple principles and is efficient for the matching of two data sets characterized by the same margin-sample plots (Tucker 1958; Chessel and Mercier 1993; Bornette et al. 1994). Two analyses were done (Table 14.3), the first on the whole data set (107 sampling plots × 6 variables, Table 14.1, 107 sampling plots × 6 variables, Table 14.2) and the second on the three upstream sampling plots of each plug (33 sampling plots × 6 variables, Table 14.1; 33 sampling plots × 6 variables, Table 14.2). In the first analysis, the rows of the two tables were weighted by the number of sampling plots of each plug in order to avoid any bias due to the higher sampling of longer plugs.

Data sets were analysed using the ADE 4.0 statistical software (Chessel and Dolédec 1996).

14.3 RESULTS

14.3.1 Characteristics of the Plugs

Plug length but also vegetation, sedimentation and soil characteristics differ greatly from one abandoned channel to another. The plug length varies from a few metres to 1500 m, for a mean length of 775 m (±437 m).

Figure 14.3 shows the results of the MCA of field variables. The first two factors of the MCA account for only 9.9% and 7.5% respectively of total inertia, demonstrating the

Table 14.3 Statistical analysis used to combine structural and field variables

Structural variables		Field variables	
		Length of plugs	6 field variables
MCA, 11 plugs × 6 variables	*Scatter value*	Projection on the factorial map*	
MCA, 107 stations × 6 variables**	*Co-inertia analysis*		MCA, 107 stations × 6 variables**
MCA, 33 stations × 6 variables	*Co-inertia analysis*		MCA, 33 stations × 6 variables

* Factorial map of the MCA 'structural variables'.
** Rows were weighted by the number of plug stations.

complex variability of vegetation, soil and sedimentation characteristics along the abandoned channel plugs of a freely meandering river. These factors identify a main gradient based notably on NSL, DOD and VEG (correlation ratios respectively: 0.81, 0.81 and 0.62 on F1 axis, and 0.64, 0.55 and 0.50 on F2 axis). The highest values on the first axis are related to coarse granulometry, thick layers of overbank sediment, thin upper sediment layers, a high number of sediment layers, and soft-wood vegetation. Semi-aquatic and hard-wood communities are grouped together. The second axis is positively correlated with hard-wood vegetation, thick layers of overbank sediment, a high number of sediment layers, and well-developed soil. Plant communities are clearly discriminated along this axis, semi-aquatic communities, alder/willow, ash/poplar and plantation being distributed along the gradient.

The plugs are distributed on the factorial map according to the heterogeneity of their sampling plot characteristics. PLA, PGA, SMO, PN1 and PN2 are mainly plotted on positive values along the first axis, which demonstrates the high heterogeneity of their sampling plots for the parameters GRA, DOD, DUL and VEG. The abandoned channels CFO, GOU, SBR, PNO and partly BRO and RIC are plotted as positive values on the second axis. Plugs are characteristically homogeneous for the parameters GRA, DOD and DUL, but more heterogeneous for VEG, HOR or NSL.

Figure 14.3 shows that sedimentological and soil characteristics of the plugs partially depend on the upstream–downstream location of the sampling plots. On most plugs (SMO, PGA, PLA, PN1 and PN2), upstream sampling plots have coarser sediment, thinner overbank sediment layers and soft-wood vegetation, whereas downstream sampling plots are colonized by hard-wood or semi-aquatic communities (dominated by *Phragmites* or *Carex* sp.). PNO, BRO, CFO and RIC sampling plots are also ranged along a gradient from upstream to downstream that mainly superimposes on the second factor. The PNO and BRO plugs show terrestrial pioneer vegetation in the inner part of the plug and mature units on developed soil and very thick overbank deposits in the upstream part, whereas CFO and RIC present an inverse gradient: their upstream sampling plots have similar characteristics to those of downstream parts of PNO and BRO. Finally, GOU and SBR do not exhibit any upstream–downstream trend and are characterized by a highly complex sediment and vegetal mosaic.

14.3.2 Characteristics of the Abandoned Channels

Plugs are plotted on the first factorial map of the MCA, combining six structural variables for 11 plugs. The first two axes account for 52.8% of total inertia and are mainly linked to AGE, VEI, HOI but also in a minor part to SIN, ANG and STY (Figure 14.4).

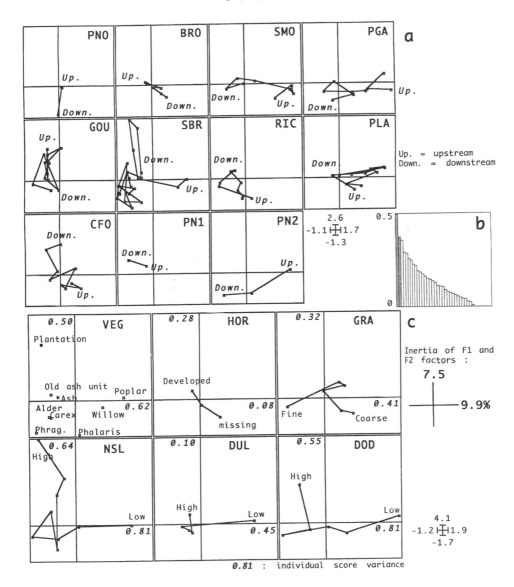

Figure 14.3 Multiple correspondence analysis of the field data: (a) distribution of the sampling plots of each alluvial plug on the F1–F2 factorial map; (b) graph of the eigenvalues; (c) projection of the modalities of the six field variables on the F1–F2 factorial map. See Tables 14.1 and 14.2 for coding

The first axis mainly discriminates alluvial plugs according to cut-off date, the older ones having positive values. This age distinction is also related to structural factors. The older plugs are located in aggrading sections, and are thus currently subject to erosion (i.e. SMO and RIC). In contrast, the younger plugs are found near channels that are incising and migrating away from them.

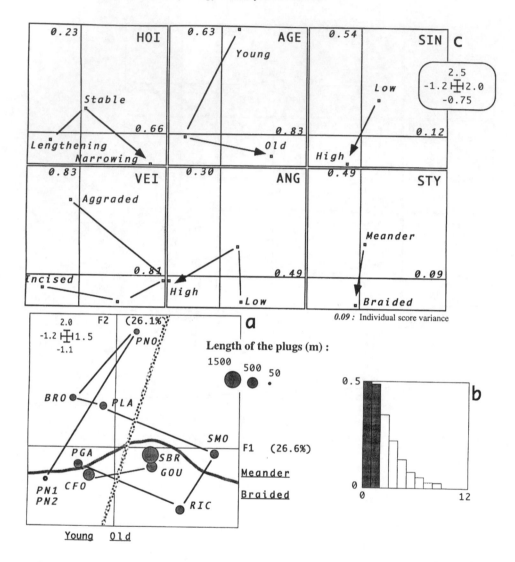

Figure 14.4 Multiple correspondence analysis of the structural data: (a) F1–F2 factorial map of the abandoned channels showing projection for the plug lengths; (b) graph of the eigenvalues; (c) projection of the modalities of the six structural variables on the F1–F2 factorial map. See Table 14.2 for coding

The second axis primarily opposes two groups according to the geomorphic pattern of their respective reaches (SIN, STY), the meanders having positive values and the braided channels negative values. Some meander cut-offs are also very recent (PNO) and are located in the most aggraded sections of the river (PNO and BRO). ANG moderately opposes meanders and braided abandoned channels. Although PN1 and PN2 were cut off from braided reaches, they have as high a cut-off angle as most meander plugs. This is because they were artificially cut off by a bridge that disconnected their upstream and downstream parts.

14.3.3   Relations between Abandoned Channel Characteristics and Plug Length

The lengths of the plugs, represented by proportionally sized circles in the factorial map, are also related to structural variables (Figure 14.4): the longer the alluvial plug, the older the abandoned channel (rho = 0.79; $z = 2.62$; $p < 0.01$). Thus, the rate of plug edification varies from 1 to 19 cm a^{-1}, the highest values being observed on the oldest abandoned channels. PN1 and PN2 plugs are anthropogenic exceptions as explained above.

The position of SMO and RIC on the first axis does not validate this general trend. Their lengths are conditioned by the recent displacement of the main river channel. Their plug lengths are the shortest of the older abandoned channels because of plug narrowing due to bank erosion of the main channel. If the main channel had remained stable, they would probably be much longer due to their age. Indeed, plug lengths measured on aerial photographs demonstrate that these plugs were longer in 1981 than in 1991 (plug length reduction of 1.5–2 m per year between 1981 and 1991 for SMO and 7–7.5 m per year for RIC).

Aggradation does not influence plug lengthening, as short plugs are observed both on incised and very aggraded reaches. Plug length also appears to be independent of angle (rho = -0.54, $z = -1.71$, $p > 0.05$), sinuosity rate (rho = 0.31, $z = 0.97$, $p > 0.05$) or geomorphological pattern (F-test value = 1.35, $p > 0.05$) even when PN1 and PN2 are excluded from the test.

14.3.4   Relations between Abandoned Channel Characteristics and Vegetation, Sedimentation and Soil Dynamics of the Plugs

Analysis of total plug length

A co-inertia analysis was used to combine structural variables and field variables in order to identify the co-structure of the two tables. The plug sampling plots correspond to the position of the end of the arrows, and the associated channel characteristics are plotted at their beginning.

As shown on Figure 14.5a, the first two axes account for 66.4% of the total inertia. Two main groups of plugs can then be distinguished according to their original geomorphic pattern (STY); the correlation ratio of STY with the first axis is 24%.

ANG, HOI and, to a certain degree, VEI, do not exhibit clear patterns on the F1–F2 map. The braided *versus* meandering gradient is also considered as partly indifferent to the age gradient because SMO, which is old, is located within the meander medium age group. Nevertheless, as shown on Figure 14.5b, the opposition of the two groups defined above is also related to age (i.e. gradient from old to medium aged). Age gradient also explains the specificity of PNO within the group of meanders or of CFO and PN1/2 within the group of braided channels.

The plugs of the older braided arms show fine granulometry, thick deposits of overbank sediment and a high number of sediment layers. Their vegetation includes both hard-wood units, some of them being planted, and semi-aquatic units such as *Phragmites* communities. Inversely, meander plugs have coarse to medium granulometry, thin deposits of overbank sediment and a low number of sediment layers. Their vegetation is dominated by soft-wood units such as willow and poplar communities.

The plugs of the youngest braided abandoned channels have characteristics similar to those of meanders: fine to medium granulometry, and moderately thick overbank sediment deposits.

Among the group of meander plugs, a secondary gradient contrasts with the main one, opposing, on one hand, young plugs with a thick upper sediment layer colonized by willows

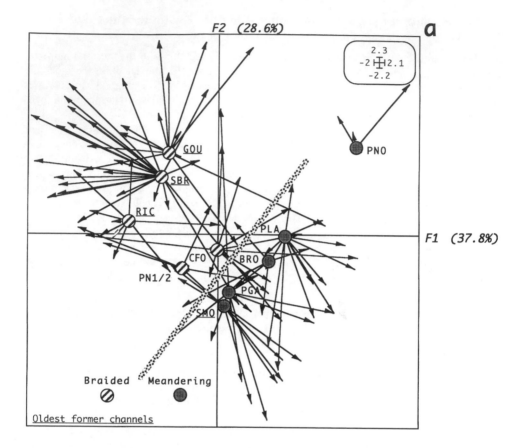

Figure 14.5 Co-inertia analysis between the field data and the structural data: (a) match of the two scatters of the first factorial map; (b) projection of the modalities of the structural variables and of the field variables on the F1–F2 factorial map. See Tables 14.1 and 14.2 for coding

or *Phalaris* sp. communities, with older ones, on the other hand, colonized by poplar units. Granulometry varies independently of this gradient, the overbank sediment deposits being very thin in all cases.

Analysis of the upstream part of plugs

The co-inertia analysis was processed in order to compare the structural variables and the field variables of the three upstream sampling plots of each plug. The first two axes account for 59.1% of the total inertia (Figure 14.6). The match of the two scatters on the F1–F2 factorial map shows that the distribution of plugs differs from the one delineated on the sampling plots for entire plug length (Figure 14.5).

The opposition between meanders and braided channels is clearer than the distinction between old and recent abandoned channels. This opposition is based on field variables: meander plugs have thinner overbank sediment accumulation but their granulometry is

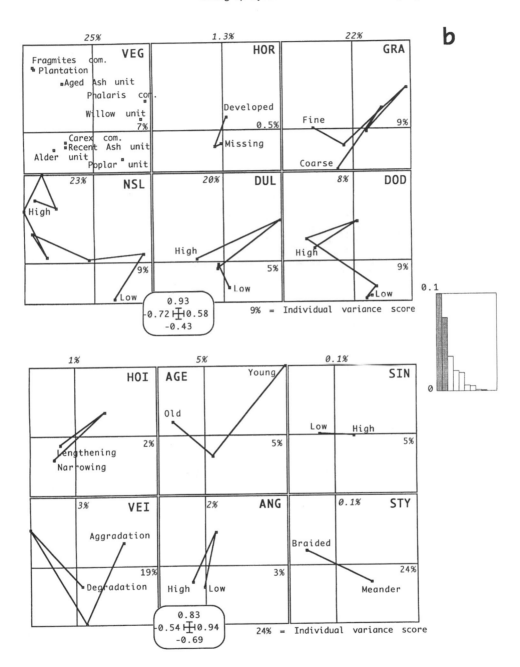

Figure 14.5 *(cont.)*

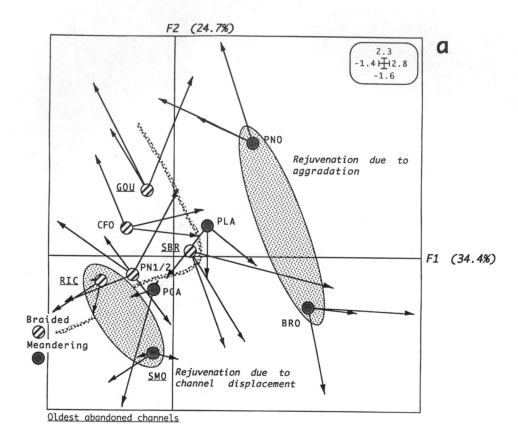

Figure 14.6 Co-inertia analysis between the field data and the structural data of the three upstream sampling plots of the abandoned channels: (a) match of the two scatters of the first factorial map; (b) projection of the modalities of the structural variables and of the field variables on the F1–F2 factorial map. See Tables 14.1 and 14.2 for coding

coarser than in braided plugs, whatever their age (i.e. scatter position of the youngest braided plugs CFO, PN1 and PN2). The upstream sampling plots of plugs do not distinguish them on the basis of age, confirming that abandoned channel geometry rules the dynamics of sediment deposition.

Figure 14.6 shows a second gradient that differs from the first one (i.e. meander–braided), isolating plugs characterized by (i) aggradation (PNO and BRO), (ii) channel displacement (i.e. eroded plugs) and/or (iii) low angle (RIC and SMO). The plugs located in aggraded reaches are characterized by both medium-sized sediments and a thick upper sediment layer, indicating rejuvenation processes. PNO is also colonized by *Phalaris* communities growing on unstable sandy substrates (Petit and Schumacker 1985). The soil of BRO is influenced by floods, as shown by overlaying or mixed-in sand in the A1 soil horizon. Moreover, the plugs located in reaches characterized by bank erosion and low angle (i.e. plug narrowing) are mainly characterized by very coarse sediment input, very thin overbank deposits and the absence of an A1 soil horizon despite their age.

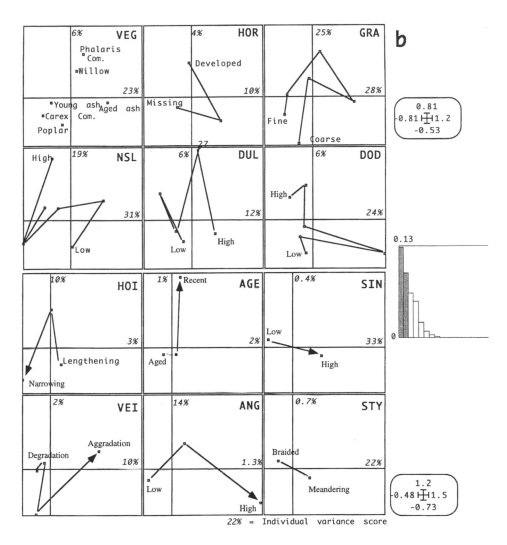

Figure 14.6 (*cont.*)

14.4 DISCUSSION

14.1.1 Causes of Vegetation, Sedimentation and Soil Differences among the Alluvial Plugs Studied

The characteristics of alluvial plugs differ greatly from one cut-off channel to another, essentially because they are influenced by numerous factors that vary in time and space. Our hypothesis was that plug development is primarily dependent upon three factors: the age of the abandoned channel, its form and location relative to the main channel, and the instability of the main river channel. Given that none of the tributaries modify suspended sediment

Table 14.4 Synthesis of results combining structural and field variables describing alluvial plugs of the Ain River

		Length	GRA	NSL	DOD	HOR	DUL	VEG
Age	from young to aged	+	−	+	+	+	0	+
Pattern	braided/meander	+	−	+	+	0	0	+
Sinuosity	low sin./high sin.	+(?)	?	+	?	0	0	+
Angle	low/high	+(?)	+(?)	0	0	−(?)	+(?)	0
Aggradation*	y/n	+(?)	+	0	0	−	+	+
Narrowing*	y/n	−	+	?	?	−	0	0

+ = Increase or change; − = decrease, 0 = no differences; ? = no response or uncertain.
* In the upstream part of the plugs (the three upstream sampling plots).

supply along the studied reach, we consider that sediment amount and calibre are the same and that sediment patterns are controlled by local factors. One exception may be PNO which is also affected by the Rhône floods.

The results show that the vegetation, the soil and the sedimentation of the alluvial plugs studied are effectively ruled by factors such as:

(i) abandoned channel geometry (braided reach abandoned channels have different characteristics from meander ones; the granulometry and the thickness of overbank deposits vary along some plugs);

(ii) cut-off age, related to the number of flood events having affected plug evolution (the youngest plugs have a sedimentary gradient from upstream to downstream whereas the oldest ones show more homogeneous sedimentation); and

(iii) channel evolution in which instability favours rejuvenation processes (growth of psammophile plants, soil regression, increase of overbank sediment granulometry and thickness).

Nevertheless, recent channel instability (i.e. aggradation or channel displacement) has only affected sedimentation dynamics in the upstream part of plugs, whereas long plugs do not show any sedimentological or soil changes in their middle or downstream sections. The influence of the structural variables on plug vegetation, soil and sediment fill are summarized in Table 14.4.

These results also indicate that attempts to explain spatial variations of plug length, soil and sedimentation along highly unstable rivers should include historical evaluation of both hydrological events (i.e. the frequency and the intensity of flooding) and floodplain hydraulics (i.e. floodplain topography, vegetation roughness, existence and orientation of overbank flow axes). The volume and the size of sediments deposited in the floodplain vary greatly in time and space in relation to these various and sometimes opposing factors, such as changes in vegetation roughness due to ecological succession or rejuvenation phenomena. Main channel mobility produces continuous evolution that may result in progressive, regressive or alternating chronosequences dominated by allogenic/autogenic processes.

14.4.2 Role of Geomorphic Pattern and Age

Differences in sediment and soil composition from one plug to another are mainly explained by channel geometry (i.e. braided or meandering channel) and the age of the abandoned channels. When analysing the entire length of plugs, it is often difficult to distinguish abandoned channels according to a unique gradient (braiding–meandering origin or age) with sediment variables because they are strongly correlated:

(i)   the oldest abandoned channels are mostly associated with a braided pattern;
(ii)   the sedimentological and soil characteristics of the youngest braided abandoned channels (i.e. CFO or PNO1/2) are similar to those of meanders;
(iii)   the considerable difference between the oldest meander cut-off channels (SMO) and old braided cut-off channels may be due to the rejuvenation of SMO by channel displacement and aggradation, rather than to a difference in channel pattern.

These geometrical differences are linked to differences in granulometric and overbank sediment thickness in the upstream part of plugs. Braided plugs integrate more clearly the dynamics of sediment inputs than the meander ones because of their location in the floodplain. Hooke (1995) previously showed on smaller systems that infill rate was much slower on chute cut-offs than on neck ones. Their probability of being located in the axis of the main channel is much greater than for meanders, whatever the present position of the main channel, because they are straight and have been abandoned along straight reaches. Pioneer species such as willow and poplar can also grow quickly and favour deposition of overbank sediment due to their high roughness (Piégay 1996). Bradley and Brown (1992) have also previously underlined that silting-up is quicker in abandoned channels close to the main river channel because overbank sedimentation rates increase with proximity to active channel margins.

If the braided abandoned channels are characterized by greater sedimentary inputs, their plugs might be longer than those of meander dead arms. The results have shown that plug length is statistically linked to plug age but not to pattern. However, three braided abandoned channels present case-specific aspects: the lengths of PN1 and PN2 are artificially short, whereas the RIC plug has been narrowed by channel displacement. Moreover, CFO, one of the youngest abandoned channels, is longer than older meander abandoned channels such as PGA, PLA, BRO or SMO. Results have also shown that the plug progression was quicker in the older abandoned channels than in the younger ones. These sedimentation rates contrast with other observations where the sedimentation rate is usually higher in the younger units than in the older ones because they are still in close relationship with the present river (Nanson and Beach 1977; Lewis and Lewin 1983; Hooke 1995; Piégay 1996). Cases studied by Hooke (1995) also highlighted that later phases of infill in the centre part of abandoned channels are much slower than at the extremities, implying that the aquatic area may be maintained for decades. Thus age is certainly not the only factor explaining the rate of plug progression. This rate may be higher in the old braided forms because they were inserted in a more active system with higher bedload transport and peak flows (Marston et al. 1995). Contemporary incision might also have reduced floodplain flooding, thereby slowing plug development. In this study, it seems that plug length depends on both age and pattern.

It should also be noted that sediment fill of older braided plugs (GOU or SBR) is finer than in the meander plugs. These old braided plugs thus seem less active, which may be explained by the fact that their difference in altitude relative to the main channel is greater than that of the meander plugs, implying that they are flooded less frequently by flows of lesser competence. Indeed, the two young braided channel plugs CFO and PN1/2 show coarser granulometry than the meander ones do.

Differences in vegetation between meander and braided abandoned channels have also been observed by Bornette et al. (in press). The meander cut-off channels are almost all characterized by species that do not tolerate flood disturbances well (Nuphar lutea, associated with *Phragmites australis* and *Carex elata*) and by the absence of flood-tolerant species whereas the cut-off braided channels are characterized by flood-tolerant species such as *Berula erecta, Callitriche platycarpa, Sparganium emersum* and *Phalaris arundinacea*. The least frequently flooded and oldest braided channel (SBR) is characterized by the occurrence of both flood-tolerant species and species intolerant to floods as mentioned above.

14.4.3   Channel Instability

Abandoned channels of the Ain River have specific sedimentation and soil dynamics due to channel instability. We do not observe a gradient away from the river of decreasing hydrological connectivity and increasing age, as Bradley and Brown (1992) did on English systems. Although we have found an upstream–downstream sedimentation gradient along some plugs, which completes the observations of Schwarz *et al.* (1996) on sediment texture of Iowa and Cedar River abandoned channels, numerous exceptions exist. The oldest plugs, GOU and SBR, are characterized by spatially homogeneous sedimentation. Their inner heterogeneity is explained by vegetation cover which is probably dependent on plug micromorphology and water table altitude. This observation may be related to local marsh conservation due to micromorphology, hillslope water supply and the presence of fine impermeable overbank deposits. Some aquatic vegetation patches are conserved in small marshes located within mature terrestrial units.

As observed by Carbiener *et al.* (1993) on the Rhine floodplain, the theoretical pattern of fluvial forms and granulometry is greatly modified by energy variation due to channel mobility. Each plug may present several morphogenic sequences that periodically disrupt the previous sedimentation trend.

Plug sedimentation characteristics may serve to identify reaches that are submitted to recent aggradation or channel horizontal mobility, as these two mechanisms increase floodplain energy and tend to show thicker and coarser recent deposits. In the case of the Ain River, aggradation does not influence plug lengthening. Short plugs are observed on both incised and very aggraded reaches. Two factors may explain this: (i) the plugs located in aggraded reaches are too young to register lengthening due to increasingly frequent flooding; and (ii) the 10-year period, during which vertical mobility has been studied, may also be too short. However, the PNO plug exhibits a very thick layer of sediment similar to braided abandoned channels, even though it is a meander cut-off plug. This is probably because of high inputs linked to channel aggradation and high roughness of upstream willow units.

Bornette and Heiler (1994) and Bornette *et al.* (in press) have demonstrated that aquatic vegetation is influenced by the vertical mobility of the main river channel on several abandoned channels along the Ain and the Rhône rivers, namely because of its influence on aquifer levels. Indeed, incision lowers the floodplain water table, which can increase the shallowness and dry up cut-off channels supplied primarily by river seepage. River incision can also reduce the temporal and spatial heterogeneity of dead-arms, and consequently lower aquatic plant richness. Nevertheless, no relationship between abandoned channel vegetation and main channel displacement has yet been demonstrated.

14.4.4   Guidelines for Abandoned Channel Restoration

Abandoned channel restoration projects must include consideration of the potential life span and the ecological potential of these aquatic zones (Henry and Amoros 1995a, b). As the silting-up of the upstream part of abandoned channels could be one of the processes which controls the life span of these wetlands, the present results may provide valuable information for managers in choosing which abandoned channels to restore. These results demonstrate that plug processes and dynamics may serve as indicators of the potential inorganic sediment input to wetland areas. Alluvial plugs may function as buffer zones which trap sediment before transit and deposit in the wetland area. Depending on their characteristics, plugs may also contribute to preserving the life span of their neighbouring aquatic zones.

The geometry of plugs influences sediment inputs and thus potentially influences the life span of aquatic zones because of the progression of the plug into the wetland area. Wetland areas of meander abandoned channels might have a longer life span than those of braided ones because they are isolated from the main channel, and because floods are usually less frequent, less effective and are filtered by upstream plug vegetation in such ecosystems. Nevertheless, other factors such as groundwater discharge or river instability can also influence the silting rates of cut-off channels. Abandoned channels may thus have very long life spans where they are rarely flooded or flooded only by well-filtered waters. Some abandoned channels which are often flooded may also conserve their aquatic zone because hillslope phreatic water supply can maintain a sufficient discharge to export sediment inputs (Bornette et al. 1994). Moreover, some braided abandoned channels can have a long life span because their forested plug is not rough enough to dissipate hydraulic energy during flooding and pristine aquatic zone geometry is maintained by surface erosion.

Consequently, along high energy piedmont rivers, the life span of wetlands depends on various geomorphological factors that occur at different temporal and spatial scales. Also, our findings on such systems do not confirm the observations of Bradley and Brown (1992) on rivers of the English Midlands showing that abandoned channels of greatest interest for conservation are often those which are located furthest from the river because they are less subject to sediment input and are protected by larger buffer zones.

Further geomorphological research is required to clarify these questions. There is thus great interest in determining what the essential morphological characteristics of plugs are (i.e. slope, micromorphology, upstream–downstream variability of gravel substrate, altitude, etc.), and how they differ in relation to other environmental factors (i.e. geomorphic pattern, age of the abandoned channel, etc.). There is a need to better understand the relationships between the sedimentological characteristics of abandoned channel wetlands and those of their plugs. Finally, the temporal dynamics of the plug progression should be elucidated, while further research should determine if this is a continuous process or rather a highly active process in the first few years that rapidly decreases in intensity over time.

ACKNOWLEDGEMENTS

The authors wish to thank Arlette Laplace Dolonde and Pierre Clément of the Rhodanian Geomorphological Laboratory (University of Lyons II) for their critical advice in the field, and Sylvain Dolédec of the Laboratory of River Ecology (University of Lyons I) for advice on statistical analysis.

REFERENCES

Amoros, C. and Petts, G.E. (eds) 1993. *Hydrosystèmes fluviaux*, Masson, Paris.
Amoros, C. and Roux, A.L. 1988. Interactions between water bodies within the floodplains of large rivers: function and development of connectivity. *Münstersche Geographie Arbeiten*, **29**, 125–130.

Babinski, Z. 1992. Hydromorphological consequences of regulating the lower Vistula, Poland. *Regulated River Research and Management*, **7**, 337–348.

Barnaud, G. 1996. Fonctions et rôles des zones humides. In *24° Journées de l'Hydraulique: L'eau, l'homme et la nature*. Société Hydrotechnique de France, Paris, 307–316.

Best, J.L. and Bristow, C.S. (eds) 1993. *Braided Rivers*. The Geological Society, London.

Bornette, G. and Heiler, G. 1994. Environmental and biological responses of former channels to river incision: a diachronic study on the upper Rhône river. *Regulated River Research and Management*, **9**, 79–92.

Bornette, G. Amoros, C. and Chessel, D. 1994. Effect of allogenic processes on successional rates in former river channels. *Journal of Vegetation Science*, **5**, 237–246.

Bornette, G., Amoros, C. and Rostan, C. 1996. River incision and vegetation dynamics in cut-off channels. *Aquatic Science*, **58**(1), 31–51.

Bornette, G., Amoros, C., Piégay, H., Tachet, J., and Hein, T. 1998. Ecological complexity of wetlands within a river landscape. *Biological Conservation*, **85**, 35–45.

Bradley, C. and Brown, A.G. 1992. Floodplain and palaeochannel wetlands: geomorphology, hydrology and conservation. In Stevens, C., Gordon, J.E., Green, C.P. and Macklin, M.G. (eds), *Proceedings of the conference conserving our landscape: evolving landforms and ice-age heritage*. English Nature, Crewe, 117–124.

Bravard, J.-P. 1986. *Le Rhône du Léman à Lyon*. La Manufacture, Lyon.

Bravard, J.-P. and Peiry, J.-L. 1993. La disparition du tressage fluvial dans les Alpes françaises sous l'effet de l'aménagement de cours d'eau (19–20ème siècle). *Zeitschrift für Geomorphologie, supplementband*, **88**, 67–79.

Carbiener, P., Carbiener, R. and Vogt, H. 1993. Relations entre topographie, nature sédimentaire des dépôts et phytocénose dans le lit alluvial majeur sous forêt du Rhin dans le fossé rhénan: forêt de la Sommerley (commune d'Erstein). *Revue Géographique de l'Est*, **33**, 297–311.

Castella, E., Richardot-Coulet, M., Roux, C. and Richoux, P. 1991. Aquatic macroinvertebrate assemblages of two contrasting floodplains: the Rhône and Ain Rivers, France. *Regulated River Research and Management*, **6**, 289–300.

Chessel, D. and Dolédec, S. 1996. ADE version 4.0: Hypercard© Stacks and Quickbasic Microsoft© Programme Library for the Analysis of Environmental Data. Université Lyon I, Villeurbanne.

Chessel, D. and Mercier, P. 1993. Couplage de triplets statistiques et liaisons espèces-environnement. In Lebreton, J.D. and Asselain, B. (eds), *Biométrie et Environnement*. Masson, Paris, 15–43.

Church, M. 1992. Channel morphology and typology. In Calow, P. and Petts, G. (eds), *The Rivers Handbook. Hydrological and Ecological Principles*. Blackwell, Oxford, 126–143.

Furness, H.D. and Breen, C.M. 1980. The vegetation of seasonally flooded areas of the Pongolo River floodplain. *Bothelia*, **13**, 217–231.

Glitz, D. 1983. Artificial channels – the 'ox-bow' lakes of tomorrow: the restoration of the course of the Wandse in Hambourg-Rahlstedt. *Garten+Landschaft* (Landscape Architecture + Planning), **2**, 109–111.

Henry, C.P. and Amoros, C. 1995a. Restoration ecology of riverine wetlands. I. A scientific base. *Environmental Management*, **19**(6), 891–902.

Henry, C.P. and Amoros, C. 1995b. Restoration ecology of riverine wetlands. II. An example in a former channel of the Rhone river. *Environmental Management*, **19**(6), 903–913.

Hooke, J.M. 1995. River channel adjustment to meander cut-offs on the river Bollin and river Dane, northwest England. *Geomorphology*, **14**, 235–253.

Kellerhals, R. and Church, M. 1989. The morphology of large rivers: characterization and management. In *International Large River Symposium*. Special Publication, Canadian Fisheries and Aquatic Science, 31–48.

Lewis, G.W. and Lewin, J. 1983. Alluvial cutoffs in Wales and the Borderlands. *International Association of Sedimentologists, Special Publications*, **6**, 145–154.

Marston, R.A., Girel, J., Pautou, G., Piégay, H., Bravard, J.P. and Arneson, C. 1995. Channel metamorphosis, floodplain disturbance, and vegetation development: Ain River, France. *Geomorphology*, **13**, 121–131.

Möller, H.M. and Wefers, K. 1983. The restoration of cut-off river channels as illustrated by the lower reaches of the Krückau in the District of Pinneberg. *Garten+Landschaft*, **2**, 107–108.

Monnier, G. and Stengel, P. 1982. La composition granulométrique des sols: un moyen de prévoir leur fertilité physique. *Bulletin Technique d'Information*, **370–373**, 503–512.

Nanson, G.C. and Beach, H.F. 1977. Forest succession and sedimentation on a meandering-river floodplain, Northeast British Columbia, Canada. *Journal of Biogeography*, **4**, 229–251.

Pautou, G. 1984. L'organisation des forêts alluviales dans l'axe rhodanien entre Genève et Lyon; comparaison avec d'autres systèmes fluviaux. In *Documents de Cartographie Écologique*, Grenoble, Université J. Fourier, **24**, 43–64.

Pautou, G. and Girel, J. 1986. La végétation de la basse vallée de l'Ain: organisation spatiale et évolution. *Documents de Cartographie Ecologique*, **29**, 75–96.

Peiry, J.L. 1990. Paléodynamique fluviale et chronologie de l'incision holocène de la basse vallée de l'Arve (Haute Savoie). *Revue Géographique de l'Est*, **30**, 77–92.

Petit, F and Schumacker, R. 1985. L'utilisation des plantes aquatiques comme indicateur du type d'activité géomorphologique d'une rivière ardennaise. *Colloque Phytosociologique*, **13**, 691–710.

Pialot, D., Chessel, D. and Auda, Y. 1984. Description de milieu et analyse factorielle des correspondances multiples. *Compte rendu hebdomadaire des séances de l'Académie des sciences*, Paris, 298, Série III, **11**, 309–314.

Piégay, H. 1996. La forêt d'inondation des rivières à forte énergie, un patrimoine écologique à gérer. *Annales de Géographie*, **590**, 347–368.

Piégay, H. and Peiry, J.L. 1997. Water line profile evolution of an intra-mountain stream and gravel load management: example of the middle Giffre River (Haute-Savoie, France). *Environmental Management*, **21**(6), 909–919.

Point, P. 1993. Valeur économique des zones humides. In *Pour un retour des poissons migrateurs*. Agence de l'Eau Adour-Garonne, Toulouse, 7.

Salvador, P.G. 1991. *Le thème de la métamorphose fluviale dans les plaines alluviales du Rhône et de l'Isère (Bassin de Malville et Ombilic de Moirans, Bas-Dauphiné)*. Thesis of geography, University of Lyon III.

Schwarz, W.L., Malanson, G.P. and Weirich, F.H. 1996. Effect of landscape position on the sediment chemistry of abandoned-channel wetlands. *Landscape Ecology*, **11**, 27–38.

Shankman, D. 1991. Botanical evidence for the age of oxbow lakes: a test of Harper's hypothesis. *Southeastern Geography*, **XXXI**, 67–74.

Shankman, D. 1993. Channel migration and vegetation patterns in the Southern coastal plain. *Conservation Biology*, **7**, 176–183.

Shields, F.D. and Abt, S.R. 1989. Sediment deposition in cutoff meander bends and implications for effective management. *Regulated River Research and Management*, **4**, 381–396.

Starkel, L. and Thornes, J.B. (eds) 1981. *Palaeohydrology of River Basins*. British Geomorphological Research Group, Technical Bulletin 28, Geo Books, Norwich.

Tenehaus, M. and Young, F.W. 1985. An analysis and synthesis of multiple correspondence analysis, optimal scaling, dual scaling, homogeneity analysis and other methods for quantifying categorical multivariate data. *Psychometrika*, **50**, 91–119.

Tsujimoto, T. and Kitamura, T. 1996. River-bed degradation influenced by growth of vegetation along a stream. In Leclerc, M., Capra, H., Valentin, S., Boudréault, A. and Côté, Y. (eds), *Second International Symposium on Habitat Hydraulics*. INRS-Eau, Québec, 389–394.

Tucker, L.R. 1958. An inter-battery method of factor analysis. *Psychometrika*, **23**(2), 111–136.

Ward, J.V. and Stanford, J.A. 1995. Ecological connectivity in alluvial rivers and its disruption by flow regulation. *Regulated River Research and Management*, **11**, 105–119.

Weihaupt, J.G. 1977. Morphometry definitions and classifications of oxbow lakes, Yukon river basin, Alaska. *Water Resources Research*, **13**(1), 195–196.

Yon, D. and Tendron, G. 1981. *Les forêts alluviales en Europe, élément du patrimoine naturel international*. Conseil de l'Europe, Strasbourg.

PART 3   Hazard and Risk

Overview: Hazard and Risk

DAVID K. CHESTER

Department of Geography, University of Liverpool, Liverpool, UK

ABSTRACT

More research on natural disasters is being carried out today than ever before. It is argued that this is due to the increasing global toll of disaster, the disproportionately high burden of losses in poor countries and the framework of specific research objectives which have been defined by the United Nations, who designated the 1990s the International Decade for Natural Disaster Reduction (IDNDR). Recent developments in hazard analysis have seen an integration of earth and social scientific perspectives, and today most research groups are focusing upon relating forecasts, based on scientific investigations, to human vulnerability. The importance of place and locality is increasingly being emphasized. New technologies, in particular remote sensing and geographical information systems (GIS), are having major impacts upon hazard studies and the potential and problems of these systems of data collection and presentation are discussed. An international consensus about how hazard research should be conducted is emerging at the beginning of the 21st century. This overview looks at the ways in which the chapters in Part 3 fit into this framework.

INTRODUCTION

The hazard analyst David Alexander (1997: 284–285) has stated that around half of all disaster research has been carried out during the past two decades, and it is not difficult to see why this has been the case. The principal reason is the sheer magnitude of disaster losses, which has prompted both increased research activity and concerted international action. In 1993 Frank Press, distinguished seismologist and humanitarian, estimated that in a typical year, i.e. one without a major catastrophe on the scale of the Bangladesh cyclone of 1991 (estimated deaths 145 000) or the Ethiopian droughts of 1985 (estimated deaths 118 000), some 250 000 people would die and losses exceeding US$40 billion would result from natural disasters (Press 1993). More recent figures for severe disasters, causing a minimum of 1000 deaths or US$1 billion in losses reviewed over the period 1977–1997, show fairly steady average annual death tolls of nearly 24 000, but annual losses of US$15 billion (Alexander 1997). What is even more worrying is that there is plenty of evidence that the world is becoming more hazardous over time and each decade sees more people being affected adversely, with monetary losses also rising rapidly. Using constant 1990 costs in US dollars, Charlotte Benson (1998) has estimated that the financial burden of all global disasters in the 1980s was *c.* $120 billion, compared with *c.* $70 billion in the 1970s and *c.* $40 billion in the 1960s. Before 1987 there was only one case where the insured loss from a disaster exceeded $1 billion; in 1995 there were 14 such instances.

Applied Geomorphology: Theory and Practice. Edited by R.J. Allison.
© 2002 John Wiley & Sons, Ltd.

Much has been written about the deleterious effects of human pressure on natural systems (e.g. Simmons 1989; Goudie and Viles 1997) and its undoubted contribution in making certain extreme events more probable is well established. Examples are numerous and include some forms of desertification (Thomas 1993) and flooding following the removal of vegetation from catchments. There is little doubt, however, that the principal reason why the world is more at risk from hazards today than in the past is increased population pressure in parts of the world which are susceptible to environmental extremes (Anon. 1993; Chester 1993: 228–244). Industrial heartlands are now to be found throughout the circum-Pacific region, low latitudes are generally more susceptible to climatic extremes than mid-latitudes and many of the world's largest cities are located on, or lie alongside, active tectonic belts (Figure 1). It has been estimated that only 22 of the world's 100 largest cities are unexposed to hazards, with some 86 of the first hundred cities in poor countries being threatened (Degg 1992, 1998; ODA 1995; Steedman 1995; Figure 1). (Many terms may be used to describe the development status of a country: *north* and *south*; *First*, *Second* and *Third Worlds*; *developed* and *underdeveloped*; *developed* and *developing*; and *economically more* and *economically less developed*. All are euphemisms for a broad division of the world into rich and poor countries and some are pejorative. For this reason the terms rich and poor will be used in this overview.) This contrast between disaster losses in rich and poor countries is the second reason why the volume of research carried out over the past two decades has been so high. It is sobering to note that since 1971 some 97% of all deaths and 99% of people affected by disasters have resided in poor countries (Twigg 1998) and, although material losses have been smaller in absolute terms because there is less of value to be destroyed, when viewed in the light of total wealth, losses have been very high indeed. For example, the 1980 Mount St Helens volcanic eruption cost the United States *c.* 0.03% of its gross national product (GNP), whereas in poor countries many disasters cost national exchequers from *c.* 15% to 40% of GNP in a disaster year (Harriss *et al.* 1985; Kates 1987; Chester 1993, 2001; Benson 1998).

In 1988 the combination of high and expanding global losses and the increasing burden on poor countries prompted the United Nations General Assembly to designate the 1990s the International Decade for Natural Disaster Reduction (IDNDR) and it is under this umbrella that much recent research has been carried out (Table 1). In the earth sciences the IDNDR defined an administrative structure of closely interlocking international scientific organizations, an agenda for hazard reduction and in some cases sources of research funding. The International Geographical Union (IGU) currently has not only a special committee to consider all matters related to the IDNDR, but also special responsibility for projects on Drought and Famine and Flood Hazard Reduction in Bangladesh, and joint responsibility for programmes in the Vulnerability of Megacities and Education for Natural Disaster Reduction (Verstappen 1998).

RECENT DEVELOPMENTS IN HAZARD ANALYSIS

Over the past two decades, especially during the last ten years and under the stimulus of the IDNDR, major lessons have been learnt about how studies of natural hazard reduction should be carried out, which have affected earth scientists, social scientists and policy makers. So great have been these changes, that they constitute a Kuhnian paradigm shift in the methodology and practice of hazard analysis (Chester 1998). Today the approach adopted contrasts markedly with that employed only ten years ago.

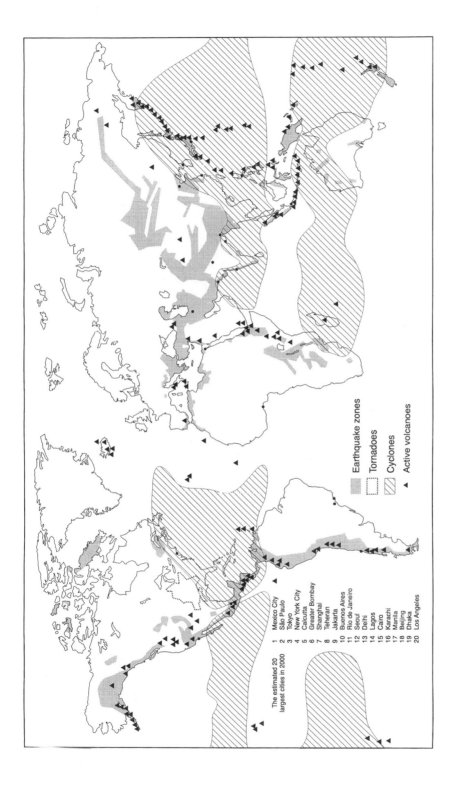

Figure 1 Hazard zones of the world (based on Munich Re, *World Map of Natural Hazards* (1998) plus information from Degg (1992) and Anon. (1993)). It should be noted that the list and rank order of megacities varies between authors, depending on differing definitions of the metropolitan area

Table 1   The principal aims of the United Nations IDNDR in 1989 (information from Eades 1998)

The General Assembly called on all governments to:
1. Formulate programmes of national disaster mitigation.
2. Take part in international action to reduce the impact of natural disasters.
3. Establish, where appropriate, national IDNDR committees, in cooperation with appropriate scientific and technological communities.
4. Encourage support from both public and private sectors
5. Take measures to increase public awareness of risk and the value of preventative measures.

Developments in Social Theory

The greatest change has been from a perspective which views disasters as being principally environmentally determined, to one grounded more securely within social theory (Hewitt 1983a, b, 1997; Chester 1993; Blaikie *et al.* 1994; Chester *et al.* 1999a, b). Until the 1990s the framework within which hazards were studied was so well established that it became known as the *dominant paradigm* (Hewitt 1983a). Although it was admitted that such factors as prior experience of hazardous events, systems of belief, psychological factors and economic and social circumstances may contribute to disaster losses, physical factors – especially the magnitude, frequency and recurrence intervals of extreme events – were viewed as their principal determinants. Stress was placed on mitigation measures based on scientific investigations of processes, forecasting, technology and planning.

The main reason why the dominant approach achieved its dominance was its success in quickly reducing losses for short-onset disasters with limited spatial impacts (e.g. volcanic eruptions, earthquakes, windstorms and floods) within rich countries, against a background of increasing hazard exposure in the four decades following the Second World War. Even when the magnitude of an event is high, losses are now generally small, even minimal, in most rich countries. Despite its large size (7.1 Richter magnitude), the 1989 Santa Cruz (or Loma Prieta) earthquake in California, for example, claimed the lives of only 63 people (Bolt 1993) and it comes as no surprise to find that from the 1950s the dominant approach began to exert a progressively pervasive influence over the research literature and the policies of national governments and international agencies, including the United Nations Disaster Relief Office. At the start of the IDNDR the dominant paradigm formed the *zeitgeist* of hazard analysis, as was clear from many policy statements which emerged early in the decade (Lechat 1990). Transfer of mitigation technology and disaster management techniques from rich to poor countries was part and parcel of achieving the dominant agenda of loss reduction.

It is because the disparity between the effects of disasters on rich and poor countries is not only wide but also widening and because technology which works well in rich countries is often inappropriate or unaffordable when transferred, that the dominant paradigm came in for increasing criticism from social theorists both in the 1980s and at a quickening pace during the course of the IDNDR. It is interesting to compare the effects of the earthquake in Armenia with that of Loma Prieta quoted above. With a similar magnitude, but occurring some ten months earlier, it killed 25 000 people and left 500 000 homeless (Degg 1993). It has been argued that most disaster losses in poor countries are not primarily a function of physical processes, but rather of deprivation and poverty (Hewitt 1983a, b, 1997; Susman *et al.* 1983; see review in Chester 1993: 228–248), with some of the so-called radical critics arguing that those who suffer are either economically and/or geographically marginalized (i.e. they are poor and/or are forced to live in disaster-prone regions such as 'megacities' or on desert margins). It is further posited that relief aid and transfers of technology often benefit those who are already wealthy, not only on the world scale but also within individual poor countries, leading to further marginalization of deprived people (Susman *et al.* 1983: 280; Chester 2001). According to more radical scholars, natural disasters are not primarily natural at all, but reflective of a process maintaining and enhancing disparities in wealth and poverty.

It is not difficult to find support for the radical critique. In the low-lying lands of Bangladesh, floods and cyclones frequently cause major loss of life. In recent decades economic factors have driven farmers to settle flood-prone lands in unprecedented numbers, because they perceive themselves to be protected by engineering works built using funds provided by programmes of foreign aid (Burton *et al.* 1978; Stoddart 1987). When cyclones strike as they did in 1970 and 1991, or when flooding occurs as it did in 1988, losses are much greater than anticipated, because inappropriate adjustment measures have encouraged the spatial concentration of poor people in hazard-prone areas. In Bangladesh rural poverty has also been responsible for the partial destruction of mangrove swamps, which formerly protected the coast, for use as firewood, and environmental resilience has been further reduced by the deforestation of many upland catchments in India and Tibet. Following the floods of 1988, Robert Hodgson (1995) found that some of the foreign aid used to rehabilitate the housing stock was still not appropriate.

By the mid-1990s there were few hazard researchers who would not broadly accept the thrust of the radical critique, that disaster exposure and poverty are linked, though this did not mean that the implications of its socialist agenda were generally accepted. It is also possible to push the relationship between wealth and disaster losses too far. It is sobering to recall that as recently as 1995 around 5000 people died in the Kobe earthquake in Japan in one of the world's richest and most technologically advanced countries (Keller and Pinter 1996: 3). Also, until major institutional changes occurred in the mid-1980s, Italy had a very poor record of responding to a range of disasters, despite being a wealthy country (Duncan *et al.* 1996: 17–18). In the case of the Kobe earthquake, many buildings dated from the economic boom years of the 1960s and 1970s and were not earthquake-proofed to a sufficiently high standard and this, combined with a deep-seated cultural desire to cope without outside assistance, contributed greatly to the high casualty rates both during and following the earthquake (Dawkins 1995; Glancey 1995). In Italy, until the political reforms of the last decade, all disaster planning took place under the heavy hand of organized crime, clientism and corruption (Chester 1993: 249–308; Chester *et al.* 1999a). In contrast some poor countries manage to cope with a range of hazards in a manner more typical of a rich country, a good example being Indonesia (Chester 1993). In Indonesia there is a deeply incultured tradition of coping with a range of natural disasters, which means a high priority has been given to hazard mitigation within overall development planning, and a mixture of technological (i.e 'western') and indigenous techniques are used (Kasumadinata 1984).

A major advance in the later 1980s and during the IDNDR has been the realization that places are unique: Mount Etna is not Mount Pintatubo; San Francisco is not Cairo; and the desert margins of the Sahel are not the same as the semi-arid lands of the USA. To be successful, hazard research and disaster planning have not only to take into account the detailed physical processes which control extreme events, but also require to be incultured within the society which is threatened. In a seminal volume (Blaikie *et al.* 1994), it was argued that what is important about a society is its *vulnerability*. This is defined as 'the characteristics of a person or group in terms of their capacity to anticipate, cope with, resist and recover from the impact of a natural hazard'. There is a distinction between hazard and risk, risk being 'a compound function . . . of the . . . natural hazard and the number of people characterized by their varying degrees of vulnerability . . . to extreme events. A disaster occurs when a significant number of *vulnerable* people experience a hazard and suffer severe damage and/or disruption . . . (to) . . . their livelihood' (Blaikie *et al*. 1994: 21 [my italics]).

This changing emphasis in disaster research may be seen both in the significant modification of the aims of the IDNDR as the 1990s ran its course, and in the output of research groups. In 1994 an IDNDR international review conference was held in Yokohama (Japan). It was clear from the proceedings that the social consequences of disaster, and the need for a more strongly incultured approach to mitigation, were emerging as major agenda items for the remainder of the decade (United Nations 1995; Eades 1998). As far as changes in the approach adopted by research teams is concerned, this may be seen in the work of the author and his colleagues. In 1980s we were concerned with volcanic hazard reduction on Etna volcano in Sicily. Stress was placed on understanding physical processes and on the basis of these investigations hazard maps were compiled (Chester *et al*. 1985). In subsequent research undertaken during the IDNDR on Furnas volcano on the island of Sao Miguel in the Azores, under the auspices of European Union's Laboratory Volcano initiative, we have not only carried out conventional hazard mapping, but also related this to the economy, culture and society of people living in the shadow of the volcano (Chester *et al*. 1999b and Table 2).

At the close of the IDNDR, hazard assessment typically involved a combination of the investigation of processes, their mapping and prediction, combined with vulnerability analysis: in short, a conflation of scientific and social scientific approaches to loss reduction.

Technological Change

It is a matter of observation that there has been a revolution in global communications over the past two decades and especially since 1990. Although David Alexander (1997: 294) rightly draws attention to the 'privileged beneficiaries' of this revolution, with most of the world's poor being distanced from it, it has nevertheless transformed hazard analysis. Today major catastrophes quickly gain their own Internet pages, albeit of varying quality, but what is more important for hazard studies is that expertise may now rapidly be shared across research communities through dedicated web sites, many of which may be accessed via the International Union of Geodesy and Geophysics (http://www.iugg.org), the International Geographical Union (http://www.igu-net.org) and Worldwide Disaster Aid and Information (http://www.disasterrelief.org).

The start of the IDNDR in 1990 coincided with the diffusion of two important new technologies through the worldwide disaster research community. The older of the two was satellite-based remote sensing and the more recent was geographic information systems (GIS), and many commentators at the time rightly assumed that they would transform hazard analysis (UNESCO 1994). In the case of remote sensing, valuable contributions have and are being made, not only to disaster preparedness studies, but also to work on prevention and relief (see Table 3 and Chapter 15). One issue of concern is the 'frustration felt by many scientists who sense the tremendous potential . . . (of remote sensing), but who cannot put it into practice, because . . . data are not appropriate for the job; . . . (or) . . . are too costly or too slow in arriving' (Wadge, 1994: 3). Although written several years ago, Geoff Wadge's remarks were still germane at the end of the IDNDR (see Table 3).

GIS and remote sensing are complementary technologies and, increasingly, the former is being used to interpret data collected by the latter (see Chapters 15 and 17). Information derived from other sources may also be incorporated into a GIS provided it is capable of spatial expression. Large volumes of geographically referenced data, which would overwhelm manual processing, may now be analysed rapidly. Excellent recent examples include:

Table 2 An example of the demographic, socioeconomic and cultural/behavioural information which is now routinely collected in hazard studies. This particular example relates to evacuation planning at Furnas Volcano on São Miguel Island in the Azores. It is argued that these factors could complicate evacuation planning based on hazard mapping and that they should be carefully considered by the Civil Defence authorities

Factors	Details
Demographic	The Furnas District had a resident population of 22 644 in 1991, accommodated in 5693 houses. These figures do not, however, capture (a) the large number of visitors, especially in summer, and (b) the under-occupancy of many houses for most of the year. These are inhabited for some of the time by weekend residents, return migrants from mainland Portugal and abroad and by tourists. Since the Portuguese census date is 15 April, a month of relatively low resident population, far more people would potentially require transporting to safety at certain times of the year than the published figures suggest. A further demographic factor which could complicate evacuation is the degree to which population is either clustered within the principal settlement of the local authority area, or widely dispersed over its area.
Socioeconomic	Traditionally an agricultural/fishing area, alternative employment opportunities are limited. This is reflected in recent decades by the permanent or temporary out-migration of many people in the economically active cohort, with the result that 'dependency ratios' (i.e. % under 15 + % over 65 in the population) range from 38 to 46% across the district. The percentage of the population classed as economically active is never greater than 36% of the total in any administrative area
Cultural/behavioural	As a result of an aged population and the fact that before the 1974 Portuguese Revolution many people did not receive even an elementary education, illiteracy ranges from 8 to 23% of the total population across the Furnas District. Behavioural factors are important and include close links between many of the inhabitants and the land, as a result of both active and traditional family-based agricultural ties. At its most simple and bearing in mind that cattle rearing and fattening dominate contemporary agriculture, large numbers of livestock – both living and dead – could block roads in the event of an eruption, whilst another close attachment, of people to village, land and farm, could cause some inhabitants to resist evacuation.

Adapted from Chester *et al.* (1999b), and based on numerous additional sources, social surveys and interviews carried out in the field area. Reproduced with permission.

(a) management of water resources in Kenya
(b) hazard and land-use planning in Senegal
(c) aspects of risk assessment in Ecuador (UNESCO 1994) and
(d) the work of the Montserrat Volcano Observatory (MVO) in the Caribbean.

In the case of the MVO scientists have been monitoring the continuing eruptions of the Soufriere Hills volcano which began in 1995 (Wadge 1998). One great advantage of GIS-based hazard mapping over conventional manual cartography is that revision and updating may be easily accomplished as new data become available and this may take place after the disaster begins as 'hard' information gradually replaces partial forecasts. It is this methodology which is currently being used by the MVO (see Clay *et al.* (1999) and references therein).

Table 3 Examples of the use of satellite-based remote sensing techniques in hazard assessment and management. The normal typeface indicates that the given application is either operational or needs little further research and development, while the *italic* typeface indicates that further research and development is required and/or improved observational capacity would be desirable and/or spatial or temporal resolution requires improvement

Hazard type	Prevention	Preparedness (warning)	Relief
Earthquakes	Mapping faults and land use	*Geodynamic measurement of strain accumulation*	Locate stricken areas, map damage, *direct rescue teams*
Volcanic eruptions	*Topographic* and land-use mapping	*Hazard mapping (i.e. general prediction), detection and measurement of gas emissions*	*Mapping of volcanic products and the damage they have caused*
Landslides	*Topographic* and land-use mapping	*Soil porosity, rainfall and slope stability mapping*	Mapping of the slide, updating maps
Flash floods	Land-use mapping	*Local forecasts of rainfall*	Mapping of the flood damage
Major floods	*Detailed maps of floodplains,* land-use maps	*Regional and local weather forecasts, estimates of evapotranspiration*	Mapping of flood damage, *directing relief teams*
Storm surge and tsunami	Land-use and vegetation maps	Sea state, *ocean surface wind velocities*	Mapping the extent of damage
Hurricanes		Synoptic weather forecasts	Mapping the extent of damage
Tornadoes		*Local weather forecasts*	Mapping amount and extent of damage
Drought		*Development of long-range climatic models*	Monitoring vegetation biomass and crops
Wildfires		Location of fires	Mapping the extent of damage, directing evacuation, controlling fire fighters

Prevention includes measures designed to provide permanent protection; *preparedness* involves activities devised to minimize loss of life and damage through warnings, predictions, forecasts, hazard mapping, evacuation, planning for rehabilitation and recovery; and *relief* consists of assistance and/or intervention during or after the disaster.
Based on Walker (1994) and updated from numerous other sources.

The Evolving International Consensus

Greater knowledge of extreme events of nature, the trenchancy of much of the radical critique, the lessons which have been learnt from the first-hand study of many disasters during the IDNDR and new technological developments, mean that new frameworks are evolving for the study of hazards. Indeed, the implication of insights from both the earth and the social sciences has been to emphasize the uniqueness of place and human vulnerability. Recently the author and his colleagues, in the context of volcanic risk studies (Chester *et al.* 1999b), have suggested that assessment should proceed sequentially: first, hazard mapping – using conventional and GIS-based techniques – should be carried out; second, interactions with other elements of the physical environment should be specified; and, finally, factors

Table 4 Evolving framework for the study of hazards and society. Close parallels with environmental impact analysis should be noted

	Technique	Features	Comments
Increasing complexity	Checklist	Lists all the factors – physical, economic, cultural and societal – which need to be considered. Cause/effect relationships are implied but not specified in detail.	Initiatives in many countries involve, either implicitly or explicitly, a checklist approach. It is the evolving norm of the IDNDR.
	Overlays	Traditionally this has relied on overlay maps showing physical, social and historical aspects of the region. Today geographical information systems (GIS) are commonly used.	There is much scope for this approach to be used in hazard analysis, because many of the variables are spatial and capable of being either mapped or incorporated into a GIS. The impact of satellite-based systems, significant at present, is likely to be much more prominent in the future.
	Matrices	Matrices are used to identify first-order cause/effect relationships.	At the present time variables are not sufficiently well specified to enable matrices and network-based studies to be carried out. There may be much scope in the future.
	Networks	Used to identify 'chains' of complex interactions. Ideally this approach requires mathematical modelling.	

Based, with permission, on a table in *Volcanoes and the Environment* (eds J. Marti and G.J. Ernst), Cambridge University Press, 2002.

which include, *inter alia*, demography, socio/economic and cultural/behaviourial characteristics of the population at risk should be considered.

A framework of more general applicability is also being adopted using a methodology which is similar to Environmental Impact Analysis (EIA) (Jones and Hollier 1997: 228–255). EIA was developed under legislation passed in many countries from the late 1960s to assess the impact of environmentally damaging projects (Mitchell 1989). In the particularity of their effects on unique places, hazards are in many ways analogous to such projects and close parallels may be drawn between EIA and the framework currently evolving for the study of hazards (Chester 2002; see also Chapter 18). As Table 4 shows, for all hazards the profusion of physical and social factors which need to be studied may be expressed in checklists, while the overlay approach – using either conventional or GIS-based cartography – is increasingly being employed to compare spatial data. In the present volume the chapters by Amadio *et al.*, Catani *et al.* and Del Monte *et al.* exemplify this approach.

NATURAL HAZARDS AND APPLIED GEOMORPHOLOGY

At the present time research is emphasizing the complex character of the relationships between hazards and society, the uniqueness of place, the innovation of new technologies and research synergies both within and between the sciences and social sciences. Indeed these themes are likely to become even more prominent as the United Nation's new approach,

International Strategy for Disaster Reduction (ISDR) (United Nations 1999), succeeds that of the IDNDR (Hamilton 1999: 306). The chapters which follow give a flavour of some of the research themes discussed in this overview.

In Chapter 15, on the role of geomorphology in landscape ecology, Vittorio Amadio and his colleagues describe the compilation of the 1:250 000 Landscape Unit Map of Italy. This map provides the Italian government with a tool to aid planning and monitor impacts of potentially damaging development projects on soil conservation and the wider natural environment. Using an updated methodology of 'land systems analysis' (Cooke and Doornkamp 1990: 20–28) and an 'overlay' approach (see Figure 15.4), both remote sensing/ aerial photographic data and GIS techniques are employed. *Inter alia* the authors collected data on morphology, lithology, vegetation and land use, and used these to classify the Italian landscape into distinctive units. Information from existing thematic maps and literature, field surveys and government reports was also incorporated. Data are presented not only in map form but also digitally, and applied to central Italy.

Shallow landslides are a major category of hazard in Italy and each year are responsible for huge economic and social losses. In Chapter 16 Mauro Casadei and Enzo Farabegoli examine the problem in the context of the northern Apennines. Because the factors which control landslides are so numerous, planners have traditionally assessed risk using just four critical slope categories (10%, 20%, 35% and 50%), as limits for different types of economic activity. In a highly innovative study the authors test several techniques of terrain modelling. Although the results are derived from one catchment, implications of the research are of more general interest, questioning the widespread assumption that in Italy slopes of 20–35% are suitable for ploughing without significant risk of landsliding.

Conservation of historic sites is usually conceived of in terms of preserving buildings and artifacts against decay, but during the IDNDR another issue emerged. Many sites are at risk from extreme natural events and in a number of countries research is progressing into assessing the threat to the archaeological and historical heritage. Filippo Catani, Riccardo Fanti and Sandro Moretti (Chapter 17) provide an interesting pioneer study of the progress that is currently being undertaken in Italy at the interface between earth sciences and archaeological conservation. Using as case studies the Vulci (Lazio region, central Italy) and Tharros (west Sardinia) archaeological sites, the authors employ GIS techniques to assess geomorphological risk factors, in particular flooding, enhanced erosion and slope instability.

As discussed earlier, environmental impact assessment (EIA) is rapidly becoming the *zeitgeist* in hazard studies and Arthur Conacher (Chapter 18) discusses some of the problems involved in applying this methodology to farming. Dryland agriculture in Australia and salinization in the western wheat belt exert complex impacts on soils, with forcing factors being not only pedological, but also hydrological and geomorphological. The author evaluates the role of EIA in this area of research, pointing out its practical difficulties and the patchy progress which has been made to date.

During the IDNDR and as discussed above, techniques using overlays of maps showing different variables (Figure 15.4), have emerged as powerful tools in examining spatial data relevant to hazard assessment. In Chapter 19 Maurizio Del Monte and his colleagues apply the technique quantitatively to geomorphological hazards within three drainage basins in Italy. By comparing overlays of factors which are known to control landscape instability they are able to demonstrate that the key variables are lithology, land use, slope gradient and drainage density. They go on to show how the relative importance of each variable may be determined quantitatively.

In volcanology one of the advances of recent years has been the realization that volcanoes are not only intrinsically unstable landforms, but that their morphology can also have a major impact on the distribution of damaging products from pyroclastic flows, pyroclastic surges and lahars, i.e. volcanic debris flows (Chester 1993). In Chapter 20 Masaru Iwamoto first reports on the 1990–1995 eruption of Unzen volcano in Japan and then discusses how geomorphology largely controlled the distribution of volcanic products and resulting hazard. The importance of geomorphology in volcanic hazard mapping is emphasized.

ACKNOWLEDGEMENTS

The author wishes to thank his colleague, Dr Katie Willis, for commenting on an earlier draft of this chapter. Artwork was expertly drafted by Mrs Sandra Mather.

REFERENCES

Alexander, D. 1997. The study of natural disasters, 1977–1997: Some reflections on a changing field of knowledge. *Disasters*, **21**(4), 284–304.

Anon. 1993. *Disaster Reduction*. UNESCO, Paris. Environment and Development Brief.

Benson, C. 1998. The cost of disasters. In Twigg, J. (ed.), *Development at Risk: Natural Disasters and the Third World*. Oxford Centre for Disaster Studies, Oxford, 8–13.

Blaikie, P., Cannon, T., Davis, I. and Wisner, B. 1994. *At Risk: Natural Hazards, People's Vulnerability, and Disasters*. Routledge, London.

Bolt, B.A. 1993. *Earthquakes*. Freeman, New York.

Burton, I., Kates, R.W. and White, G. 1978. *The Environment as Hazard*. Oxford University Press, New York.

Chester, D.K. 1993. *Volcanoes and Society*. Edward Arnold, London.

Chester, D.K. 1998. The theodicy of natural disasters. *Scottish Journal of Theology*, **51**(4), 485–505.

Chester, D.K. 2002. Volcanoes and society. In Marti, J. and Ernst, G.J (eds), *Volcanoes and the Environment*. Cambridge University Press, Cambridge.

Chester, D.K., Duncan, A.M., Guest, J.E. and Kilburn, C.R.J. 1985. *Mount Etna: The Anatomy of a Volcano*. Chapman and Hall, London.

Chester, D.K., Duncan, A.M., Dibben, C., Guest, J.E. and Lister, P.H. 1999a. Mascali, Mount Etna, Sicily: An example of Fascist planning during the 1928 eruption and its continuing legacy. *Natural Hazards*, **19**, 29–46.

Chester, D.K., Dibben, C., Coutinho, R., Duncan, A.M., Guest, J.E. and Baxter, P.J. 1999b. Human adjustments and social vulnerability to volcanic hazards: The case of Furnas Volcano, Sao Miguel, Azores. In Firth, C. and McGuire, W.J. (eds), *Volcanoes in the Quaternary*. Geological Society, London, Special Publication, 161, 189–207.

Clay, E., Barrow, C., Benson, C., Dempster, J., Kokelaar, P., Pillai, N. and Seaman, J. 1999. *An Evaluation of HMG's Response to the Montserrat Volcanic Emergency*. Department for International Development, London, Evaluation Report EV635 (two volumes).

Cooke, R.U. and Doornkamp, J.C. 1990. *Geomorphology in Environmental Management* (2nd edition). Clarendon Press, Oxford.

Dawkins, W. 1995. Faith in the Authorities Fades. *Financial Times* (London), 21 January, 9.

Degg, M. 1992. Natural disasters: Recent trends and future prospects. *Geography*, **77**, 198–209.

Degg, M. 1993. Earthquake hazard, vulnerability and response. *Geography*, **78**, 165–170.

Degg, M. 1998. Natural hazards in the urban environment: The need for a more sustainable approach to mitigation. In Maund, J.G. and Eddleston, M. (eds), *Geohazards in Engineering Geology*. Geological Society, London, Engineering Geology, Special Publication, 15, 329–337.

Duncan, A.M., Dibben, C., Chester, D.K. and Guest, J.E. 1996. The 1928 eruption of Mount Etna volcano, Sicily, and the destruction of the town of Mascali. *Disasters*, **20**(1), 1–20.

Eades, T. 1998. The International Decade for Natural Disaster Reduction. In Twigg, J. (ed.), *Development at Risk: Natural Disasters and the Third World*. Oxford Centre for Disaster Studies, Oxford, 20–22.

Glancey, J. 1995. Stricter rules proved their Worth. *The Independent* (London), 20 January, 12.

Goudie, A. and Viles, H. 1997. *The Earth Transformed: An Introduction to Human Impacts on the Environment*. Blackwell, Oxford.

Hamilton, R. 1999. Natural disaster reduction in the 21st century. In Ingleton, J. (ed.), *Natural Disaster Management*. Tudor Rose, Leicester, 304–307.

Harriss, R.W., Hohenemser, C. and Kates, R.W. 1985. Human and non-human mortality. In Kates, R.W., Hohenemser, C. and Kasperson, J.X. (eds), *Perilous Progress: Managing the Hazards of Technology*. Westview Press, Boulder, 129–155.

Hewitt, K. 1983a. The idea of calamity in a technocratic age. In Hewitt, K. (ed.), *Interpretations of Calamity*. Allen and Unwin, London, 3–32.

Hewitt, K. (ed.) 1983b. *Interpretations of Calamity*. Allen and Unwin, London.

Hewitt, K. 1997. *Regions of Risk: A Geographical Introduction to Disasters*. Addison Wesley Longman, Harlow.

Hodgson, R. 1995. Housing improvements: Disaster response or hazard mitigation? Examples from Bangladesh. *Built Environment*, **21**(2/3), 154–163.

Jones, G. and Hollier, G. 1997. *Resources, Society and Environmental Management*. Paul Chapman Publishing, London.

Kasumadinata, K. 1984. Indonesia. In Crandall, D.R., Booth, B., Kasumadinata, K., Shimozuru, D., Walker, G.P.L. and Westercamp, D. (eds), *Source-book for Volcanic Hazard Zonation*. UNESCO, Paris, 55–60.

Kates, R.W. 1987. Hazard assessment and management. In McLaren, D.J. and Skinner, B.J. (eds), *Resources and World Development*. Wiley, Chichester, 741–753.

Keller, E.A. and Pinter, N. 1996. *Active Tectonics: Earthquakes, Uplift and Landscape*. Prentice Hall, New Jersey.

Lechat, M.F. 1990. The International Decade for Natural Disaster Reduction: background and objectives. *Disasters*, **14**(1), 1–6.

Mitchell, B. 1989. *Geography and Resource Analysis*. Longman, Harlow.

Munich Re, 1998. *World Map of Natural Hazards* (2nd edition). Munich Reinsurance Company, Munich.

ODA. 1995. *Megacities: Reducing Vulnerability to Natural Hazards*. Overseas Development Administration and The Institution of Civil Engineers, Thomas Telford Press, London.

Press, F. 1993. The need for action. In *Disaster Reduction*. Environment and Development Brief, 5, UNESCO, Paris, 2–4.

Simmons, I.G. 1989. *Changing the Face of the Earth: Culture, Environment and History*. Blackwell, Oxford.

Steedman, S. 1995. Magacities: The unacceptable risk of natural disaster. *Built Environment*, **21**(2/3), 89–94.

Stoddart, D.R. 1987. To claim the high ground: geography for the end of the century. *Transactions Institute of British Geographers* NS **12**, 327–336.

Susman, P., O'Keefe, P. and Wisner, B. 1983. Global disasters, a radical interpretation. In Hewitt, K. (ed.), *Interpretations of Calamity*. Allen and Unwin, London, 263–283.

Thomas, D.S. 1993. Sandstorm in a teacup? Understanding desertification. *Geographical Journal*, **159**(3), 318–331.

Twigg, J. 1998. Disasters, development and vulnerability. In Twigg, J. (ed.), *Development at Risk: Natural Disasters and the Third World*. Oxford Centre for Disaster Studies, Oxford, 2–7.

UNESCO. 1994. *New Technologies*. Development Brief 3, UNESCO, Paris.

United Nations. 1995. *Yokohama Strategy and Plan of Action for a Safer World: Guidelines for Natural Disaster Prevention, Preparedness and Mitigation*. United Nations, Geneva.

United Nations. 1999. *International Decade for Natural Disaster Reduction: Successor Arrangement*. United Nations, New York.

Verstappen, H.Th. 1998. Special committee for the International Decade for Natural Disaster Reduction (SC/IDNDR). *Bulletin International Geographical Union*, **48**(2), 121–126.

Wadge, G. 1994. Preface. In Wadge, G. (ed.), *Natural Hazards and Remote Sensing*. The Royal Society and The Royal Academy of Engineering, London, 7–12.

Wadge, G. 1998. *A Hazard Evaluation System for Montserrat*. IDNDR Flagship Programme – Forecasts and Warnings No. 3, Thomas Telford, London.

Walker, L.S. 1994. Natural hazard assessment and mitigation from space: The potential of remote sensing to meet operational requirements. In Wadge, G. (ed.), *Natural Hazards and Remote Sensing*. The Royal Society and The Royal Academy of Engineering, London, 7–12.

15 The Role of Geomorphology in Landscape Ecology: the Landscape Unit Map of Italy, Scale 1:250 000 (Carta Della Natura Project)

VITTORIO AMADIO, MARISA AMADEI, ROBERTO BAGNAIA, DANIELA DI BUCCI, LUCILLA LAURETI, ANGELO LISI, FRANCESCA R. LUGERI AND NICOLA LUGERI

Italian Department for National Technical Services (DSTN), Rome, Italy

ABSTRACT

The first step of the Carta della Natura Project is the realization of a Landscape Unit Map of Italy at a scale of 1:250 000, which permits generalized territorial phenomena to be focused upon. In this context, the study presented here consists of the definition of the conceptual approach and of the methodologies developed for the characterization and mapping of the landscapes of Italy. The aim is the identification of land settings that show homogeneity at the study scale. These land settings were identified following an approach based on an integrated study and synthesis of the elements forming the landscape structure, analysed according to their composition and pattern. The elements considered were the morphological, lithological, vegetational and land-use features. This allowed a number of landscape types to be identified and described, each with a characteristic physiognomy, and the landscape units of central Italy to be mapped, described and codified.

Remote sensing methods played a primary role in the realization of the map, allowing the pattern that marks each landscape unit to be defined.

The main results are discussed in the light of considerations that emphasize the importance of geomorphology in landscape ecology studies. In fact, at the regional scale our findings suggest that it is the physiography which best approximates the results of a landscape classification performed following a holistic approach.

15.1 INTRODUCTION

The Carta della Natura Project (Italian Law 394/91) aims to evaluate the condition of the natural environment in Italy, identifying the natural assets and the environmental vulnerability of the country. The project was conceived to provide the government with a tool to aid territorial planning, which takes into account the natural and environmental aspects (interventions of national interest, environment and soil conservation).

In order to carry out land management functions, this tool must be updatable, multiscalar, and it has to contain a wide range of information concerning the environment, i.e. the physical, biotic and anthropic aspects and their interrelations. From this point of view, a geographic information system is needed.

In the realization of the Carta della Natura Project, we started from the identification and classification of the landscape regional setting and of the ecosystemic units (i.e. the patches

Applied Geomorphology: Theory and Practice. Edited by R.J. Allison.

forming the landscape). Consequently, two main scales were chosen: the first, 1:250 000, for a regional level; the second, 1:50 000, for a more detailed analysis.

Since the Carta della Natura Project involves mapping the entire national territory, the first stage of our work consisted of (1) defining the conceptual approach and (2) developing the methodologies to be followed in the characterization and mapping, at a regional scale (1:250 000), of the landscapes of Italy; the aim was to identify land settings that show homogeneity at this scale.

The landscapes were identified following a holistic approach based on an integrated study and synthesis of the composition and pattern of the elements that constitute the physiognomy of each type of landscape, using remote sensing.

The main results and problems are discussed in the light of considerations that point to the importance of geomorphology in landscape ecology studies. Indeed, our finding that land-form is related to landscape at the regional scale has implications for the semi-automatic identification of landscapes by morphometric data.

15.2   CONCEPTUAL APPROACH

Our theoretical approach to landscape study and classification starts from the modern conception of Nature in the sciences, which is well summed up in the following definition of the Universe: 'not homogeneous, dynamic, multiscaled and hierarchically organised' (Prigogine and Stengers 1984).

Nature is considered to be a system that can be studied and described by taking into account its structural features (i.e. components with their characteristics and spatial distribution), its functional features (interrelations among spatial elements), and its spatial–temporal hierarchical organization.

This conception is particularly appropriate for such a complex reality as the landscape, which is the result of interaction among physical, biotic and anthropic phenomena acting in different spatial–temporal scales (Forman and Godron 1986).

Landscape can be defined from various points of view, depending upon the disciplinary approach adopted (Meinig 1979; Vink 1983). In accordance with the view of Nature described above and with the aims of the Carta della Natura Project, we followed a conception of landscape well expressed by Zonneveld (1979): '. . . (a) part of the space on the Earth's surface, consisting of a complex of systems, formed by the activity of rock, water, air, plants, animals and man and that by its physiognomy forms a recognizable entity' and more recently expressed by Forman and Godron (1986): 'a heterogeneous land area composed of a cluster of interacting ecosystems that is repeated in similar form throughout'.

Therefore, considering landscape to be a complex system, an eco-systemic model of interpretation was chosen (O'Neill *et al.* 1986; May 1989). According to this model the landscape is a system which is organized in scale-dependent, hierarchical levels of complexity (O'Neill 1989). To address this organized complexity, the use of strictly analytical methods is not appropriate: instead a holistic approach is necessary, which considers all the different factors characterizing the land globally and integrally. Therefore, a systemic approach, typical of the ecological sciences, was chosen.

In order to classify and map the landscapes, we thought that this holistic and integrated approach could be attained in practice by studying the elements which form the landscape structure (e.g. physiographic components and land cover; Table 15.1), since these are the visible expression of the endogenous and exogenous phenomena, as well as of the human activity, generating the landscape. This study consisted of the identification of such elements,

Table 15.1 Landscape types of central Italy

Name	Landscape general structure	Elevation (a.s.l.)	Relief energy	Main lithologies	Drainage pattern	Physiographic components	Main landcover
PC Coastal plain	Flat or subflat area bordered by a low coast. The shape of the area is generally elongated in the direction of the coastline.	Lower than 100 m	Low	Clay, silt, sand, sandstone, gravel, conglomerate.	Parallel and subparallel. Meandering streams and artificial channels are present.	Shoreline, beach, dune, back-dune, lagoon-pond-marsh, fossil dune, river delta, marine terrace. Subordinately: channel, reclaimed land, river plain, terrace and flat alluvial fan.	Agricultural land, urban areas, infrastructures, wetlands.
PA Valley floor plain	Flat or subflat area inside a river valley. The shape of the area is elongated in the direction of the river stream.	Not significant	Low	Clay, silt, sand, sandstone, gravel, conglomerate, travertine.	Generally characterized by a single principal stream, often meandering. Artificial channels are present.	Stream, natural levee, floodplain, lake-pond-marsh, river terrace. Subordinately: travertine plateau, channel, reclaimed land, flat alluvial fan, river delta.	Agricultural land, urban areas, infrastructures, wetlands.
PP Open plain	Flat, subflat or undulate, wide area, with variable geometry; not limited to a river valley.	Few tens of metres – about 400 m	Low	Clay, silt, sand, sandstone, gravel, conglomerate, travertine.	Well developed, parallel and subparallel. Meandering streams and artificial channels are present.	River and marine terrace, stream, natural levee, floodplain, lagoon-pond-marsh, travertine plateau. Subordinately: reclaimed land, flat alluvial fan, river delta, little hill.	Agricultural land, urban areas, infrastructures, wetlands.

continues overleaf

Table 15.1 (*cont.*)

Name	Landscape general structure	Elevation (a.s.l.)	Relief energy	Main lithologies	Drainage pattern	Physiographic components	Main landcover
VI Inner valley	More or less wide valley, generally with elongated shape and bordered by ridges of mountains on both sides. Depressed belt intra- or forechain, which presents many kinds of land shapes (flat to hilly). Typically, the outcropping lithologies are more affected by the erosion than those present on the surrounding relief.	Not significant, generally less than 1000 m	Low, mean	Clay, silt, sand, gravel, conglomerate, sandstone, marl. Subordinately: travertine, volcanics.	Dendritic, trellis, pinnate. Generally characterized by a single principal stream, often meandering.	River terrace and plain, flat alluvial fan, travertine plateau, terrigenous hills with various top-shape (flat to smoothed; crests), V-shaped valley, flat valley-floor.	Agricultural land, infrastructures, urban areas, forest.
CI Intra-montane basin	Closed and depressed area, surrounded by relief, characterized by a morphology that is flat to slightly undulate.	Not significant	Low	Clay, silt, sand, sandstone, gravel, conglomerate, travertine.	Centripetal, dendritic. Meandering streams and artificial channels are present.	River and lacustrine terrace, alluvial plain, lake-pond-marsh, reclaimed land, channel, flat alluvial fan, travertine plateau.	Agricultural land, urban areas, infrastructures, wetlands.
TC Carbonate plateau	Flat rocky area bordered by low escarpments, mainly formed by limestone.	0–500 m	Low	Limestone, dolomitic limestone, marly limestone.	Scarcely developed, conditioned by karst.	Carbonate plateau, escarpment, slope talus, all karst morphology elements.	Agricultural land, grassland, scrub, infrastructures, urban areas.
CV Volcanic hills and plateaux	Hilly reliefs with conic to tabular shape, formed by present or past volcanic activity.	Up to about 1000 m	Mean, high	Lava, pyroclastics. Subordinately: travertine, clay, silt, sand.	Centrifugal, parallel, dendritic.	Plateau, smoothed hill top, cone, caldera, crater, gorge, V-shaped valley, flat valley-floor. Subordinately: sub-circular lake or alluvial plain in caldera or crater depressions, travertine plateau, badlands, cliff.	Forest, agricultural land, grassland, scrub.

CA Clay hills	Hills with smoothed to flat top (occasionally crests) and graded to flattened slopes, mainly formed by clay.	Some tens of metres – about 700 m	Mean	Mainly clay, silt, sand. Subordinately: gravel, conglomerate, travertine, volcanics.	Dendritic, parallel, pinnate.	Smoothed and/or flat top, crest, graded to flattened slope, V-shaped valley, flat valley-floor, badlands, 'biancane', 'crete', widespread landslide and accelerated erosion phenomena. Subordinately: travertine plateau, sandstone or conglomerate plateau, river plain and terrace, alluvial fan.	Agricultural land, scrub, grassland.
CT Siliciclastic hills	Mainly siliciclastic hills forming wide parts of the forechain.	Some hundred metres	Mean	Sandstone, marl, clay. Subordinately: calcarenite, conglomerate, evaporite.	Dendritic, pinnate, meandering streams are present.	Smoothed top, crest, graded slope, V-shaped valley or flat valley-floor, badlands, widespread landslide and accelerated erosion phenomena. Subordinately: river terrace and plain, alluvial fan.	Agricultural land, forest, grassland, scrub.
CC Carbonate rock hills	Hills mainly composed by carbonate rocks, forming parts of the chain and the forechain.	Some hundred metres	Mean, high	Limestone, dolomitic limestore, dolostone, marly limestone.	Scarcely developed, linked to karst morphology; trellis, parallel, rectangular.	Crest, smoothed top, steep slope, V-shaped incised valley, gorge, all karst morphology elements, small closed depression filled with incoherent sediments, slope talus. Subordinately: river terrace and plain, alluvial fan.	Agricultural land, grassland, scrub, forest, exposed rocks.
CE Heterogeneous hilly landscape	Hilly landscape characterized by a strong lithologic and morphologic heterogeneity and, consequently, by a typical internal inhomogeneity.	Up to about 1000 m	Variable, generally from low to mean	Highly variable lithotypes.	Complex, compound.	Terrigenous and rocky hills with variable shapes (smoothed, flattened, with crests), graded to flattened slope, V-shaped valley, flat valley-floor, slope talus, river terrace, alluvial plain and fan.	Agricultural land, forest, grassland, scrub.

continues overleaf

Table 15.1 (*cont.*)

Name	Landscape general structure	Elevation (a.s.l.)	Relief energy	Main lithologies	Drainage pattern	Physiographic components	Main landcover
PI Areas with isolated hummocks	Group of isolated and defined hills with smoothed top and variable slope dip, generally low. The hummocks are separated by winding valleys, often with flat floor, but sometimes locally incised.	Up to about 1000 m	Mean	Strongly heterogeneous. Mainly clay, marl, limestone, sandstone, sand, volcanics, ophiolitic rocks, travertine.	Dendritic, centrifugal, parallel, irregular. Main streams often meandering.	Hills with smoothed or locally flat top, V-shaped valley or flat valley-floor, alluvial fan, graded slope.	Agricultural land, forest, scrub, grassland, infrastructures, urban areas.
CP Terrigeneous hills with isolated rocky ridges	Reliefs forming entire parts of the chain and the forechain, characterized by the strong morphologic evidence of rocky ridges and spurs, which abruptly rise with respect to the surrounding areas, which present smoothed and topographically depressed morphologies.	Some hundred metres – less than 2000 m	Mean, high	Clay, marl. Subordinately: calcarenite, conglomerate, sandstone, radiolarian chert, evaporite.	Dendritic, pinnate. Main streams often meandering.	Rocky ridge and/or spur with cliffs and sharp crests, V-shaped valley or flat valley-floor, widespread landslide and accelerated erosion phenomena. Subordinately: travertine plateau, river terrace and plain, alluvial fan, slope talus.	Agricultural land, forest, grassland, scrub, exposed rocks.
RC Isolated coastal relief	Isolated rocky relief, facing the sea, surrounded by low coastal plains and, in part, by the coastline. Generally it constitutes a promontory, or it is linked to the mainland by a tombolo.	0 to some hundred metres	Mean, high	Rocky, lithotypes.	Scarcely developed; centrifugal, parallel.	Rocky or sandy shore, mountainous coastline, shore cliff, little beach, steep slope, crest. Subordinately: dune, slope talus.	Exposed rocks, scrub, forest.
RI Isolated rocky relief	Single rocky relief, isolated into a topographically lower area and bordered by abrupt breaks of slope.	Some hundred metres – about 1000 m	High	Rocky lithotypes.	Scarcely developed; centrifugal, parallel, trellis.	Crest, steep slope, cliff. Subordinately: peak, slope talus.	Forest, exposed rocks, scrub, grassland.

MV Volcanic mountains	Mainly cone-shaped mountains, formed by present or past volcanic activity.	Over about 1000 m	Mean, high	Lava, pyroclastics.	Centrifugal. Subordinately: parallel.	Caldera, crater, cone. Subordinately: subcircular lake and/or plain in caldera or crater depression, plateau, gorge, hill, cliff, slope talus.	Forest, grassland, scrub, exposed rocks.
MT Siliciclastic mountains	Mainly siliciclastic mountains, often structured in ridges, forming wide parts of the chain and the forechain.	Some hundred metres – about 2500 m	Mean, high	Sandstone, marl, clay. Subordinately: calcarenite, conglomerate, evaporite.	Dendritic, pinnate. Main streams often meandering.	Smoothed top, crest, graded or steep slope, V-shaped valley, flat valley-floor, widespread landslide and accelerated erosion phenomena, badlands, slope talus. Subordinately: river terrace and plain, alluvial fan.	Forest, scrub, grassland.
MC Carbonate rock mountains	Mountains mainly composed by carbonate rocks, structured in ridges and massifs, forming wide parts of the chain and the forechain.	Up to about 3000 m	High	Limestore, dolomitic limestone, dolostone, marly limestone.	Scarcely developed, conditioned by karst; trellis, parallel, rectangular.	Crest, peak, steep slope, cliff, V-shaped incised valley, gorge, U-shaped valley, all glacial morphology elements, karst high plain, all karst morphology elements, small closed depression filled with incoherent sediments, slope talus. Subordinately: river terrace and plain, alluvial fan.	Exposed rocks, forest, grassland, scrub.
MX Crystalline massif	Group of mountains with a strong morphologic contrast, characterized by well defined crests and valleys with steep slopes.	500–2000 m	High	Crystalline rocks (metamorphic and intrusive).	Subparallel, dendritic.	Sharp crest, peak, steep slope, V-shaped narrow valley, accelerated erosion phenomena, slope talus. Subordinately: alluvial fan.	Forest, grassland, scrub, exposed rocks, agricultural land.

continues overleaf

Table 15.1 *(cont.)*

Name	Landscape general structure	Elevation (a.s.l.)	Relief energy	Main lithologies	Drainage pattern	Physiographic components	Main landcover
IS Small islands	Islands extended less than 500 square kilometres, with coastlines strongly developed with respect to the island area.	Up to about 1000 m	Not significant	Not significant.	Scarce developed; centrifugal.	Low and/or mountainous coastline, rocky or sandy shore. Subordinately: coastal plain, dune, relief, cliff, volcano.	Scrub, grassland, agricultural land, exposed rocks.
AM Metropolitan area	Wide area, totally covered by buildings except some little zones, anyway completely surrounded by the urban texture.	Not significant	Not significant	Not significant.	Not significant.	Not significant.	Urban areas, infrastructures.

of their spatial relationships and of the patterns in which they are organized. Because each landscape has a clearly identifiable pattern, it was possible to distinguish any particular landscape from surrounding ones.

15.3 CLASSIFICATION OF LANDSCAPE TYPES AND UNITS

Among the elements forming the landscape, only those that can be directly observed at the regional scale were considered in our landscape classification. These are the physiognomic–structural features of the land, and treating them as a set allows the different landscape settings to be distinguished as structurally homogeneous portions of the Earth's surface. The integrated study of these elements reveals an arrangement of physical, biotic and anthropic features to which well defined functional processes correspond. At the regional scale the main elements considered were the morphological, lithological, vegetational and land-use features, analysed according to their composition and pattern.

It is important to note that the definition of the landscape structure is a prerequisite to the understanding of its functional relationships and changes, and therefore it is essential to the future development of the Carta della Natura Project.

On the basis of the conceptual approach outlined above, a number of landscape types were identified. These are typical associations of land features recognizable at a regional scale and therefore they do not refer to a specific geographically defined portion of the land, e.g. the type 'coastal plain' can be found in many different places along the Italian coast.

The criteria used to distinguish and describe the different landscape types are summarized in Table 15.1, which presents the types identified in central Italy. As noted earlier, the parameters which were considered were in the main related to the morphologic (elevation, relief energy, drainage pattern, physiographic components), geologic and landcover settings (see also SIGEA 1994–1995). Moreover, in order to define some types, it was important to observe their relationships with the surrounding landscape shapes. Such observations inform many of the descriptions of the landscape's general structure in Table 15.1, and they can be clarified by referring to the example of the plain areas. At a regional scale, plains are often very similar from a litho-morphologic and landcover point of view, but in reality they may correspond to different landscapes. If they occupy a valley floor between ridges ('valley floor plain' type) they will have a different configuration compared to those plain areas not limited by the sides of a river valley ('open plain' type) or to those developed along the sea coast ('coastal plain' type).

Identifying and delimiting the landscape type that characterizes one defined geographical area, we obtain a 'landscape unit', and this represents the fundamental unit for the Carta della Natura system at the 1:250 000 scale. Landscape units differ from landscape types in that every landscape unit is marked by a precise geographical location, which gives its name to the unit; units are, therefore, unique. Their identification and mapping was achieved by identifying the local and unique features of specific geographically defined portions of the landscape.

We emphasize that the landscape units are part of a hierarchical, multiscalar system which is organized on a geographical basis. Types, instead, are outside such a hierarchical system, but they nevertheless represent an important tool for its definition. Any specific landscape unit is characterized by one landscape type only; however, each landscape type may be used to classify several landscape units.

At the 1:250 000 scale, the landscape units presented here can be related to the land systems class in the land classifications applied by some international institutes and organizations

(Christian and Stewart 1953, 1968; CFS: Lacate 1969; FAO 1974; CSIRO: ITC: Van Zuidam and Van Zuidam Concelado 1978; Zonneveld 1979; Tricart and Kilian 1985; Giordano 1989). However, in our classification the term 'landscape system' is used to define an ensemble of landscapes that can be detected with a more generalized approach (e.g. Northern Apennines, Po Plain landscape systems).

Our method is founded on the same basic principles as those of the above-cited land classifications, but it takes into particular account an important aspect, fundamental for the aim of the Carta della Natura Project: the transformation of the landscape through time. Thus our system is devised to take into account the possibility that it will be updated through time, following the dynamics of the landscape units and their related functional processes. The relevance of this to the definition of the natural values and of the environmental vulnerability of our country is evident, because these will change through time too.

15.4 METHODS

The Landscape Unit Map of Italy was realized principally by use of remote sensing, and mainly through interpretation of aerial photographs. This method is the most appropriate for landscape analysis at a regional scale: it permits a synoptic study of the land which is consistent with a holistic approach and the immediate reading of the elements forming landscape structure. The general principles for the application of remote sensing methods to a holistic approach to the landscape can be found in Naveh and Lieberman (1984). The *Volo Italia* aerial photographs (Ferretti 1989), at the nominal scale of 1:75 000, were used because they cover, homogeneously, the whole national territory at an appropriate scale for our study.

Interpretation of the aerial photographs allowed the identification of the landscape units, and was accompanied by their mapping on Istituto Geografico Militare Italiano (IGMI) topographic maps at the scale 1:250 000. Details of each unit were also noted on worksheets (not presented here). These worksheets contain the following: a general description of the landscape structure, both geographic and related to the landscape type attributed to the unit; the altitude arrangement and the relief energy class; the observed physiographic components; the lithology; the drainage pattern; and the land cover, classified following the *CORINE Land Cover* suggestions (CEE 1993).

Moreover, the results were integrated with field surveys (suitably performed), and compared to the national and international scientific literature (*Carta geologica d'Italia* 1:100 000 and 1:500 000 (Servizio Geologico Nazionale); Sestini 1963; Boccaletti 1982; Forman and Godron 1986 (and references therein); Bonardi *et al*. 1988; Pedrotti 1991; Reichenbach *et al*. 1992; Pignatti 1994 (and references therein); Accordi *et al*. 1988). Previous studies by public administrations were also taken into account (e.g. Regione Emilia Romagna 1987; Regione Marche 1991; Regione Toscana 1994). In particular, careful reference was made to the published thematic maps.

We emphasize that the landscape unit map presented in this chapter has been produced as a direct synthesis of all the factors characterizing the landscape considered together; it is not the result of the superimposition of different analytic layers.

The map and the related tables, converted to digital format, form a database that is part of a geographical information system (GIS); in fact, a GIS is considered to be the most logical choice for structuring an ecological database (Risser and Treworgy 1985; Haines-Young *et al*. 1993, and references therein). The landscape unit map is in vector format, and the description of types and units is stored in tables linked to the map by the built-in Relational Data Base Management System (RDBMS) of the software used (Arc/Info 1997). In our case, this way of managing the data was selected and designed with a view to evaluating the quality and the vulnerability of the environment (Haines-Young *et al.* 1993). In order to do this, in the same database many other layers by different authors have also been stored, including the digital elevation model (DEM), the lithological map and the land-cover map. Each layer is accompanied by a number of tables in an associated database. All these layers are available in the Italian Department for National Technical Services (DSTN) database, and they cover the whole national territory.

From what has been described so far, we can sum up the principal steps of the work undertaken (Figure 15.1). (1) Identification and description of the landscape types for the Italian territory. From this classification, applied to specific and unique geographic zones, and completed with the description of more local features, we obtain: (2) the mapping, description and codification of the landscape units. These units were mapped on IGMI topographic maps at the 1:250 000 scale. (3) The creation of a database organized in a GIS, designed *ad hoc*, was required for management of all the data and maps. In the GIS a number of thematic maps are also present as digital layers; these can be easily compared to our own map.

15.5 RESULTS AND DISCUSSION

The methodology that has been outlined was applied and fine-tuned to central Italy. The first result of this work is given by the 1:250 000 scale map presented here in a reduced scale (Figure 15.2).

In the upper part of Figure 15.2 the identified landscape types are represented. There are 21 for central Italy and their description can be found in Table 15.1; some types extend over wide and continuous areas, whereas other types show a fragmented distribution, with relatively reduced extension.

In the lower part of Figure 15.2, the unit boundaries are superimposed onto the digital elevation model. There are 285 units in the study region (Table 15.2); they show extremely variable shape and extension, and this can be partially related to the shape and extension of the types represented in the upper part of the figure.

For some types an internal distinction into units was easily made, whereas for other types the pattern was so homogeneous that this was impossible. The consequence is that the dimensions of the units with respect to the relative types are not random. For example, we notice that the average unit area for 'open plains' is 2047 km^2, while that for 'small islands' is 27 km^2 (Table 15.2). The average area for all units is 307 km^2.

Comparing the landscape unit map with the DEM (lower part of Figure 15.2) we can observe that the boundaries between two units, whether pertaining to the same type or not, generally correspond to morphologic discontinuities, such as valleys, breaks of slope or coast lines. An exception is the 'metropolitan area' type, which is also an exception from another point of view: urban development is mainly governed by human activity and city boundaries are not necessarily controlled by morphological features.

For a critical evaluation of our methodology, two main aspects can be considered. The first concerns the subjectivity of the method, which is a common problem in qualitative studies like this, where interpretation plays a primary role. To estimate the subjectivity of the method, we compared maps which were independently created by different researchers on the same test areas. The comparison showed a good correspondence among the unit boundaries drawn by the different researchers. This is also the case when we mosaic the sheets – made by different authors – that make up the landscape unit map. These observations indicate that the procedure is sufficiently defined to allow different people to obtain comparable results.

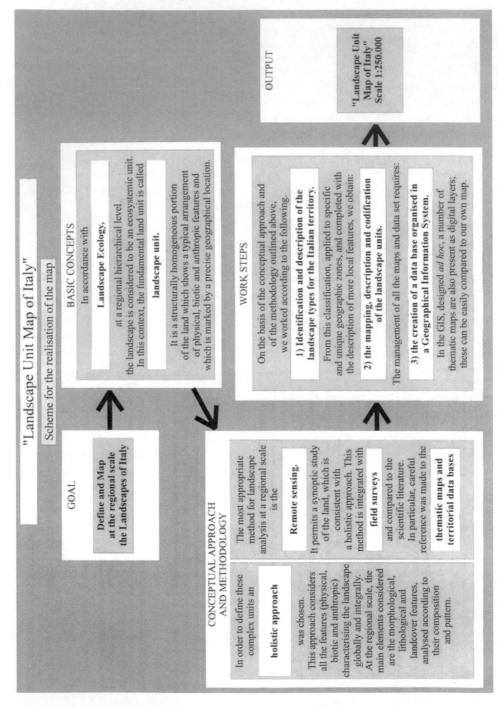

Figure 15.1 Scheme for the realization of the landscape unit map of Italy

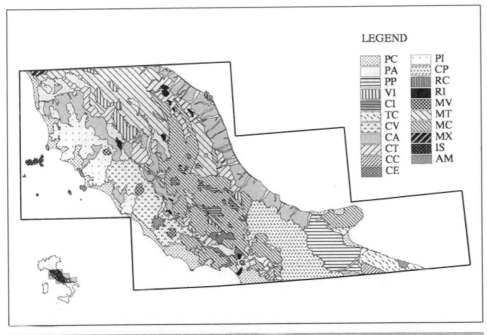

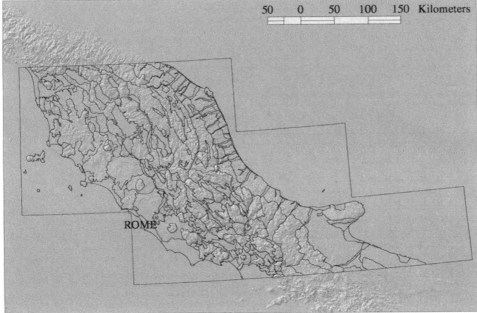

Figure 15.2   Landscape types (upper part) and units (lower part) maps of central Italy. For explanation of the legend, see Table 15.1

Table 15.2 Landscape units areal data

Landscape types	Number of units	Total area (km^2)	Average unit area (km^2)
AM	2	350	175
CA	42	12 210	291
CC	4	655	164
CE	4	714	179
CI	15	1074	72
CP	10	7378	738
CT	16	5687	355
CV	10	5737	574
IS	10	273	27
MC	41	16 373	399
MT	16	10 476	655
MV	4	453	113
MX	1	155	155
PA	49	5342	109
PC	22	5367	244
PI	10	6628	663
PP	2	4094	2047
RC	5	248	50
RI	13	508	39
TC	3	1826	609
VI	6	1870	312
Total	285	87 417	307

The second point concerns the flexibility of the method. In fact, by using a feedback process in the mapping of the landscape units, we progressively detailed our definition of the types and, consequently, modified the units already identified. The possibility of revising the map and database according to new information, coupled with use of the GIS tool, together form an open and dynamic system which was remarkably effective in improving the quality of the final product.

In order to understand which elements are the most important in determining the landscape units identified under our methodology, we compared our map with other thematic maps that depict various analytical features for the same area (Figure 15.3).

We chose Rome and its surroundings as a key area, because many different landscape types were detected there. In particular, the information layers related to the elements that are largely responsible for structuring the landscape were considered, i.e. relief features (studied on a DEM with a square grid of $200 \times 200 \, \text{m}^2$), lithology (Lithological Map of Italy, Geological Survey of Italy), hydrographic network (Italian Hydrographic and Oceanographic Service) and land cover (in which the vegetational aspects are also present; *CORINE Land Cover* (CEE 1993) of Italy). All these maps are available in digital format in the DSTN database. In Figure 15.3, the lithological and land-cover maps have been reproduced (in grey scale), which allows the different patches and their distribution pattern to be compared.

The comparison of these layers with our landscape unit map shows that generally there is a good correspondence between the landscape unit map and the other thematic maps. In fact, each area defined by a landscape unit usually corresponds to a characteristic pattern of the elements represented in the other maps, as can be observed in Figure 15.3 (see also Figure 15.4).

An important comment can be made about the dimensions of the homogeneous areas drawn in each thematic map. These are generally less extended than in the landscape unit

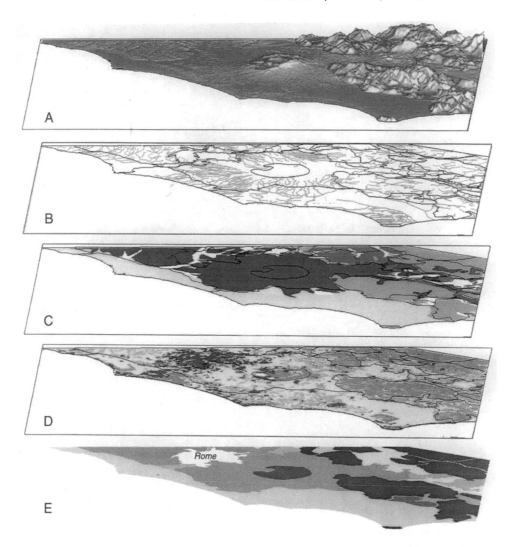

Figure 15.3   Main structural elements that characterize the landscape, represented as thematic layers (A, B, C, D), compared with the Landscape Unit Map (E): detail of Rome and its surroundings. (A) Relief features (DEM); (B) hydrographic network (hydrographic map); (C) lithology (lithological map); (D) land cover (land-cover map). (B, C, D) The boundaries of the landscape units are superimposed. It can be observed that a characteristic pattern of the elements represented in the other maps corresponds to each area defined by a landscape unit. The good correspondence between the landscape unit (E) and the relief features (A) is evident: the boundaries between the units correspond to morphologic discontinuities and each unit shows a striking homogeneity in the arrangement of relief forms

map, but even though homogeneous areas of small extension may differ from one another at a local level, when grouped together from the perspective of the regional level they never-theless (as a group) show differences that are distinct enough from surrounding areas to allow them to be classified as corresponding to a particular landscape unit.

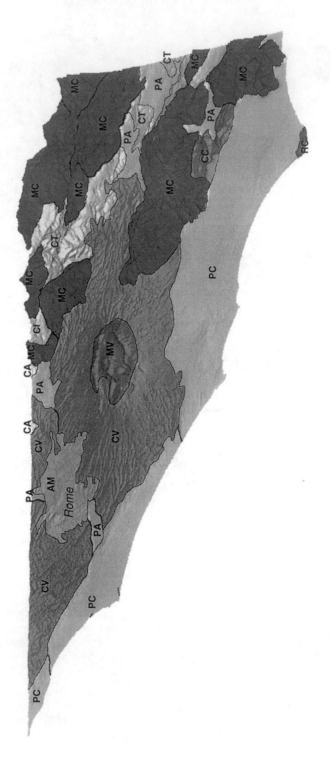

Figure 15.4 The Landscape Unit Map: the same land area as depicted in Figure 15.3 is superimposed on the related DEM in a perspective view. The lines represent the limits between the landscape units, and the grey tones distinguish the different landscape types for each unit, marked by the relative code (for an explanation of the landscape types see Table 15.1)

The best correspondence between our map and the thematic maps can be observed for the distribution of the landforms in the physiography, as highlighted by the DEMs in Figures 15.3 and 15.4. An exception is the 'metropolitan area' type, whose boundaries are not necessarily related to physiographic discontinuities. This confirms what was said earlier, because cities are artificially imposed on one or more previously existing natural landscapes (Figure 15.4).

Thus, at the regional scale, our findings suggest that the physiography best approximates the results of a landscape study which is performed following a holistic approach. From a geological point of view, this could have been expected *a priori*, since the physiography results from the interrelations between features (lithology, structural setting) that exert a passive control on the landscape, and processes (exogenous and endogenous natural processes, human activities) that exert an active control on the Earth's surface. Many natural science approaches to defining landscapes – outside the Earth sciences – tend to attribute more relevance to vegetational aspects (see, for example, Pignatti 1994). In this light our work contributes a new perspective to landscape ecology studies.

15.6 CONCLUSIONS

The development of a methodology that allows the Italian territory to be divided into landscape units provides the basis for attributing a value to the quality and the vulnerability of environmental conditions, which is the final goal of the Carta della Natura project. In this context, the realization of the map has been accompanied by a comparison between holistic and analytical approaches. Taking as its point of departure the assumption that precise emergent properties can be recognized for each scale of observation, our study has shown that at the hierarchical level corresponding to a regional scale, it is the physiography which best approximates the results of a landscape study performed following a holistic approach. As a matter of fact, the landscape unit map shows that, at a general scale, the main features of the landscape are strictly linked to the physiographic patterns. Therefore, geomorphology is the most important tool for the study of the landscape at the regional hierarchical level, allowing the morphological pattern that marks each type, and to which the vegetation and the functional relationships are strictly related, to be defined.

This qualitative result could provide the basis for a quantitative method for the definition, starting from numerical parameters, of the 'morphometric signature' that typifies each landscape type. Therefore, the kind of work presented here is a significant step towards the application of semi-automatic methods to the recognition of landscape features over wide areas.

ACKNOWLEDGEMENTS

Thanks are due to Renato Ventura for his kind help.

This chapter is dedicated to Beatrice Crescenzi, whose contributions were crucial.

REFERENCES

Accordi, G., Carbone, F., Civitelli, G., Corda, L., De Rita, D., Esu, D., Funiciello, R., Kotsakis, T., Mariotti, G. and Sposato, A. 1988. Carta delle litofacies del Lazio-Abruzzo ed aree limitrofe. Note illustrative. *Quaderni de 'La ricerca scientifica'*, **114**(5), 223.

Arc/Info, 1997. Version 7.1.1. Copyright 1982–1997. Environmental Systems Research Institute, Inc.

Boccaletti, M. (ed.) 1982. Carta strutturale dell'Appennino settentrionale. Scala 1:250 000. CNR, *P.F. Geodinamica*, **429**.

Bonardi, G., D'Argenio, B. and Perrone, V. (eds) 1988. Carta geologica dell'Appennino meridionale. Scala 1:250 000. *Memorie dela Società Geologica Italiana*, **41**.

Christian, C.S. and Stewart, G.A. 1953. *General report on survey of Catherine-Darwin region, 1946*. Land Research Series, **1**, CSIRO, Canberra.

Christian, C.S. and Stewart, G.A. 1968. Methodology of Integrated Surveys. *Proceedings of the Conference on Aerial Surveys and Integrated Studies*, UNESCO, Toulouse, 233–280.

CEE, 1993. *CORINE Land Cover*. Guida tecnica, CORINE Programme.

FAO, 1974. Approaches to land classification. *FAO Soils Bulletin*, **22**.

Ferretti (ed.) 1989. *Volo Italia*. Aerial photographs of Italy, nominal scale 1:75 000.

Forman, R.T.T. and Godron, M. 1986. *Landscape Ecology*. Wiley, Chichester.

Giordano, A. 1989. *Il telerilevamento nella valutazione delle risorse naturali*. Rivista di Agricoltura Subtropicale e Tropicale, **83**(1), Istituto Agronomico per l'Oltremare, Firenze.

Haines-Young, R., Green, D.R. and Cousins, S.H. (eds) 1993. *Landscape Ecology and GIS*. Taylor & Francis, London.

Lacate, D.S. 1969. *Guidelines for Bio-physical Land Classification*. Canadian Forest Service Publication, 1204.

May, R.M. 1989. *Ecological Concepts*. Blackwell, Oxford.

Meinig, D.W. 1979. *The Interpretation of Ordinary Landscapes*. Geographical Essays, Oxford University Press, Oxford.

Naveh, Z. and Lieberman, A.S. 1984. *Landscape Ecology Theory and Applications*. Springer-Verlag, Berlin.

O'Neill, R.V. 1989. Perspectives in hierarchy and scale. In Roughgarden J., May R.M. and Levin S.A. (eds), *Perspectives in Ecological Theory*. Princeton University Press.

O'Neill, R.V., De Angelis, D.L., Waide, J.B. and Allen, T.F.H. 1986. *A Hierarchical Concept of Ecosystems*. Princeton University Press.

Pedrotti, F. (ed.) 1991. *Carta della vegetazione reale d'Italia. Scala 1:1.000.000*. Ministero dell'Ambiente, Relazione sullo stato dell'ambiente.

Pignatti, S. 1994. *Ecologia del paesaggio*. UTET, Torino.

Prigogine, I. and Stengers, I. 1984. *Order Out of Chaos: Man's New Dialogue with Nature*. Bantam, New York.

Regione Emilia Romagna, 1987. *Piano Territoriale Paesistico regionale*. Regione Emilia Romagna, Bologna.

Regione Marche, 1991. *L'ambiente fisico delle Marche*. Ass. Urbanistica-Ambiente, Ancona.

Regione Toscana, 1994. *I sistemi di paesaggio della Toscana*. Firenze.

Reichenbach, P., Acevedo, W., Mark, R.K. and Pike, R.J. 1992. A new landform map of Italy in computer shaded relief. *Bollettino di Geodesia e Scienze Affini*, **52**, 21–44.

Risser, P.G. and Treworgy, C.G. 1985. Overview of ecological research data management. In *Research Data Management in the Ecological Sciences*. University of South Carolina Press, Columbia.

Sestini, A. 1963. *Il paesaggio*. Collana: Conosci l'Italia, T.C.I., Milano, **7**.

SIGEA (Società Italiana Geologia e Ambiente), 1994–1995. I paesaggi geologici italiani. *Materiali Verde Ambiente*, nos 2, 3, 4, 5, 6 (1994), 1, 2, 3 (1995).

Tricart, J. and Kilian, J. 1985. *L'ecogeografia e la pianificazione dell'ambiente naturale*. Franco Angeli, Milano.

Van Zuidam, R. and Van Zuidam Concelado, F.I. 1978. *Terrain Analysis and Classification using Aerial Photographs*. ITC Textbook, International Institute for Aerospace Survey and Earth Science, Enschede, **8**(6).

Vink A.P.A. 1983. *Landscape Ecology and Land Use*. Longman, New York.

Zonneveld, S.J. 1979. *Land Evaluation and Landscape Science*. ITC Textbook of Photointerpretation, International Institute for Aerospace Survey and Earth Science, Enschede, **7**.

16 Estimation of the Effects of Slope Map Computing on Shallow Landslide Hazard Zonation: a Case History in the Northern Apennines (Italy)

MAURO CASADEI
Department of Earth and Planetary Science, University of California, Berkeley, CA, USA

AND

ENZO FARABEGOLI
Dipartimento di Scienze della Terra e Geo-ambientali, University of Bologna, Bologna, Italy

ABSTRACT

Topographic slope represents a key factor in the evaluation of shallow landslide potential by any computer-based techniques. In the Italian Apennines several planning tools still rely on a discrete slope scale based on four critical slopes (10%, 20%, 35%, 50%) derived from geotechnical and agricultural considerations. In order to evaluate the performances of contour-based (CBM), digital elevation model (DEM) and triangular irregular network (TIN) terrain modelling techniques, the Centonara catchment, a small watershed (2.1 km²) in the northern Apennines was analysed. Since the different digital terrain models (DTMs) are very sensitive to the quality of input data, the 10 m elevation spaced contours from a 1:5000 scale topographic map were used.

The performances of TIN and DEM were assessed after a comparison with the reference CBM results, and further investigated taking into account the general morphometric features and the observed distribution of landslides, which cover more than 44.5% of the catchment. Slope frequency distribution computed over the whole catchment bears the signature of different calculation methods: DEM underestimates the extent of slopes steeper than 40%, overestimating slopes less steep than 35%; TIN yields results more consistent with CBM, diverging slightly within the slope ranges 15–40% and >60%. Both TIN and DEM fail to represent correctly the morphology within gullied hillslopes, proving unsuitable for the assessment of the hazard related to debris flows and mud flows, which typically affect such areas.

In view of the widely accepted assumption that hills with 20–35% slope are suitable for ploughing without great risk of landsliding, agricultural practice in Italy is commonly extended to slopes as steep as 35%. The high frequency of earth slides and earth flows within the 15–40% slope range emerges independently of the chosen DTM and underlines the need to reconsider the critical slope ranges.

Shallow landslide hazard zonation would reflect both the underestimation of the steepness of critical areas and the use of obsolete slope ranges: the amount of areas misclassified as suitable for agricultural practice exceeds 10 ha in the Centonara catchment.

Applied Geomorphology: Theory and Practice. Edited by R.J. Allison.
© 2002 John Wiley & Sons, Ltd.

16.1   INTRODUCTION

Shallow landsliding is the most widespread cause of landscape development in the Italian northern Apennines, and is often responsible for catastrophic events leading to huge economic and social losses.

The set of parameters governing the stability of hillslopes include the geotechnical parameters of rock and soil (friction angle ϕ, cohesion c, bulk density γ), the local hydrological and morphological setting (geometry of water table, slip surface and topography), and local climatic and environmental variables (precipitation, soil thickness, vegetation, presence of plant root and animal burrows: Dietrich and Dunne 1993).

The lack of well distributed geotechnical and hydrological data has not previously allowed reliable extrapolation of slope stability analyses from single landslide to catchment scale. Hence, much effort has been made in the last few decades to reduce the number of variables needed to assess shallow landslide potential at catchment scale, with reasonable success.

Some physically based hydrological approaches have been proposed since the early 1970s for forecasting slope evolution and shallow soil slips (Carson and Kirkby 1972; Smith and Bretherton 1972; Ahnert 1976) and have been developed with variations to the contemporary models (Willgoose *et al.* 1990; Howard 1994; Dietrich and Montgomery 1998). These models rely on at least two major parameters: (a) local slope, whose effect is well known over slope stability; (b) upslope contributing area per unit contour length, which estimates the rate of run-off and shallow subsurface flow, involving eventual failure due to pore pressure increase.

During the same period, most Italian Apenninic catchments have been studied mainly on a statistical basis rather than by physical models in order to define some threshold slopes within homogeneous geological domains. This topic has been widely discussed since the end of the 1960s (Valentini 1967; Lucini 1969). The development of these concepts and the subsequent applications (Amadesi and Vianello 1978, 1982) relied on thematic mapping of mass wasting, land use, lithology and topographic slope, often integrated with many other parameters (hillslope aspect, structural setting, etc.) (see Canuti and Casagli (1996) for a review).

The advances in geographical information system (GIS) technology in the 1980s provided a widespread tool for landslide hazard zonation (Carrara 1983; Varnes and IAEG 1984; Brabb 1984; Hansen 1984; Carrara *et al.* 1991; Clerici *et al.* 1993; Canuti and Casagli 1996), which however still relies primarily on the identification of critical slope angles. Figure 16.1 shows the most common slope classification schemes chosen by Italian authors for slope stability purposes.

The importance of the above-mentioned frameworks lies in the fact that they are still being used as environmental planning tools ('slope stability maps') in some Italian regions (Coltorti *et al.* 1987; Spagna and Cabriel 1987; Marchionna *et al.* 1988; Clerici *et al.* 1993, Canuti and Casagli 1996).

A crucial phase of computer-based work is data collection whose accuracy should be compatible with the detail of the analysis. The three-dimensional representation of hillslope morphology has been increasingly adopted in order to calculate topographic features more readily, but the choice of technique for producing the model is still under discussion. As a matter of fact, the calculation of such quantities turns out to be highly sensitive to the choice of the digital terrain model (DTM) scheme (Band *et al.* 1995; Dietrich and Montgomery 1998).

This issue may be even more crucial in areas featuring highly irregular shapes such as the 'calanchi' (Farabegoli and Agostini 2000), badland-like landforms typical of many catchments in the Italian Apennines (Figure 16.2), which require a very fine-scale analysis.

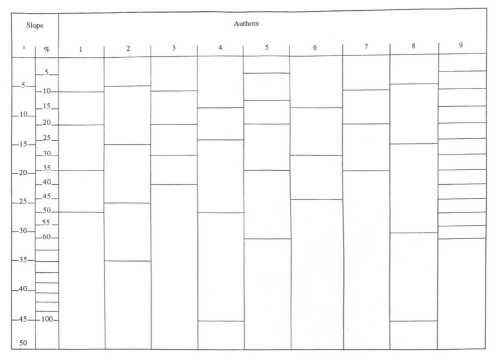

Figure 16.1 Slope angle discrete scales previously adopted in various Italian works. Reference key: 1, Amadesi and Vianello (1982); 2, Carrara (1983); 3, Coltorti *et al.* (1987); 4, Corniello *et al.* (1980); 5, Costantini (1987); 6, D'Alessandro and Pantaleone (1987); 7, Marchionna *et al.* (1988); 8, Spagna and Cabriel (1987); 9, present work

This chapter tries to verify at a reasonable scale (10 m elevation spaced contours) the relationships between digital terrain models calculated by three different techniques and the most significant threshold slopes highlighted by shallow landslide distribution in a small sample catchment from the Italian Apennines.

16.2 GENERAL INFORMATION AND GEOLOGICAL–GEOMORPHOLOGICAL SETTING

The Centonara catchment is a watershed about 2.1 km^2 in area located near Bologna, in the northern Apennines (Italy). The mean monthly temperature ranges from about 0°C in the cold season to over 25°C during the hot season. Annual rainfall precipitation ranges from 500 mm to 1400 mm, mostly concentrated in a few storm events per year, which in the past have reached over 300 mm in three days.

From the geological viewpoint this sector is one of the most complicated in the Po Plain Apenninic margin (Figure 16.3) (Farabegoli *et al.* 1994). Here the autochthonous Romagna series is tectonically covered by the Miocene–Lower Pliocene chaotic complex (Argille Scagliose *auct.*). In the south the chaotic complex is covered unconformably by marine Lower Pliocene terrains. In the north the Miocene evaporitic sequence (Gessoso Sc[...] Colombacci Formations) is covered with a transgressive erosional unconformity b[...]

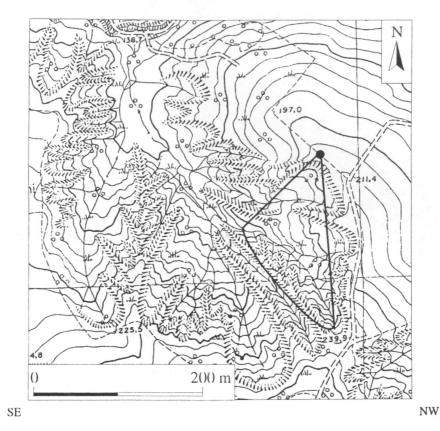

SE   NW

Figure 16.2 Calanchi landforms on the southern flank of the Centonara catchment. The unvegetated chaotic bedrock features a highly irregular topography at fine scale. Contour elevation spacing is 5 m. Relief of lateral gullies ranges between 20 and 40 m

Yellow Sands Fm (Lower Pleistocene). These units are dissected by strike-slip and thrust faults. Moreover, the area features complex geomorphological characters owing to the variable lithology and the marginal position with respect to the main Apenninic valley incision.

The spatial distribution of the processes is differentiated according to bedrock lithology:

- the western upstream part of the basin undergoes strong badland erosion over clayey bedrock; the topography is very irregular, featuring very fine contour crenulations and very steep hillslopes; however, in spite of an almost impermeable bedrock, the drainage network is not as dense as expected, since the overall landscape is dominated by mud flows and debris flows (Figure 16.4) rather than fluvial processes;
- the eastern Plio-Pleistocene terrains provide gentler soil-mantled slopes, mostly concerned with slow shallow landslides (earth flows, earth slides), overland flow erosion and fluvial depositional landforms (alluvial fans and terraced deposits) near to the catchment's outlet.

The overall frequency of landslides amounts to 44.5% of the catchment. The geomorphological setting has several points in common with the landscape outlined by Dietrich and

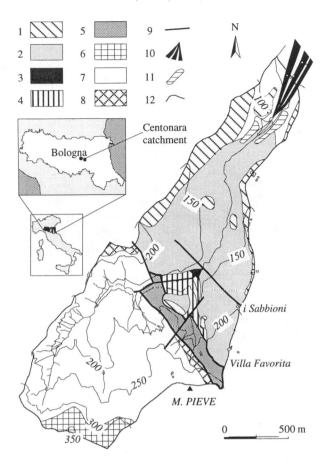

Figure 16.3 Geological sketch of the Centonara catchment. Key: 1, alluvial terraces (Pleistocene); 2, Imola Yellow Sands Fm.: conglomerate, sandstone, clay (Lower Pleistocene); 3, Gessoso-solfifera Fm.: conglomerate, sandstone, clay (Messinian); 4, Gessoso-solfifera Fm.: gypsum, clay (Messinian); 5, marl, clay (Tortonian–Messinian); 6, sandstone, clay (Pleistocene); 7, Chaotic complex including: 8, siliceous and silty marls (Lower to Middle Miocene); 9, fault; 10, alluvial fan; 11, artificial lake; 12, watershed boundary (after Farabegoli and Forti 1997)

Montgomery (1998), who describe soil-mantled unchannelled valleys commonly featuring the occurrence of shallow landslips sliding along the soil–bedrock interface. At the toes of such valleys, the colluvium deposited within the hollows is often further mobilized as debris flow.

16.3 GENERAL INFORMATION ON TERRAIN MODELLING

A large amount of work has been produced during the last two decades, mainly relating to three different approaches (e.g. Mark 1975; Carlà *et al.* 1986; Carrara 1988; Band *et al.* 1995). Each method performs differently in terms of accuracy, consistency and computational time,

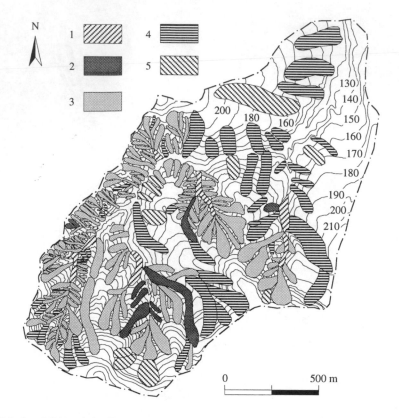

Figure 16.4 Landslides of the Centonara catchment. Key: 1, debris fan; 2, debris flow; 3, mud flow; 4, earth flow; 5, earth slide. Contour elevation in metres

and the choice largely depends on the detail required by the morphometric analysis. We summarize below the advantages and pitfalls relevant for geomorphological purposes.

16.3.1   Contour-based Methods (CBM)

The very first manual approach in three-dimensional hillslope modelling is the contour-based method (CBM), which captures very effectively the hydrological flow net. Slope is computed perpendicular to the contour lines, which are usually extracted from existing topographic maps.

The CBM was used with success in hydrological (O'Loughlin 1986; Moore *et al*. 1988) and geomorphological (Montgomery and Dietrich 1989; Prosser and Abernethy 1996) catchment-scale investigations thanks to the development of computer programs (e.g. TOPOG: Moore *et al*. 1988) which build the catchment's hydrological flow net given contour lines, breaklines and a threshold for elementary flow tube width.

On the other hand, this approach is time-consuming and is sometimes discarded as unnecessarily slow for most purposes. Another common argument against the use of CBMs relies on the fact that the resulting spatial structure is oversampled in one direction (along contours) while being undersampled in the other (flow path) (Band 1993; Weibel 1997).

However, the field evidence from the sample area and from other Apenninic catchments, as well as other works (Band *et al.* 1995; Prosser and Abernethy 1996; Dietrich and Montgomery 1998), shows that this method handles the topographic data set more realistically, producing fewer artefacts than other methods.

16.3.2 Rectangular Grids (DEM)

Raster-based catchment modelling schemes are widespread and well developed (O'Callaghan and Mark 1984; Band 1986, 1993; Lammers and Band 1990; Willgoose *et al.* 1990). Such a representation greatly simplifies the computation of hillslope morphometric and hydrological features by reducing the complexity of the topology. However, the resolution required for a detailed geomorphological analysis (1:5000 in this work) leads either to a huge amount of data or to a poor representation according to the detail of the resolution (Mark 1975). Even in the case of a well-defined grid, the techniques for computing hydrological and geomorphological parameters are still under discussion (Costa Cabral and Burges 1994; Dietrich and Montgomery 1998).

Local slope can be computed in various ways: for instance, by choosing the mean of the values in the four possible directions, or by choosing their maximum value. The former algorithm (averaging) leads to a generalized smoothing of the data, which inhibits the recognition of threshold slopes; the latter solution, while being probably the lesser evil, might be sensitive to the orientation of the grid.

The same grid artifacts also influence the determination of other important hydrological parameters (upslope-contributing area and local convexity) (Wolock and Price 1994; Dietrich and Montgomery 1998), together with the choice of the flow tracking technique (Costa Cabral and Burges 1994; Wolock and Price 1995).

16.3.3 Triangular Meshes (TIN)

Triangular irregular network (TIN) techniques (Puecker *et al.* 1978) feature widely in three-dimensional surface modelling, as well as finite element preprocessing for several applications. The adoption of such a scheme allows a better fit of irregular spatial structures, if compared to grid-based frameworks. The increased storage and computing complexity due to the topological scheme is compensated by the reduction of data point requirements (Mark 1975). However, as many authors have pointed out (e.g. Weibel and Brändli 1995), if the data source is contour lines some inconsistencies usually occur (artificially flat triangles); the resulting misleading representation concerns concave/convex zones, which are of vital importance for hydrological (Smith and Bretherton 1972; Beven and Kirkby 1979) and geomorphological (Montgomery and Dietrich 1989; Dietrich *et al.* 1992) applications. As yet, only specific constrained triangulations seem to be able to tackle the problem, at the cost of inputting all of the significant breaklines – streams, divides and cliffs – which can be very numerous in irregular areas (for instance, calanchi landforms). Some authors have been trying to overcome this major problem by means of adaptive interpolation (Weibel and Brändli 1995; Schneider 1995).

16.4 CRITICAL SLOPE RANGES

Two further aspects should be discussed in order to explain the relationships between topographic slope and geomechanical and environmental planning constraints in the Italian Apenninic context.

Table 16.1 Geomechanical properties

β (%)	ϕ (°)
50	45.0
35	35.0
20	21.8
10	11.3

β = topographic gradient; ϕ = friction cycle

16.4.1 Geomechanical Properties

The available data from bibliographic references (Colosimo 1978; Lancellotta 1987; Casadei and Farabegoli 1997) show that the peak friction angle for Apenninic clay rocks ranges from 35° down to 25°, when the residual ϕ_r may drop to 12–5°. The critical topographic gradients associated with these friction angles can then be estimated by applying the simple infinite slope stability model (Skempton and DeLory 1957). By approximating the bulk density of the material $\gamma \approx 2\,t\,m^{-3}$ and a water table at ground level, the analysis yields the values reported in Table 16.1.

16.4.2 Environmental Planning Issues

In the case of thematic mapping produced by the Emilia-Romagna Cartographic Survey, the choice of the slope classes was determined mainly on the basis of agricultural practice and access for mechanical devices. The resulting classification (Amadesi and Vianello 1978; see Figure 16.1) turns out to be equivalent to the above scale obtained by the application of the infinite slope model.

Usually, in areas steeper than 35% mechanical ploughing is not allowed. The same threshold (or sometimes 30% steepness) is commonly adopted as a maximum slope for building. The above criteria converge in identifying five critical domains bounded by four slope thresholds (10%, 20%, 35%, 50%). The detailed slope scale (13 categories) adopted in this work was chosen in order to observe the differences from the different DTMs at intermediate slopes and verify that the boundaries are reflected by the different representations.

16.5 DTM MODELLING OF THE SAMPLE AREA

The most detailed available topographic map (produced by the Emilia-Romagna Carto-graphic Survey) provides the set of contours with 5 m elevation spacing at 1:5000 scale collected after aerophotogrammetric survey. The magnitude of the error in elevation within non-forested and moderately steep areas is around 1 m. Previous experience with such maps has shown (Figure 16.5) that in small crenulated areas the representations derived from contours at 5 m and 10 m elevation spacing feature a sufficient convergence, as opposed to 25 m spacing. For this reason we selected 10 m contours as a good compromise between accuracy and speed. This choice agrees with the recent advances in small catchment modelling, which require a topographic detail not inferior to 10 m elevation spacing (e.g. Zhang and Montgomery 1994).

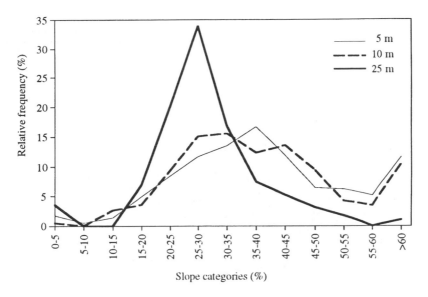

Figure 16.5 Slope frequencies computed by TIN from planar regular hillslopes (triangular facets) given different contour elevation spacing, in areas similar to the Centonara catchment

The slope maps were computed with CBM, DEM and TIN techniques (Figure 16.6), adopting a discrete scale selected bearing in mind the above-mentioned critical slope ranges.

(1) CBM: the contour elevation lines were processed using an application developed at the Department of Earth Sciences of the University of Bologna. The program is interactive, in order to overcome some problems intrinsic in the nature of the matter: the user has the option of 'suggesting' a realistic flow path at runtime, which proves very useful in particular conditions where the visual inspection of the map evidently shows that downslope flow is not strictly normal to the digitized contour lines (presence of break-lines, blind structures, concave/convex shapes; see Figure 16.7). Such difficult conditions were met very often because of the high contour crenulations within the calanchi areas. The number of output polygons is 7434.

(2) DEM: a kriging of the contour vertices data set was performed over a grid of 20 m × 20 m cell spacing. The choice of the spacing was made in order to obtain a most accurate grid without extrapolating the data source too much, and avoiding the creation of common artefacts such as stepped profiles. For each cell the slope angle was computed along steepest descent rather than by four direction averaging. This choice depends on the need to recognize some threshold slopes: by choosing the maximum slope it was intended to limit the intrinsic smoothing of DEMs derived by interpolation. The number of output cells is 5373.

(3) TIN: a triangulation from the contour vertices data set was performed; the slope of each TIN cell was calculated along the projection of the normal to the facet. The main streams and ridges were used as breakline constraints to the Delaunay-based algorithm. The geomorphological setting and the scale of the analysis showed a large occurrence of such physical boundaries which could not be fully supplied by the operator to the triangulation algorithm, thus leading to some well-known inconsistencies (artificially flat

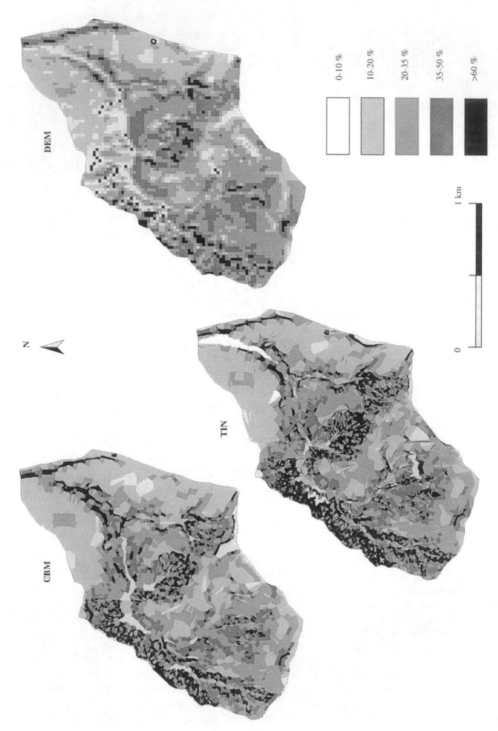

Figure 16.6 Slope maps of the Centonara catchment computed from the three DTMs. For simplicity only the five critical slope ranges are shown

0-10 %

10-20 %

20-35 %

35-50 %

>60 %

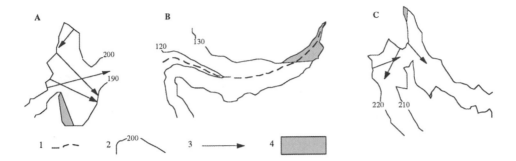

Figure 16.7   Three critical configurations unresolvable by two contours without interactive assessment. (A) Concave/convex structure; (B) stream valleys; (C) sharp morphological changes. Key: 1, channel; 2, contour line; 3, inconsistent flow path due to geometric irregularities; 4, blind stuctures: contiguous contours cannot be directly compared

triangles). The number of output triangles is 19 804: this increase in the number of polygons compared to the contour-based method is due to the fact that in the CBM homogeneous contiguous flow cells were implicitly grouped into single polygons featuring the same slope angle class.

16.6   DATA AND DISCUSSION

Since the contour-based framework is widely regarded as the most realistic representation of slope (O'Loughlin 1986; Band *et al*. 1995; Prosser and Abernethy 1996; Dietrich and Montgomery 1998), it was used as a reference for comparison of the DEM and TIN schemes.

16.6.1   Overall Slope Angle Frequency Distribution

The slope frequency distribution computed over the whole area highlighted the following features (Figure 16.8).

For TIN versus CBM, the two methods yield similar results, differing very slightly except for the steepest range which suffers from some intrinsic TIN artifacts:

(a)   15–40% – underestimated by TIN by 3.3%
(b)   40–50% – overestimated by TIN by 11%
(c)   >60% – overestimated by TIN by 65.1%.

For DEM versus CBM, the two curves are strongly divergent within the following slope ranges, which are important for practical purposes (as we shall point out later):

(a)   20–35% – overestimated by DEM by 12.5%
(b)   >40% – underestimated by DEM by 35.6%.

In practice, the presence of critical thresholds makes the environmental planning process very sensitive to small slope changes, even when at first sight the differences might appear to be of relatively minor importance, as in the present case. If we look at the results from the three

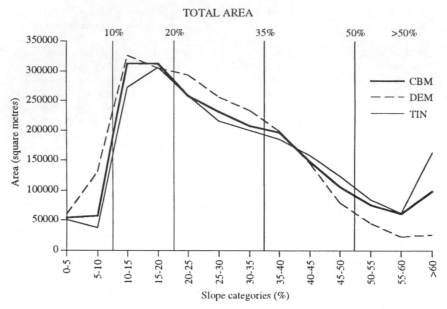

Figure 16.8 Slope frequency distribution computed over the whole area. Critical slopes are also highlighted

Table 16.2 Discrepancies in critical slope ranges between the three methods

Slope (%)	CBM	DEM	TIN
0–10	110968.2	189836.7	86957.0
10–20	623180.2	629458.8	576294.8
20–35	692818.7	779143.1	671554.4
35–50	447474.6	418046.2	463986.0
>50	233966.5	91923.2	309616.0

methods under the critical slope ranges outlined above (Figure 16.6), the previously observed discrepancies increase (Table 16.2).

It is worth noting that the DEM-based slope classification would underestimate the extent of hillslopes steeper than 35% by 8% to 12% of the whole territory as compared to the other schemes. This might be ascribed to the grid spacing (20 m), which, despite being finer than the standard 30 m mesh, should probably be refined to at least 10 m, as suggested for similar areas by Zhang and Montgomery (1994).

16.6.2 Slope Angle versus Hillslope Morphometry

In order to verify the sensitivity of the DTM methods to the topographic configuration, the total area was further divided into two morphometric subdomains (Figure 16.9): ungullied areas, gently sloping, and mainly subjected to unchannelled sediment transport; and ungullied areas, whose main sediment transport is controlled by gullying and landsliding.

Figure 16.9 Landform types of the Centonara catchment. Key: A, ungullied; B, gullied. Contour elevation in metres

Ungullied areas

The three methods (Figure 16.10) feature almost convergent results. A slight difference can be observed within the 10–20% slope range, where TIN and DEM underestimate the related area (respectively by 9.2% and 12.8%): this might be ascribed to common misclassifications of slope breaks at hillslope toes. Bearing in mind the slight inconsistencies, we will not discuss in detail the differences between each pair of curves.

Gullied areas

The three slope frequency distributions (Figure 16.10) show great differences, and demonstrate that the inconsistencies found in the results from the total area are strongly related to gullied landforms.

The TIN and CBM diverge significantly in the middle range of slope angle and at the steepest interval:

(a)   15–40% – underestimated by TIN by 14.7%
(b)   40–60% – slightly overestimated (5%) by TIN
(c)   >60% – overestimated by TIN by 74.5%.

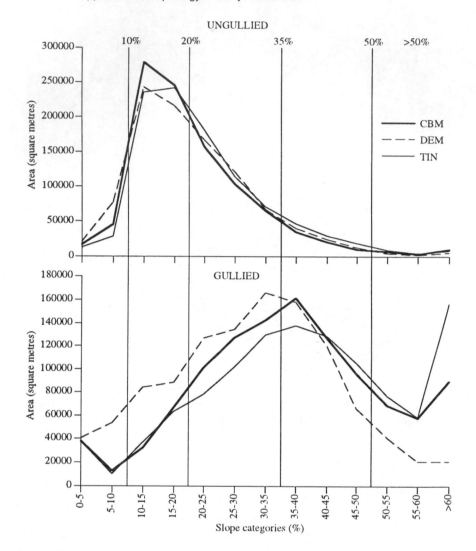

Figure 16.10 Slope frequency distribution compured over ungullied and gullied hillslopes. Critical slopes are also shown

For DEM versus CBM the divergence is striking over the whole slope range, while the threshold value of slope between the underestimation and overestimation still lies in the 35–40% slope class:

(a)   20–35% – overestimated by DEM by 15.4%
(b)   >40% – underestimated by DEM by 38.9%.

The fact that most of the misclassifications lie within gullied areas underlines the risk of underestimating the steepness of hillslopes which represent the most likely source areas for

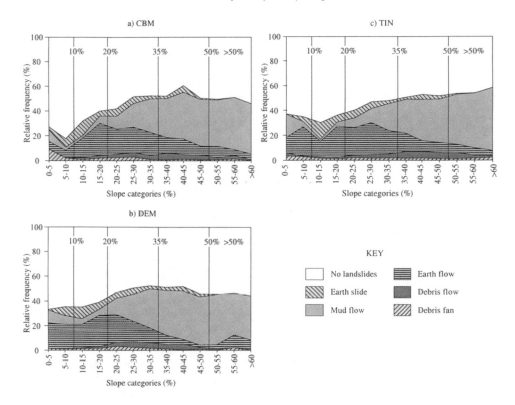

Figure 16.11 Relative frequency of landslide occurrence within each slope category, computed by: (a) CBM; (b) DEM; (c) TIN. Critical slopes are also shown

debris flows. While the differences are substantial, it is worth noting that they are much smaller than those obtained under similar conditions by treating contour lines with 25 m elevation spacing (Casadei 1997).

16.6.3 Slope Angle versus Landslide Types

The overlay of the slope maps obtained by the three DTMs with the landslide map allowed three plots to be made of slope angle versus landslide type (Figure 16.11).

TIN and CBM yield similar results, differentiating at the tails of slope classes, where the frequency of earth flows is underestimated by TIN. A further small difference, which emerges weakly from the figure, is related to the slope distribution of the axial debris fans, whose tendency to accumulate at very small gradients (0–5%) is better captured by the CBM.

For DEM versus CBM as a whole, the distribution of mass wasting within the critical geotechnical domains is comparable. Nonetheless, the percentage of earth flows is greatly underestimated (12.7% over homogeneous slope rather than 31.4%) within the 35–50% range, up to 55% steepness.

As was the case with TIN, the low-angle distribution of debris fans is better outlined by CBM. The relative frequency of mud flows obtained by DEM reaches over 40% of the 40–50% steepness range, as opposed to the lower CBM-derived frequency.

It is worth noting that the maximum bias emerges for earth flows, which are regarded as being more closely related to local slope than debris flows (Casadei 1997). On the other hand, both the CBM and TIN agree in detecting the occurrence of debris fans below angles as high as 30%, although the latter method fails to appreciate the high concentration within almost flat areas (0–10%). Such gently sloping areas should feature a high hazard related to debris impact and accumulation when located at the outlet of the sub-basins.

The same analysis performed separately on the erosion, transportation and deposition landslide areas (not performed in this work) yielded even greater differences in other analogous Apenninic catchments (Casadei 1997).

16.7 LANDSLIDE HAZARD IMPLICATIONS

If we recall that topographic slope angle is treated as one of the most important parameters in any landslide hazard zonation technique (see Section 16.1), we can draw a few important suggestions from the above data.

16.7.1 Earth Flows and Earth Slides

The maximum occurrence of earth flows and earth slides is within the 15–40% slope range, when the relative frequencies reach over 30% of the area. It cannot be excluded that such a widespread phenomenon is related to current and past agricultural practice, as others have claimed (Sidle *et al.* 1985; Alexander 1992; Casadei 1997).

The maximum relative differences between DEM results and the other methods occur within the >10% and 35–50% slope intervals, and they cannot be neglected as they range between 10% and 15% of the available area. The straightforward conclusion is that while CBM and TIN give a consistent representation of these phenomena, the 20 m spaced DEM performs poorly, and should be avoided for earth flow (and earth slide) hazard zonation.

16.7.2 Debris Flows and Mud Flows

Debris flows and mud flows develop through steep (>35%) gullies. At such slopes, this kind of mass wasting occupies roughly 40% of the area. The frequency decreases at lower gradients to almost disappear below slopes of 20% steepness, whereas the final deposits accumulate at slopes of less than 10%. The same tendency for debris fans is not clearly highlighted by TIN (as opposed to CBM) because of its typical difficulties in handling convex shapes.

The location of the area should be taken into account for hazard assessment, as even gently sloping areas are prone to debris impact and accumulation, when located at the outlet of the sub-basins.

The morphometric analysis has shown that neither DEM nor TIN perform well in handling gullied areas and therefore cannot be recommended for debris flow (mud flow) hazard zonation.

16.7.3 Land Use

Although the Centonara catchment falls within a country area, the above results underline some problems relevant from the economic viewpoint.

The DEM method overestimates the extent of hillslopes of 20–35% steepness; therefore areas steeper than 35% would be wrongly classified as suitable for building and agricultural practice, allowing the mechanical ploughing of potentially unstable hillslopes. In the Centonara catchment the amount of misclassified areas exceeds 10 ha.

Bearing in mind that the classification bias is concentrated within gullied areas and that the availability of unstable areas for agricultural practice is widely regarded as a strong factor in the initiation and evolution of gullying processes in the Mediterranean areas (see Alexander (1992) for a discussion), the above results are extremely relevant to shallow landslide hazard zonation and land-use planning.

16.8 CONCLUSIONS

The morphometric analysis of the Centonara catchment has been performed by different techniques. The complication of local landscape, featuring widespread landsliding (44.5% of the catchment) and highly crenulated calanchi landforms has enhanced some of the inconsistencies already known from previous literature about gridding and triangulation techniques.

1. DTMs: the calculation of slope angle from TIN and CBM yielded consistent results in regular areas, although quite different from the DEM representation. Both DEM and TIN inconsistencies become significant within steep and gullied areas, critical in the assessment of threshold-based processes. The observed discrepancies between the methods relying on 10 m spaced contours underline the need to review many of the planning documents used in Italy, which are based on less detailed data sources (50 m and 25 m spaced contour elevation lines), probably not consistent with many critical small-scale landforms.
2. Mass wasting: earth flows and earth slides occur predominantly within the 15–40% slope range, when the relative frequency reaches over 30% of the area. Mud flows start to occur significantly on slopes steeper than 15%, associated with the residual friction angle range for normally consolidated clays (18–20°). Debris fans tend to accumulate within almost flat areas, but their occurrence starts just below slopes as high as 30%. Therefore, these gently sloping areas feature a high hazard related to debris impact and accumulation, when located at the outlet of the sub-basins.

As a whole, the three DTMs suggest that areas steeper than 15% or 20% have to be considered vulnerable to shallow landslides in the short to medium term. The widespread agricultural exploitation of such hillslopes plays a relevant role in these mass-wasting processes, increasing water infiltration and loosening soil by ploughing; this last point suggests a more rational exploitation of the Apenninic hillslopes, in order to avoid soil loss and desertification.

REFERENCES

Ahnert, F. 1976. Brief description of a comprehensive three-dimensional process-respose model of landform development. *Zeitschrift für Geomorphologie N.F.*, Suppl.Bd. **25**, 29–49.

Alexander, D. 1992. On the causes of landslides: human activities, perception and natural processes. *Environmental Geology Water Science Bulletin*, **20**(3), 165–179.

Amadesi, E. and Vianello, G. 1978. Nuova guida alla realizzazione di una carta della stabilità dei versanti. *Mem. Soc. Geol. It.*, **19**, 53–60.

Amadesi, E. and Vianello, G. 1982. Metodologia per la realizzazione di una carta della stabilità. *Geol. appl. ed Idrogeol.*, **17**, 33–37.

Band, L.E. 1986. Topographic partition of watersheds with digital elevation models. *Water Resources Research*, **22**, 15–24.

Band, L.E. 1993. Extraction of channel networks and topographic parameters from digital elevation data. In Beven, K. and Kirkby, M.J. (eds), *Channel Network Hydrology*. Wiley, Chichester.

Band, L.E., Vertessy, R. and Lammers, R.B. 1995. The effect of different terrain representation and resolution on simulated watershed processes. *Zeitschrift für Geomorphologie N.F.*, Suppl.Bd. **101**, 187–199.

Beven, K.J. and Kirkby, M.J. 1979. A physically based variable contributing area model of basin hydrology. *Hydrological Science Bulletin*, **24**, 43–69.

Brabb, E.E. 1984. Innovative approaches to landslides hazard mapping. *Proceedings of IV International Symposium on Landslides*, **1**, Toronto: 307–324.

Canuti, P. and Casagli, N. 1996. Considerazioni sulla valutazione del rischio da frana. *Proceedings of Symposium C.N.R.-G.N.D.C.I. Fenomeni Franosi e Centri Abitati*. Bologna, 29–129.

Carlà, R., Carrara, A., Detti, R., Federici, G. and Pasqui, V. 1986. Geographical Information Systems in the assessment of flood hazard. *Transactions of the International Conference on Arno Project*, Florence.

Carrara, A. 1983. Cartografia tematica, stoccaggio ed elaborazione dati. *Transactions of Symposium C.N.R. on Project Soil Conservation-Subproject Landslide Phenomena*, Bari.

Carrara, A. 1988. Drainage and divide networks from high-fidelity digital terrain models. In Chung, C.F., Fabbri, A.G. and Sinding-Larsen, R. (eds), *Quantitative Analysis of Mineral and Energy Resources*, Kluwer, Dordrecht, 561–597.

Carrara, A., Cardinali, M., Detti, R., Guzzetti, F., Pasqui, V. and Reichenbach, P. 1991. GIS techniques and statistical models in evaluating landslide hazard. *Earth Surface Processes and Landforms*, **16**, 427–445.

Carson, M.A. and Kirkby, M.J. 1972. *Hillslope Form and Process*. Cambridge University Press, Cambridge.

Casadei, M. 1997. *Analisi quantitativa delle relazioni fra i parametri morfologici ed il dissesto nell'Appennino Settentrionale*. PhD Thesis, Bologna.

Casadei, M. and Farabegoli, E. 1997. Geomorphological analysis of a badland-type watershed for a solid urban waste disposal site location in the northern Apennines. *Proceedings of IAEG Symposium 97*, Athens, 1669–1674.

Clerici, A., Cuccuru, G., Trambaglio, L. and Lina, F. 1993. La realizzazione di una carta della stabilità dei versanti mediante l'uso di un sistema d'informazione geografica. *Geologia tecnica e Ambientale*, **4**, 25–40.

Colosimo, P. 1978. Comportamento di argille Plio-Pleistoceniche in alcuni versanti instabili dell'anconetano. *Mem. Soc. Geol. It.*, **19**, 215–224.

Coltorti, M., Nanni, T. and Rainone, M.L. 1987. Il contributo delle scienze della terra nell'elaborazione di un piano paesistico: l'esempio del Monte Conero (Marche). *Mem. Soc. Geol. It.*, **37**, 629–647.

Corniello, A., De Riso, R. and Lucini, P. 1980. La franosità potenziale del bacino del F. Tammaro (Campania). *Mem. Note Ist. Geol. Appl. Napoli*, **15**, 1–35.

Costantini, E.A.C. 1987. Cartografia tematica per la valutazione del territorio nell'ambito dei sistemi produttivi. *Ann. Ist. Sper. Studio e Difesa del Suolo*, **18**, 23–74.

Costa Cabral, M.C. and Burges, S.J. 1994. Digital elevation model networks (DEMON): A model of flow over hillslopes for computation of contributing and dispersal areas. *Water Resources Research*, **30**, 1681–1692.

D'Alessandro, L. and Pantaleone, A. 1987. Caratteristiche geomorfologiche e dissesti nell'Abruzzo sud-orientale. *Mem. Soc. Geol. It.*, **37**, 805–821.

Dietrich, W.E. and Dunne, T. 1993. The channel head. In Beven, K. and Kirkby, M.J. (eds), *Channel Network Hydrology*. Wiley, Chichester, 175–219.

Dietrich, W.E. and Montgomery, D.R. 1998. *SHALSTAB. A digital terrain model for mapping shallow landslide potential*. NCASI Technical Report.

Dietrich, W.E., Wilson, C.J., Montgomery, D.R., McKean, J. and Bauer, R. 1992. Erosion thresholds and land surface morphology. *Geology*, **20**, 675–679.

Farabegoli, E. and Agostini, C. 2000. Identification of 'Calanco', a badland landform in the northern Apennines, Italy. *Earth Surface Processes and Landforms*, **25**, 307–318.

Farabegoli, E. and Forti, P. 1997. Geomorphic evolution of karst and fluvial basins in the surroundings of Bologna. *Suppl. Geogr. Fis. Dinam. Quat.*, **3**(2), 205–213.

Farabegoli, E., Rossi Pisa, P., Costantini, B. and Gardi, C. 1994. Cartografia tematica per lo studio dell'erosione a scala di bacino. *Riv. Agron.*, **23**(4), 356–363.

Hansen, A. 1984. Landslide hazard analysis. In Brundsen, D. and Prior, D.B. (eds) *Slope Stability*. Wiley, Chichester, 523–602.

Howard, A.D. 1994. A detachment-limited model of drainage basin evolution. *Water Resources Research*, **30**(7), 2261–2285.

Lammers, R.B. and Band, L.E. 1990. Automatic object representation of drainage basins. *Computation and Geoscience*, **16**(6), 787–810.

Lancellotta, R. 1987. *Geotecnica*.

Lucini, P. 1969. Un metodo grafico per la valutazione della franosità. *Mem. Note Ist. Geol. Appl. Napoli*, **2**, 1–14.

Marchionna, G., Sacchi, L., Silvi, A., Ventura, R. and Visicchio, F. 1988. Proposta di un modello statistico di franosità dei versanti finalizzato alla realizzazione di una carta di stabilità. *Boll. Serv. Geol. It.*, **107**, 253–312.

Mark, D.M. 1975. Computer analysis of topography: a comparison of terrain storage methods. *Geografiska Annaler*, **57A**, 179–188.

Montgomery, D.R. and Dietrich, W.E. 1989. Source areas, drainage density and the problem of landscape scale. *Water Resources Research*, **25**, 1907–1918.

Moore, I.D., O'Loughlin, E.M. and Burch, G.J. 1988. A contour-based topographic model for hydrological and ecological applications. *Earth Surface Processes and Landforms*, **13**, 305–320.

O'Callaghan, J.F. and Mark, D.M. 1984. The extraction of drainage networks from digital elevation data. *Comp. Vision, Graphics and Image Processing*, **28**, 323–344.

O'Loughlin, E.M. 1986. Prediction of surface saturation zones in natural catchments by topographic analysis. *Water Resources Research*, **22**, 794–804.

Prosser, I.P. and Abernethy, B. 1996. Predicting the topographic limits to a gully network using a digital terrain model and process thresholds. *Water Resources Research*, **32**(7), 2289–2298.

Puecker, T.K., Fowler, R.J., Little, J.J. and Mark, D.M. 1978. The triangulated irregular network. *Proceedings of ASP-ACSM Symposium on Digital Terrain Models*, St Louis.

Schneider, B. 1995. Adaptive interpolation of digital terrain models. *Proceedings of ICC 1995*. Barcelona, 2206–2210.

Sidle, R.C., Pierce, A.J. and O'Loughlin, C.L. 1985. *Hillslope Stability and Land Use*. Water Resource Monograph Series 11. American Geophical Union, Washington.

Skempton, A.W. and DeLory, F.A. 1957. Stability of natural slopes in London Clay. *Proceedings of 4th International Conference on Soil Mechanics and Foundation Engineering*, London, **2**, 378–381.

Smith, T.R. and Bretherton, F.P. 1972. Stability and conservation of mass in drainage basin evolution. *Water Resources Research*, **8**, 1506–1529.

Spagna, V. and Cabriel, M. 1987. Cartografia geologico-tecnica per gli strumenti urbanistici: dal piano territoriale regionale di coordinamento (P.T.R.C.) al piano regolatore generale (P.R.G.). *Mem. Soc. Geol. It.*, **37**, 669–688.

Valentini, G. 1967. Un modello statistico nello studio della franosità nel quadro morfologico, geologico e geotecnico nella media valle del F. Fortore. *Geol. appl. ed Idrogeol.*, **2**, 197–227.

Varnes, D.J. and IAEG Commission on Landslides, 1984. *Landslide hazard zonation – a review of principles and practice*. UNESCO, Paris.

Weibel, R. 1997. Generalization of spatial data: principles and selected algorithms. In Van Kreveld, M., Nievergelt, J., Roos, T. and Widmayer, P. (eds), *Algorithmic Foundations of Geographic Information Systems*. Springer, Berlin, 99–152.

Weibel, R. and Brändli, M. 1995. Adaptive methods for the refinement of digital terrain models for geomorphometric applications. *Zeitschrift für Geomorphologic*, Suppl. Bd. **101**, 13–30.

Willgoose, G., Bras, R.L. and Rodriguez-Iturbe, I. 1990. A coupled channel network growth and hillslope evolution model. 1. Theory. *Water Resources Research*, **27**, 1671–1684.

Wolock, D.M. and Price, C.V. 1994. Effects of digital elevation model map scale and data resolution on a topography-based watershed model. *Water Resources Research*, **30**(11), 3041–3052.

Wolock, D.M. and Price, C.V. 1995. Comparison of single and multiple flow directions algorithms for computing topographic parameters in TOPMODEL. *Water Resources Research*, **31**(5), 1315–1324.

Zhang, W. and Montgomery, D.R. 1994. Digital elevation models grid size, landscape representation and hydrologic simulation. *Water Resources Research*, **30**(4), 1019–1028.

17 Geomorphologic Risk Assessment for Cultural Heritage Conservation

FILIPPO CATANI, RICCARDO FANTI AND SANDRO MORETTI

Dipartimento di Scienze della Terra, Università di Firenze, Firenze, Italy

ABSTRACT

This chapter describes the recent research activities carried out by the Earth Sciences Department of Florence University (Italy), in the field of geographical information systems applied to geomorphological hazard assessment. The research has been undertaken in two study areas in Italy (Vulci and Tharros) and constitutes an example of a methodological application for the assessment of geomorphologic hazard and risk in relation to landslide, flood and erosion processes for historical sites.

The creation of the cultural heritage database is presented and the analysis of the geomorphologic processes affecting it has been performed. Arc/Info software (© ESRI Inc.) has been employed to create the database in order to organize and georeference all the information obtained from the preliminary survey analysis. The research work includes a risk evaluation following the United Nations Educational, Scientific and Cultural Organization (UNESCO) procedure. On the basis of UNESCO's strategies all the geomorphologic processes influencing the cultural heritage stability (both on slope and floodplain) were analysed. This procedure allows the risk distribution in the considered areas to be highlighted, giving the opportunity to identify more susceptible zones where prevention or conservation activities should be applied.

The two case studies represent different levels of detail in which the final products are the definition of a risk map of a whole archaeological site addressing the management of cultural heritage (Vulci case study) and a detailed risk analysis addressing specific conservation measures (Tharros case study).

17.1 INTRODUCTION

International efforts for the evaluation and conservation of cultural heritage are mainly pursued by UNESCO, which is promoting several initiatives especially directed at specific monuments and archaeological sites. These initiatives are often concerned with restoration or conservation but rarely with prevention of damage. This is mainly due to the fact that there is almost no link between knowledge of environmental impacts and cultural heritage. Italy is one of the richest countries worldwide with respect to cultural heritage. Yet, this fact involves problems of a different nature such as storage, classification and protection from both incorrect exploitation and natural disasters. In an attempt to devise procedures and methodologies for solving conservation and damage prevention problems, the Italian National Research Council developed a Special Project for the Safeguard of Cultural Heritage. Its main

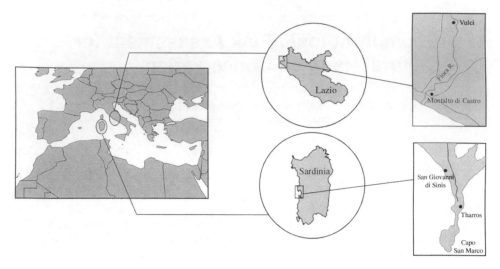

Figure 17.1 Location map of test sites

purpose is to obtain a more comprehensive and interconnected framework among the several research fields working on this subject. Recently the Earth Sciences Department of Florence University was fully involved in this project, and particular attention was given to research in the field of geomorphologic risk assessment.

The evaluation of archaeological heritage that may be at risk often depends on the survey scale. In fact, most data are extremely inhomogeneous to such an extent that problems in map reporting arise. Furthermore, an effective portrayal of heritage information can only be achieved when other important aspects are taken into account, that is, geomorphology and its morphodynamics.

Using a geographical information system (GIS) these problems can be solved since it allows one to represent several features with different scales and properties in the same integrated digital environment. The procedure presented in this chapter takes into account the problems previously expressed and the capability of a GIS to solve them in the context of two Italian case studies: Vulci (northern Lazio) and Tharros (Sardinia).

17.2 LOCATION OF TEST SITES

17.2.1 Vulci

The Vulci archaeological site is located on a gently sloping plateau of the Viterbo province (Lazio region, central Italy, Figure 17.1) and is characterized by a smooth undulating morphology. Geological deposits outcropping in the area are representative of the post-Miocene sedimentary sequence which is prograding up to the Pleistocene period. These deposits are overlying, with unconformity, the allochthonous deposits of the Tuscan Series lying on its metamorphic basement. On this geologic framework the plateau, albeit irregularly shaped, has very sharp boundaries because of the incision of the Fiora River and its affluents which are probably influenced by faults or tectonic structures. Such morphological settings were particularly favourable in the past for defensive purposes, when the plateau was almost entirely occupied by an important urban centre.

The archaeological records (Ministero LL.PP 1996) provided sufficient information for the remains of the populated settlement of this territory dating from the Copper Age to the Bronze Age to be retraced. In fact, the archaeological observations show that from the 9th century BC the protovillanovan villages of the middle Fiora Valley were abandoned and population concentrated on the plain of the future city. The Etrurian town of Vulci had its maximum expansion from the 8th century until the 5th century BC mainly due to the commercial trade favoured by the Tyrrhenian domain. In 474 BC the Cuma's defeat by the Syracusan Greeks resulted in the loss of trade on the Tyrrhenian Sea. In the 4th century land property was the new source of wealth and power but in 280 BC Vulci was once more defeated by the Romans. Under the Romans Vulci decayed and became a small province until it was completely abandoned during the 9th century. In the early imperial age there were traces of good building activity, which is recognizable in the Tempio Grande and in the reconstruction of the Domus del Criptoportico (Moretti 1982).

17.2.2 Tharros

The promontory of Capo San Marco (Oristano province, west Sardinia island, Figure 17.1) is constituted by pelitic and calcareous sediments of Messinian–Pliocene age. Plio-Pleistocene basalts which, due to selective erosion, form a hilltop plateau, overlie these deposits. A sequence of Pleistocene deposits, both marine and continental, testifies to several eustatic fluctuations and a thick deposit of aeolian sandstones, related to the Würm glaciation, closes this sequence. Recent dune and beach deposits also bury large sectors of the peninsula (Cherchi *et al.* 1978).

Since the 1950s several campaigns of archaeological excavations in the Capo San Marco peninsula have brought to light the ruins of Tharros, a village of Phoenician origin (8th century BC). In the course of its history Tharros was ruled by the Carthaginians (6th century BC) and later the Romans (3rd century BC) until it was abandoned in the 9th century (Pesce 1966; Zucca 1993). The ruins of two necropolises and a temple, related to the Phoenician–Punic period, can be found near the urban area. Instability processes of different types, related to the particular geological and geomorphologic conditions, affect all these archaeological sites.

17.3 METHODS

Risk assessment is performed according to the UNESCO procedure (Varnes and Iaeg 1984) with further modification (Hartlén and Viberg 1988; Canuti and Casagli 1994; Fell 1994; Cruden and Fell 1997), in which *hazard* (H) is defined as the probability that a potential destructive phenomenon will take place, for a defined period and area. *Vulnerability* (V) represents the expected degree of loss, for a defined element or group of elements, resulting from the occurrence of a natural phenomenon of a certain intensity. It is scaled from 0 (no loss) to 1 (total loss). *Specific risk* (R_s) is a combination of hazard and vulnerability according to the degree of the expected damage caused by a natural phenomenon of a defined intensity. Hence it can be defined as:

$$R_s = HV \tag{17.1}$$

To calculate the *total risk* (R), defined as the expected value of loss (human, property, etc.) due to a natural phenomenon, the *elements at risk* (E) have to be considered, such as people, property, economic activity and public services which are at risk in a given area. Total risk is therefore represented as:

$$R = HVE = R_s E \tag{17.2}$$

Einstein (1988) espouses that in order to characterize potential destructive phenomena in a study related to risk such as the one described here, a complete analysis should include the following 'risk definition steps' (IUGS WGL/CRA, 1997):

(i)   hazard identification and analysis
(ii)   description of individual elements at risk
(iii) analysis of vulnerability
(iv)   risk analysis
(v)   risk assessment and management.

To apply this procedure in an archaeological area, each of these steps has to be adequately defined.

(i)   This includes the following aspects: the survey of all the geomorphologic processes affecting the area's stability (both on slope and floodplain); the database and related geomorphologic maps representing the forms on which processes are acting; the analyses of processes for identifying their intensity, established on the basis of their main physical characteristics (e.g. for a landslide its velocity and dimension; for a flood the probable area of occurrence); the time and space hazard distribution which have relevance with respect to intensity.
(ii)   The number and characteristics of archaeological artifacts are determined. This step is usually based on existing archaeological surveys.
(iii) The assessment of the expected degree of damage to archaeological heritage due to the interaction with geomorphologic processes is performed.
(iv)   The synthesis of the information obtained from previous steps through the overlay of geomorphologic hazards and archaeological 'elements at risk' data is carried out. The calculation of risk is a computation of the interaction among economic value and vulnerability of 'elements at risk' and hazard following Equation 17.2.
(v)   The identification of the structural intervention for mitigating risk or reducing hazard in respect of each element at risk is accomplished. A detailed characterization of geomorphologic processes is required and the computed risk has to be compared with the acceptable risk.

The main task of this research, however, is not directed towards its complete application but rather towards the validation of its principles in the field of conservation of archaeological monuments. In fact, the complete methodology would presume a detailed knowledge of the single characteristics, such as vulnerability and 'element at risk' value, which are not always available or easily identifiable. Therefore, in the two case studies, the procedure has been applied at a qualitative level leading to the development of an inventory of the distribution of geomorphologic processes, distinguished in typology classes. The overlay of this information with archaeological sites has been performed. In this case, due to the characteristics of the archaeological heritage, a value of 1 was chosen for vulnerability, on the assumption that the involvement of an archaeological artifact in a geomorphologic process leads to its complete destruction. This assumption can be considered acceptable in most cases, because the economic cost of restoration is often comparable with the practical economic value of the object itself. Considering the value of the elements at risk in the Vulci area, a three-class subdivision has been created as a function of the effective archaeological relevance of the area. This subdivision has been devised by the archaeologists working on the site. Cross-referencing the information related to hazard and the elements at risk leads to the portrayal of a map that should highlight risk areas for which a subsequent detailed study, based on the complete definition of each single geomorphologic process, can extract the data needed for the risk management procedure.

In this research the application of the methodology up to the identification of areas at risk (step iv) is described for the Vulci test site, whilst the detailed geomorphologic characterization of several single risk situations, fundamental for step v, is provided in the Tharros case study. On the one hand, the Vulci case approach represents a generally small-scale study in which a specific geographic location and its cultural heritage are surveyed and classified according to the described methodology. This work has proved useful in producing documents (e.g. maps of the elements at risk or maps of risk areas) which are necessary for planning purposes, in a framework of sustainable management, protection and prevention of damage. On the other hand, the Tharros case study, focusing especially on steps iv and v, singles out each site where there is evidence of risk, facilitating the identification of specific prevention or mitigation actions for each surveyed risk element.

In this context, especially in order to account for the interactions among variables and different objects, both geomorphologic and archaeological, a tool capable of managing and elaborating many different kinds of data at the same time is required. The extensive use of a GIS was found to be the most effective procedure for an integrated analysis of the site information and of the intrinsic and external condition of potential instability (Goodchild 1992; Laurini and Thompson 1992). A GIS is a computer-based system capable of storing information about geographic features in a database. By maintaining the spatial location of features, as well as their attribute information, the GIS is able to perform complex analyses efficiently and rapidly. In particular, this tool is useful in the management of geographic and descriptive data and in their reciprocal overlay analysis necessary to obtain a complete description and comprehension of risk issues in the study area. The GISs applicable in this study are Arc/Info 7.0 and ArcView 3 (both © ESRI Inc.).

17.4 DISCUSSION

17.4.1 Vulci Case Study

In the Vulci case study, a synoptic level of investigation, in which the procedure is finalized to identify the general conditions of hazard, vulnerability and risk, is considered. This procedure can then be employed in an environmental management plan for risk forecast and prevention.

In the Vulci area, the analysis kept to the previously outlined risk definition steps with the exception of step iii, in which vulnerability has been considered constant for each risk element and equal to a value of 1 (100% loss), accounting for the lack of structural protection for the archaeological artifacts and for the previous assumptions about restoration costs.

The first step deals with the traditional geomorphological survey followed by a reorganization of data in order to make them suitable for a GIS. On the one hand, the traditional concept of surveying is based on the representation of geomorphological forms by means of lines and symbols, seldom using areal entities. On the other hand, the most interesting capability of a GIS is the processing of bidimensional areal data, such as polygons or clusters of pixels in either vector or raster format. This allows one to carry out many kinds of analyses, which are impossible to obtain with traditional cartographic or geomorphologic methods. In particular, the use of GIS overlay operators has enabled this research to identify geomorphologic areal characteristics, giving rise to the classification of 'process influence zones' in the study area. In this framework, a 'process influence zone' is defined as an area where one or more morphodynamic agents act on the landscape equilibrium. For this reason, the traditional geomorphological map has to be redrawn considering all the linear and point information as areas. This can be done by analysing each form in its real influence area and entering all its properties on a relational database in a GIS. This concept, regarding the influence distribution of a process, is quite difficult to apply to all the geomorphological processes at small and medium scale (1:50 000, 1:25 000), but proves easier for larger scale maps (e.g. 1:10 000 or 1:5000), in which each single process can be described as having a surface distribution. Following this mapping phase, according to the field and photo-interpretation work, each process influence zone has been assigned a suitable hazard class. Such classes were previously defined in order to fit with the relative hazard value with respect to the total area. Relative hazard has been utilized instead of absolute hazard because a detailed determination of the exact probability of occurrence of phenomena was found to be outside the aims and objectives of this work.

As already noted, steps ii and iii were performed in the Vulci test site with the traditional archaeological survey and the subsequent assumption of vulnerability equal to 1. In order to account for the value of the elements at risk, three classes have been defined as low, medium and high theoretical economic value. Different objects were assigned to these classes according to the archaeologists working on the site (Sgubini Moretti 1980, 1993; Ricciardi 1988).

The most important archaeological monuments in the area are: the François Tomb, a big underground structure composed of four rooms excavated from tephrite rocks; the Tempio Grande, the remains of a large Roman temple; the Domus del Criptoportico, a huge Roman imperial age country house whose principal remains are the mosaic floorings and the criptoportico, a beautiful and unique underground porch.

Another very relevant structure is the urban road system, mainly composed of the two traditional Roman streets, the Decumanus (in the W–E direction) and the Cardo (in the N–S direction). In Vulci these structures are particularly valuable because of their well-preserved nephrite paving. Thus, step iv was finally realized with the application of Equation 17.2, where V is equal to 1 and H and E are parametric values related to different classes of hazard and economic value, leading to the production of a relative risk map (Figure 17.2).

At the end of the procedure, with the help of the GIS overlay and forecast operators, three cases of serious risk situations have been found: erosion, landslides and flood processes are identified in areas where archaeological artifacts (of different levels of importance) are present (Figures 17.2 and 17.3). They can be described as follows.

A. Accelerated erosion of a little creek in an advanced stage of downcutting affecting an ancient wall structure near the north gate of the city (Figure 17.3A).
B. Accelerated sheet erosion near the east end section of the Decumanus, where a gentle slope begins just out of the ancient city boundary. Here, a great portion of the nephrite paving is at risk, much of which has already been destroyed (Figure 17.3B).
C. Rock falls and toppling phenomena in the rock cliff where the famous François Tomb is located. Here, a detailed structural survey revealed a risk to the west side of the underground structure (Figure 17.3C).

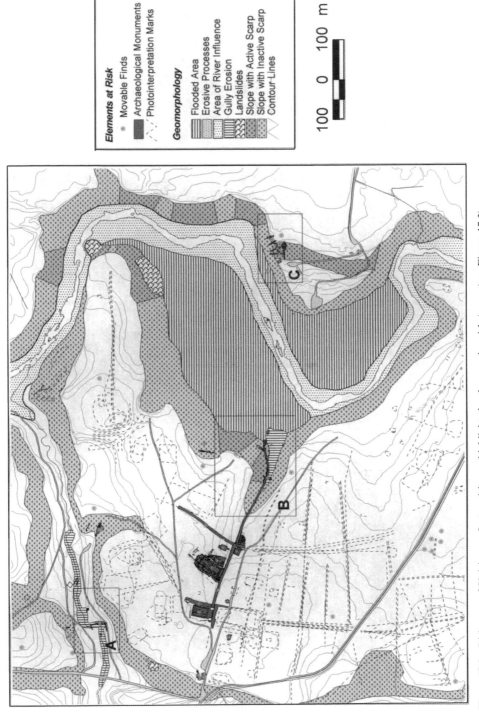

Figure 17.2 Risk map of Vulci area. Lettered boxes highlight the three major risk issues (see Figure 17.3)

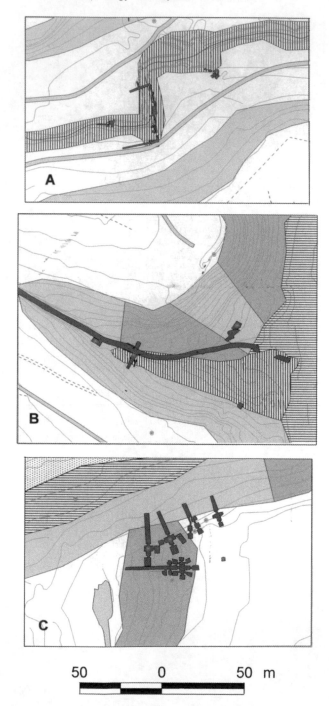

Figure 17.3   Enlargements of the three risk zones highlighted by the GIS overlay and forecast analysis. (A) Downcutting erosion of a creek close to a portion of an ancient wall structure; (B) sheet erosion by the cast-end portion on Decumano; (C) rock fall phenomena near the François tomb. See Figure 17.2 for legend

Especially for cases A and C, the GIS integrated analysis via comparison between process influence zones and elements at risk, as well as the spatial forecasting of the process evolution, allowed the discovery of potential risk situations which otherwise would probably have been neglected.

17.4.2 Tharros Case Study

For the Tharros case study the detail of investigation is dealing with a single type of geomorphological process, that is, landslide. In this case, the detailed analysis has to take into account all the information related to the phenomena acting on the cultural heritage, including geotechnical properties. Thus, the GIS has more detailed information that cannot be extrapolated to other areas without a precise analysis of the site. Therefore, with reference to the above-mentioned step procedure, each risk situation is analysed on the basis of information deriving from the implementation of the first four steps: thus, a complete geomorphologic description is achieved and this is very important for risk assessment and management (step v).

In the Tharros archaeological area four risk situations have been identified and characterized: the two necropolis cases (the San Giovanni di Sinis and Capo San Marco areas), the westward side of the urban zone and the *tempietto rustico* (Phoenician–Punic temple) site. The two necropolises are formed by hypogean chamber tombs consisting of a simple rectangular ditch of variable depth (0.5–1.5 m) excavated in the aeolian sandstones: some are provided with a staircase entrance (*dromos*), but most (especially in the San Giovanni area) are open on the upper side, as a removed stone slab formed the covering (Acquaro 1978).

In the northern necropolis, located on the seaward side of San Giovanni di Sinis village, the aeolian sandstones overlie a Quaternary sequence composed of shell breccias, conglomerates and sandstones which reflects the eustatic oscillations of the last glaciations (Pecorini 1972; Carboni and Lecca 1985). This particular geomorphologic situation determines the instability of the area on which the necropolis is located, because strong marine erosion acts in a selective way determining the creation of overhanging sandstone blocks bounded at the back by the chamber sides: the block downfall causes tomb destruction (Figure 17.4).

In the second necropolis, located on the south of the Capo San Marco promontory, the aeolian sandstones overlie a pelitic sequence of Messinian age. The contrast in competence between the two formations as well as recent tectonics are responsible for the presence of a natural joint network which, in combination with the tomb cuts, determines a complex geometry of unstable blocks. The site is also affected by an earth slump, with the rupture surface located within the pelitic basal sequence. As a consequence of its progressive retrogression it reached the aeolian sandstones of the necropolis (Figure 17.5). This complex geomorphologic and structural setting determines the progressive destruction of the necropolis via the fall of blocks of variable size (0.5–10 m).

Even the urban archaeological area of Tharros is marginally affected by an earth slump that involves Miocene pelitic deposits, partially buried under a recent dune field (Figure 17.6). These dunes mask almost all the landslide deposits, making its recognition quite difficult; but sharp traces of failure are evident along the slide flanks where the Roman Decumanus is displaced by several centimetres and the access road to the archaeological area is continuously damaged. The possible retrogression of the landslide would threaten the stability of the Tofet, the sacred area of Tharros, dating from the Phoenician period.

The Phoenician–Punic temple is located at the southern end of Capo San Marco on the border of a basaltic plateau with a dense joint system deriving from the cooling phase. The

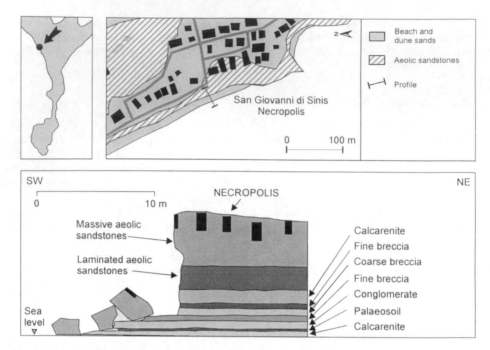

Figure 17.4 The San Giovanni di Sinis necropolis

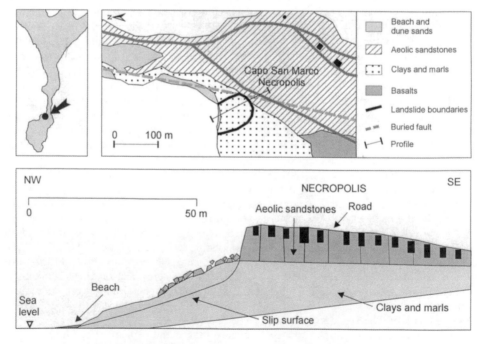

Figure 17.5 The Capo San Marco necropolis

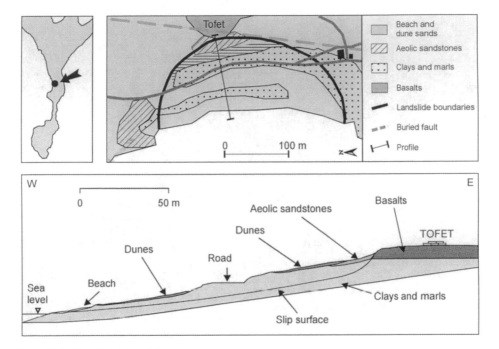

Figure 17.6 The Tofet landslide

area is also affected by several normal faults, which displaced the contact between basalts and their substratum by several metres (Figure 17.7). The fault systems are partially reactivated by gravity and the seaward blocks slowly slide apart as is shown by the presence of two main trenches filled by debris. The hypothetical slip surface is probably gently inclined and located inside the pelitic sequence. The retrogression of failure is causing the regression of the cliff crest and it is at present very close to the temple.

17.5 CONCLUSIONS

The analysis of the two case studies has shown that a complete application of the risk assessment procedure up to step iv, in order to make a reliable analysis of potentially dangerous situations, must be based on a detailed inventory of the risk elements followed by a small-scale geomorphologic survey. In particular, knowledge about cultural heritage should concern every object of any importance and should extend to the assessment of its economic value and the identification of its geographic location.

Furthermore, acting as an operational link between archaeological and geomorphological frameworks, the utilization of a GIS has been demonstrated to be highly beneficial because of its powerful geographic data management and analysis tools.

Regarding the Vulci case study, results showed that archaeological monuments of varying economic value are present in areas with geomorphological hazards (flood, erosion and landslide). A general map of relative risk was elaborated and consequently highlighted at least three important risk areas. The Tharros case study showed four different cases of heritage at risk due to landslides caused by different geomorphological processes: rock failure by marine

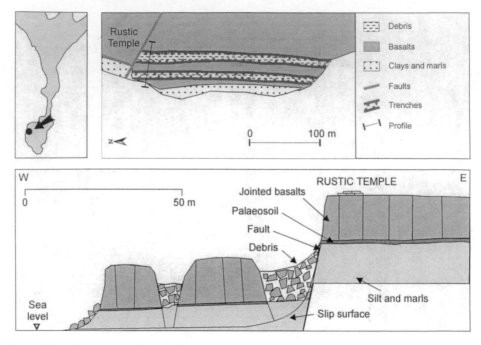

Figure 17.7   The Capo San Marco Phoenician–Punic temple

erosion, landslide due to difference in rock competence and slump involving both marine influence and lithology.

In summary, the integrated GIS system allowed us not only to identify these risk issues, but also to assess the correct operational steps to be accomplished for either the mitigation or the complete prevention of the expected damage.

ACKNOWLEDGEMENTS

This work has benefited from grants by the Italian National Research Council, Special Project 'Safeguarding Cultural Heritage', and the National Group for Geo-Hydrological Catastrophes Prevention, U.O.1.46.

REFERENCES

Acquaro, E. 1978. La necropoli meridionale di Tharros: appunti sulla simbologia funeraria punica in Sardegna. *Orientis Antiqui Collectio*, **XIII**, 111–113.

Canuti, P. and Casagli, N. 1994. *Considerazioni sulla valutazione del rischio di frana (Remarks on landslide risk evaluation)*. Fenomeni franosi e Centri Abitati, CNR-GNDCI, publication no. 846.

Carboni, S. and Lecca, L. 1985. Osservazioni sul Pleistocene medio-superiore della penisola del Sinis (Sardegna occidentale). *Bollettino Società Geologica Italiana*, **104**, 459–477.

Cherchi, A., Marini, A., Murru, M. and Robba, E. 1978. Stratigrafia e paleoecologia del Miocene superiore della penisola del Sinis (Sardegna occidentale). *Rivista Italiana Paleontologia*, **84**, 973–1036.

Cruden, D. and Fell, R. (eds) 1997. *Landslide Risk Assessment. Proceedings of the International Workshop on Landslide Risk Assessment, Honolulu, 19–21 February 1997*. Balkema, Rotterdam.

Einstein, H.H. 1988. Special lecture: Landslide risk assessment procedure. *Proceedings of 5th International Symposium On Landslides, Lausanne*, **2**, 1075–1090.

Fell, R. 1994. Landslide risk assessment and acceptable risk. *Canadian Geotechnical Journal*, **31**(2), 261–272.

Goodchild, M.F. 1992. Geographical data modelling. *Computer and Geosciences*, **18**(4), 401–408.

Hartlén, J. and Viberg, L. 1988. General report: evaluation of landslide hazard. *Proceedings of 5th International Symposium on Landslides, Lausanne*, **2**, 1037–1058.

IUGS WGL/CRA Working Group on Landslides, Committee on Risk Assessment, 1997. Quantitative risk assessment for slope and landslides – The state of the art. In Cruden, D. and Fell, R. (eds), *Landslide Risk Assessment. Proceedings of the International Workshop on Landslide Risk Assessment, Honolulu, 19–21 February 1997*. Balkema, Rotterdam, 3–12.

Laurini, R. and Thompson, D. 1992. *Fundamentals of Spatial Information Systems*. The Apic Series, Academic Press, London.

Ministero LL.PP. 1996. *Itineraries in Vulci. Contributi turistici e didattici*. Cooperativa Archeologia, Rome.

Moretti, M. 1982. *Vulci*. Novara, Italy.

Pecorini, G. 1972. La trasgressione pliocenica nel Capo S. Marco (Oristano, Sardegna occidentale). *Bollettino Società Geologica Italiana*, **91**, 365–372.

Pesce, G. 1966. *Tharros*. Rome, Italy.

Ricciardi, L. 1988. Vulci. *Studi Etruschi*, **LIV**.

Sgubini Moretti, A.M. 1980. Vulci. Scavi e scoperte. *Studi Etruschi*, **XLVIII**, 547–550.

Sgubini Moretti, A.M. 1993. *Vulci e il suo territorio*. Rome.

Varnes, D.J. and IAEG, 1984. *Commission on Landslides – Landslide Hazard Zonation – a review of principles and practice*. UNESCO, Paris.

Zucca, R. 1993. *Tharros*. G. Corrias, Oristano.

18 Geomorphology and Environmental Impact Assessment in Relation to Dryland Agriculture

ARTHUR J. CONACHER
Department of Geography, University of Western Australia, Nedlands, Australia

ABSTRACT

Geomorphologists and geomorphology have not made a conspicuous contribution to the methodology or implementation of environmental impact assessment (EIA). Yet the impacts of some human activities, such as the acceleration of erosion by water and wind in response to dryland agricultural practices, are manifestly amenable to geomorphological research. Secondary salinization in Western Australia's wheatbelt is another example where geo morphologists have made a significant contribution in researching the complex processes responsible for the problem and in proposing and trialing remedial measures.

However, EIA and even cumulative impact assessment and strategic environmental assessment either are not appropriate to these kinds of problems, or are constrained by practical difficulties. Management (manipulation of the hydrological, pedological and geomorphic processes involved) requires coordination amongst farmers, land agencies and local governments and therefore a catchment or regional approach, which in turn indicates the need for an integration of environmental assessment with planning. Despite some encouraging developments, progress in this regard has been patchy.

18.1 INTRODUCTION

Since 1969, development proposals have been subject to an environmental impact assessment (EIA) process, first in the USA, and then progressively in most other countries: the main exceptions are in Africa and the Middle East (Harvey 1998). In Australia, EIA or its equivalent are required in all jurisdictions, commencing with the Australian Commonwealth's Environmental Protection (Impact of Proposals) Act in 1974. Even though the original environmental impact matrix devised by the noted geomorphologist L.B. Leopold and his colleagues in 1971 was adopted widely in various forms, geomorphology and geomorphologists appear to have played little part in further developing the methodology of EIA. Amongst many texts, reference may be made to the excellent evaluation of EIA by Munn (1979), the papers on methods in PADC Environmental Impact Assessment and Planning Unit (1983), or the more recent treatment of EIA procedures by Thomas (1996): of the latter book's 241 references, and despite a reasonable sprinkling of geographers, none appears to deal with a geomorphologically related topic.

Applied Geomorphology: Theory and Practice. Edited by R.J. Allison.
© 2002 John Wiley & Sons, Ltd.

In response to the organization of a session on Geomorphology and Environmental Impact Assessment at the International Association of Geomorphologists' fourth conference in Bologna in 1997, this chapter offers some thoughts as to why this is so.

The types of problem investigated by geomorphologists are perhaps not really appropriate to EIA; neither is EIA appropriate for all environmental problems. For example, drawing on the work of Canadian researchers, Conacher (1988) identified the following overlapping typology of environmental problems:

1. sectoral or regional problems (such as land degradation);
2. project-specific problems;
3. numerous, dispersed, discrete actions that have a cumulative impact on the environment;
4. environmental issues (characterized by conflict and politicization).

Types 1 and 3 overlap and may be relevant to geomorphologists; but many cumulative impacts, such as air pollution arising from numerous individuals' use of motor vehicles, clearly are not. Type 2 problems are most clearly associated with EIA and, with type 4, in the public mind with 'environmental' problems. But the geomorphic impact of proposed developments may in many cases be minor, though not necessarily so. The need for mediation (conflict resolution) in type 4 means that those major environmental battlegrounds are often not appropriate to EIA methodologies and certainly not relevant to geomorphology.

Land degradation is a major environmental problem – debatably *the* major problem in Australia – but it is not really amenable to EIA. The latter is a fairly formalized sequence of procedures used by an environmental protection agency in response to a development proposal. But land degradation results from the actions of numerous farmers and is regional in extent (and sectoral in nature), and is generally not a response to a particular development proposal. On the other hand, land degradation is an environmental problem which most emphatically is relevant to geomorphology. Thus it can be argued that the context and skills of EIA and geomorphology by and large do not coincide.

This chapter develops the above argument by considering the environmental impact of broadscale, dryland agriculture in Australia, largely because aspects of its spatial scale and nature are particularly relevant to applied geomorphology. In a sense, the question then becomes not one of why geomorphologists are not playing a role in understanding, evaluating, predicting and even managing the effects of agricultural practices on environmental processes – which they clearly are – but whether EIA is appropriate in such a context.

Dryland agricultural practices set in train a very complex array of direct and indirect effects on the environment. Many of these effects take place in the soil or affect the soil, with the driving processes being pedological, hydrological and geomorphological in nature. Clearing of pre-existing vegetation in order to plant crops or pasture affects the way in which rainfall reaches the soil surface, by modifying interception, canopy drip (throughfall), stemflow, storage in the biomass, drop impact, and evapotranspiration. Pedo-geomorphic processes of rainwash (rainsplash plus overland flow), infiltration and throughflow, and the translocation of soil materials by those processes and mass movements, are also influenced by vegetative changes as well as by the range of dryland agricultural practices. These practices include ploughing and seedbed preparation, weeding, seeding, fertilizing, spraying and harvesting as well as soil disturbances caused by the movements and grazing activities of stock. Much excellent geomorphology has been done on aspects of these impacts in many countries, especially erosion by water and wind.

Table 18.1   Extent of the secondary, dryland soil salinity problem in Australia

State/territory	Area salt-affected in 1982 (ha)	Area salt-affected in 1996 (ha)	Potential area at equilibrium (ha)
Western Australia	264 000	1 804 000	6 109 000
South Australia	55 000	402 000	892 000
Victoria	90 000	120 000	unknown
New South Wales	4000	120 000	5 000 000
Tasmania	5000	20 000	unknown
Queensland	8000	10 000*	74 000
Northern Territory	unknown	minor	unknown
TOTAL	426 000	2 476 000	>12 075 000

Source: Anon. (1996; table 1.1; EPA SA 1998, p. xvi)
* Only severe salinity

18.2   THE EXAMPLE OF SECONDARY SALINIZATION IN THE WESTERN AUSTRALIAN WHEATBELT

Secondary, dryland salinization of soils and water in Australia is one of the environmental consequences of dryland agricultural practices (Table 18.1). The problem is defined as having developed following the widespread adoption of annual crops and pastures, and excludes areas of primary salinity (such as salt lakes) which pre-date European settlement and land use, and areas under irrigation. The Western Australian wheatbelt is the area worst affected, with more than 11% of the area cleared for agriculture rendered unproductive. That percentage is predicted to increase to over 30% by 'equilibrium' (Table 18.1), which could occur within ten years in higher rainfall western parts of the wheatbelt, but possibly not for another 200 years in the eastern wheatbelt (Anon. 1996a). Not surprisingly, the problem is already having profound environmental, economic and social impacts. The mechanisms which are responsible for the problem show the need for an appreciation of both geomorphology and hydrology. Intensely and deeply weathered soils, deep and shallow aquifers, overland flow and streamflow play crucial roles and are illustrated by data from the Western Australian wheatbelt.

Two examples come from work by Richard George, a geomorphologist employed as a research officer with the Western Australian agricultural agency and also convener of the State Salinity Plan Working Group responsible for producing the 1996 'Situation Statement' on salinity (Anon. 1996a). The first example, from a small (12 ha), intensively instrumented catchment near Narrogin (mean annual rainfall 520 mm) in the Western Australian wheatbelt, shows that a combination of overland flow, throughflow, return flow and seepage from a deeper aquifer have all contributed to the formation of a small, saline seep at the base of the catchment (Figures 18.1, 18.2 and 18.3; Table 18.2) (George 1991; George and Conacher 1992a, b). Although throughflow contributes a significant amount (about 50%) of water to the seep, it is relatively fresh (about 160 mg l^{-1} chlorides). Most (up to 98%) of the soluble salts are derived from the deeper aquifer (with chloride concentrations averaging 6000 mg l^{-1}), which is present in 'saprolite grits' below the kaolinized 'pallid zone' of the deep, intensely weathered, 'lateritic' soils. Mixing of the two aquifers causes waterlogging at the seep and spreads the soluble salts (predominantly sodium chloride) through the wet soils, the combination of which kills the vegetation. A dolerite dyke intruded into the granitic basement partly delimits the catchment.

Figure 18.1 Narrogin catchment shown in Figure 18.2. View across the stream at the base, facing up the catchment. Instrumentation at the gauging weir located in Figure 18.2 is visible, as are some piezometers. The subdued topography of this low energy landscape is clearly evident. Photograph taken by the author

The second example, from a drier part (mean annual rainfall 355 mm) of the Western Australian wheatbelt near Belka, is more complicated (George 1991, 1992a; George and Conacher 1993). A small, hillside seep with relatively low salinity concentrations is caused by the emergence at the soil surface of throughflow. The throughflow aquifer is perched at the base of relatively deep (<8 m), wind-derived sandplain materials above the clayey, 'mottled zone' of the deep-weathered profile (Figure 18.4). Such 'sandplain seeps' are responsible for up to 60% of the secondary soil salinity problem in the eastern wheatbelt (and for 10% of the problem over the entire region). However, in this instance the seep also comprises a *recharge* zone to the deeper, regional aquifer in the saprolite grits (Figure 18.5). The saline (<9000 mg l^{-1}; elsewhere >30 000 mg l^{-1} total soluble salts) groundwater in this deep aquifer moves beneath the topographic interfluve backing the sandplain seep, emerging at the surface in one of the broad valley systems which characterize much of the wheatbelt. Prior salt lakes (areas of primary salinity) are often present in these valley systems, as at this location, where a geological fault delimits the edge of two small lakes. However, throughout the wheatbelt these broad valley floors are becoming extensively salinized and waterlogged following the removal of the native vegetation by Europeans for agriculture (secondary salinity).

George's calculations from intensive monitoring here and elsewhere (George 1992b) show that recharge of the deep, saline, regional aquifer in the drier (<350 mm annual rainfall) parts

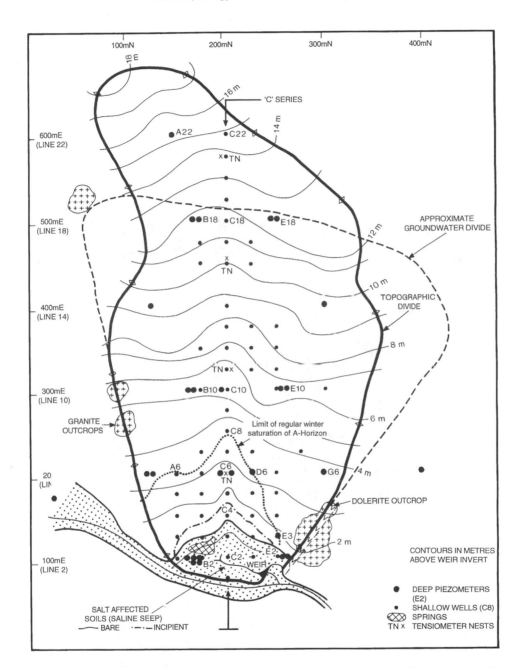

Figure 18.2 Map of George's 12 ha experimental catchment near Narrogin, Western Australian wheatbelt, showing instrumentation, contours and salt-affected area. Modified from George (1991)

322

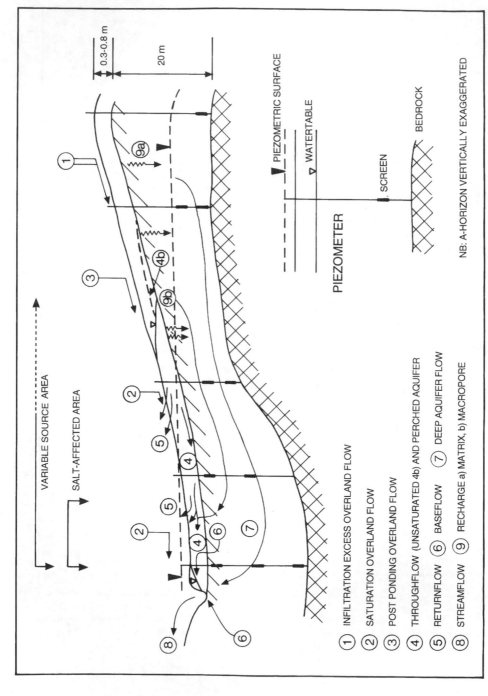

Figure 18.3 Longitudinal section down the Narrogin site (Figure 18.2), showing the hydrological components of the catchment. Modified from George (1991)

Table 18.2   The contribution of various mechanisms of water movement to run-off from a salt-affected area at the base of a small hillslope near Narrogin, Western Australia: data from nine intensively monitored rainfall events

Storm no.	Run-off generation mechanisms					Total flow (m³)	Duration days	Rainfall (mm)
	Throughflow (%)	Returnflow (%)	Saturation (%)*	Overland (%)**	Baseflow (%)***			
Winter								
3	60	(90)	35	–	5	410	6.9	13.0
8	49	(69)	38	–	13	28	1.9	3.8
9	41	(67)	39	–	20	18	1.3	2.8
10	52	(73)	38	–	10	42	1.8	7.1
11	48	(82)	40	–	12	52	2.0	8.5
13	62	(91)	32	–	6	427	9.0	30.0
Mean	52	(78)	37	–	11	163	3.8	10.8
Summer								
1	18	(16)	15	67	n.a.	60	4.0	18.0
14	15	(11)	17	68	n.a.	11	0.4	6.0
17	13	(5)	17	70	n.a.	12	0.2	6.0
Mean	15	(11)	16	69	n.a.	28	1.6	10.0

From George and Conacher (1992b; table I)
* Saturation overland flow induced from the variable source area (including the saline seep).
** Overland flow (infiltration excess) is derived solely from outside the area affected by the shallow aquifer.
*** Winter baseflow rate is 3 m³ day⁻¹. No assessable (n.a.) amount of baseflow occurs in summer.
Values in parentheses for returnflow are included in the throughflow percentages.

of the wheatbelt now varies from 6 to 15 mm a^{-1}, but with much higher rates (20–60 mm a^{-1}) beneath sandplain seeps. In contrast, the rate of horizontal displacement of the groundwater in the deep aquifer is only 0.05–0.3 mm a^{-1}, and was presumably in equilibrium with recharge under the native vegetation before clearing. Thus water is now entering the regional aquifer more rapidly than it can discharge from the system. As a result, groundwater tables in these eastern wheatbelt areas are rising at rates of up to 0.3 m a^{-1}, and more rapidly (up to 1.5 m a^{-1}) in higher rainfall areas (>500 mm a^{-1}) (Anon. 1996a). As at the Narrogin site, these deeper groundwaters are or will become responsible for most of the salts affecting the salinized soils and also a significant proportion of the water. However, the role of the perched aquifers in contributing water and mixing with more saline groundwaters needs to be kept in mind, as well as surface water which periodically moves down the valley systems during flood events (Figure 18.6).

18.3   THE RELEVANCE OF GEOMORPHOLOGY TO THE ENVIRONMENTAL PROBLEM OF SECONDARY SALINIZATION

Geomorphology is relevant both for understanding the processes and for correctly 'reading' landscapes in order to predict the future occurrence of secondary salinity. In this latter context, landscape interpretation is being assisted by remote sensing, both in the traditional

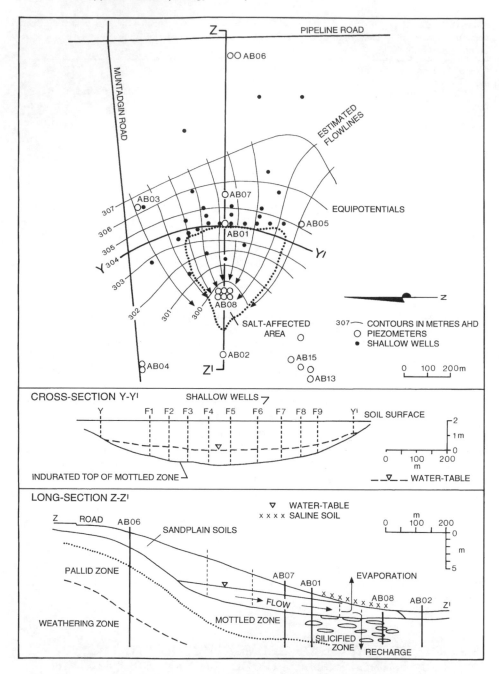

Figure 18.4 Sandplain seep at Belka (location of the seep is at 'AB PUMP' in Figure 18.5(A), showing plan view, groundwater contours, instrumentation and locations of the sections (top), cross-section Y–Y' through the seep (middle) and longitudinal section Z–Z' (base). Both the plan and the longitudinal section show the salt-affected area, with the middle and lower sections also showing the throughflow aquifer at the base of the sandplain soils, perched on the mottled zone of the deep, intensely weathered soil. Modified from George (1991)

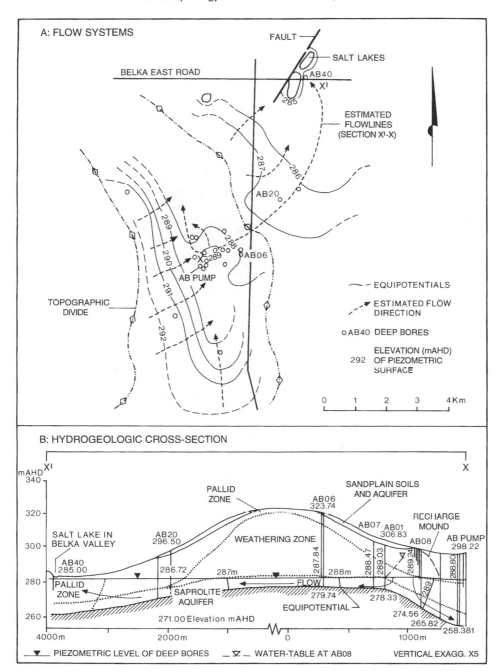

Figure 18.5 Relationship between the perched, sandplain seep shown in Figure 18.4 and the regional, groundwater aquifer at Belka. (A) Plan view with groundwater contours, topographic divides and instrumentation; (B) Cross-section from X at the sandplain seep to X' at the salt lakes in the salt-affected, major valley floor of the Belka valley. The sandplain seep recharging the deep aquifer is indicated by the recharge mound; the lack of coincidence of the regional groundwater divide (beneath the sandplain seep) with the topographic interfluve is also shown. Bedrock at both sites (Narrogin and Belka) is part of the deeply and intensely weathered granitic shield of the Yilgarn block. Modified from George (1991)

Figure 18.6 Flooding of a broad valley system east of Meckering in the Western Australian wheatbelt. Low gradients result in widespread flooding following storm events. These valley floors are often severely salt-affected: they are underlain by highly saline (<100 000 mg l⁻¹) groundwaters approaching either the capillary fringe or the surface following clearing of the natural vegetation, although naturally occurring salt lakes are also present in some of the valleys. Salts accumulate at the surface during summer and are flushed downvalley by events such as that shown here, thereby contributing salts (and water) to sites downvalley. The shrubby vegetation seen here is halophytic. Photograph taken by the author

imaging sense and by the use of airborne sensors to measure electrical conductivity and magnetic anomalies, often used in association with a geographical information system (GIS). Airborne sensing indicates that geological structures influence the movement of groundwater in deeper aquifers. These influences occur in two forms: first, the presence of 'shatter zones' (sometimes identified as faults), which appear to direct the movement of groundwater in the deeper, weathered aquifers over distances of tens of kilometres; and second, the presence of rock barriers, often in the form of dolerite dykes intruded into the granitic basement, which form barriers to the lateral displacement of groundwater, sometimes forcing it to the surface. Granite 'highs', reflecting the very uneven weathering base of the deeply weathered soils, are also important. Thus surface topography may be a poor or even irrelevant indicator of the direction of groundwater movements, and groundwater catchments may bear no relation to surface catchments. Thus geohydrology (or hydro-geomorphology) rather than 'pure' geomorphology (if there ever was such a thing) is involved.

18.4 IMPLICATIONS FOR ENVIRONMENTAL IMPACT ASSESSMENT

There are several implications arising from the salinity work in relation to the role of geomorphology in environmental impact assessment. They include the inadequacy of most current EIA methodologies in the agricultural context for accurately identifying and understanding processes. The environmental impact matrix and its numerous derivatives, for example, are particularly inappropriate. As shown in the preceding discussion, secondary salinization in dryland agricultural areas, as is undoubtedly true for most impacts of agricultural practices on the environment, is not a simple cause–effect situation. Other EIA methods, such as overlay maps, networks, systems diagrams and analysis, simulation, mathematical modelling and ecological analyses, *may* be more relevant than matrices and check lists for analysing processes (modified from Harvey 1998). Geomorphologists have certainly made important contributions to systems analysis (since Chorley and Kennedy 1971) and some of the mathematical approaches to understanding processes (since Scheidegger 1961), in particular. But none of these geomorphologists or their colleagues or more recent counterparts appear to have contributed to the development of EIA methodologies *per se*, as already noted from the absence of geomorphologists from the list of references in the EIA text by Thomas (1996). Further, the salinity research outlined above requires a considerable period of field-based measurements. Hydrological data, in particular, need to be obtained over several seasons. EIA, on the other hand, is usually carried out by consultants operating within tight time constraints, often a matter of weeks, and therefore relies almost entirely on compiling existing information. Undertaking a time-consuming, field-based research programme is not normally feasible in the EIA context.

Second, EIA is project-specific and largely irrelevant to the types of regional problems of land degradation posed by agriculture – not only secondary salinization, as above, but accelerated erosion by water and wind, loss of soil structure, soil acidification, depletion of nutrients, accumulation of fertilizer and pesticide residues, eutrophication and sedimentation of waterways, and so on (Conacher and Conacher 1995). Even cumulative impact assessment (Sonntag *et al.* 1987) is difficult to apply practically in these regional contexts. It is also difficult to envisage how the currently fashionable strategic environmental assessment (Buckley 1997) can be applied in practice to the regional, agricultural situation – with exceptions in all cases for large-scale, new proposals such as major irrigation projects or the conversion of tropical forests to agriculture. These types of inadequacies of EIA have been discussed by Marshall *et al.* (1985) and many others and more recently by Conacher and Conacher (2000).

18.5 IMPLICATIONS FOR MANAGEMENT

A further implication concerns management. Accurate identification and quantification of processes are essential if effective remedial works are to be developed. Again with reference to the secondary salinity problem, for many years the standard remedial measures recommended to farmers were to fence off the affected area (in order to control grazing by stock) and establish salt-tolerant vegetation. This approach offers the farmer some tangible reassurance that something is being done about the salt problem. Additionally, it may improve the aesthetics of the site, reduce erosion by water and wind, and provide at best some fodder for carefully controlled light grazing. But it does not and cannot deal with the causes of the problem. As has been seen, the causes are related to land-use practices in the catchments of the salt-affected areas, connected to the latter by hydro-geomorphic processes. Thus it is in the catchments that the processes must be manipulated by modifying land-use practices in order to rehabilitate the salt-affected areas. But as noted above, groundwater catchments may not coincide with the catchments of the surface drainage network. The ability of EIA to provide the necessary information, context, managerial recommendations, infrastructure and funding is questionable, to say the least.

Manipulation of land-use practices and water movements takes place by essentially engineering-type approaches (Figure 18.7) and agronomic measures (Figure 18.8), both designed to control water movements on the slopes. Transpiration, deep infiltration, through-flow and overland flow are the main water movements targeted. In turn, if manipulation is successful, the depth to groundwater tables from the soil surface in salt-affected areas at the base of the slopes is increased, mixing of saline groundwaters with relatively fresh perched throughflow waters reduced or prevented, translocation of saline water to the surface by capillarity and suction reduced, accumulation of soluble salts in the root zone prevented, and leaching of the soil by fresh rainwater in the same zone enhanced – resulting in complete rehabilitation (Figure 18.9).

However, many of these solutions are beyond the capacity of individual farmers, who lack the required skills, knowledge and resources, not to mention funds. In Australia (and Western Australia in particular, which has produced an Action Plan to deal with the problem of secondary salinity: Anon. 1996b), management is increasingly taking place at a catchment or even a regional level. These Integrated Catchment Management groups (or Landcare or Land Conservation Districts, to name some of the alternatives) comprise farmers working together to deal with a common problem, with the active assistance of local government and the state's agricultural agency, supported by federal financing. Increasingly there is also a regional planning element (for example, State Planning Commission 1994). Although these approaches are all subject to a number of difficulties of varying degrees of severity, these kinds of integrated, cooperative solutions are nevertheless essential in order to overcome the serious limitations of the narrowly focused EIA approach to environmental problem solving.

Figure 18.7 An engineering approach to manipulation of water movements in catchments of salt-affected land. This farmer is constructing interceptor banks to prevent throughflow from reaching the salt-affected land at and below the area with the dead trees. Soils in the wheatbelt typically have sand-textured surface horizons with relatively high clay-sized content subsoils. These interceptors are excavated to the subsoil, with the clayey materials pushed up against the bank on the downslope side in order to prevent the intercepted throughflow from leaking through the bank. This approach has had varied success and is largely ineffective where deeper aquifers contribute water and salts to the salt-affected areas (as is indeed the case at this site), although preventing the mixing of relatively fresh throughflow with deeper, saline groundwaters often leads to some improvement by reducing the spatial extent of waterlogging. Photograph taken by the author

330

Figure 18.8 An agronomic approach to manipulation of water movements in catchments of salt-affected land. This is an experimental site near Narrogin where CSIRO conducted research in order to determine the efficacy of planting eucalypts upslope from a salt-affected site (between the gauge and the trees) in order to lower the groundwater table and dewater the site. The research was also designed to identify the transpiration characteristics of various species. It is ironic that the cause of secondary salinization is too much water, in a semi-arid environment where the major limiting factor to crop and pasture growth is lack of water. Photograph taken by the author

(a)

(b)

Figure 18.9   Saline seep dewatered by tree planting in the catchment of the seep near Bridgetown in the relatively high rainfall (about 1000 mm mean annual rainfall) south western region of Western Australia. (A) The recently dewatered seep (with sedges), photographed in June 1996, in the foreground, with eucalypts (*Eucalyptus globulus*) in the background (also in B). The trees cover 30% of the seep's catchment. Initially, the farmer could graze sheep between the rows (B); but at the stage shown (age of trees ten years) this was no longer possible. On the other hand, the farmer was already thinning some of the trees and obtaining some income – but insufficient to compensate for the loss of grazing. It seems likely that trees planted at lower densities – figures of 10% have been mentioned – will not be successful. It is also probable that a combination of engineering (including drainage and possibly pumping) and agronomic approaches will be more successful than either method alone. Photograph taken by the author

REFERENCES

Anon. 1996a. *Salinity: a situation statement for Western Australia. A report to the Minister for Primary Industry, Minister for the Environment.* Prepared by the Chief Executive Officers of Agriculture Western Australia, Department of Conservation and Land Management, Department of Environmental Protection, and the Water and Rivers Commission, Government of Western Australia, Perth.

Anon. 1996b. *Western Australian salinity action plan.* Prepared by the Chief Executive Officers of Agriculture Western Australia, Department of Conservation and Land Management, Department of Environmental Protection, and the Water and Rivers Commission, for the Government of Western Australia, Perth.

Buckley, R. 1997. Strategic environmental assessment. *Environmental and Planning Law Journal*, **14**, 174–180.

Chorley, R.J. and Kennedy, B.A. 1971. *Physical Geography: a Systems Approach.* Prentice Hall, London.

Conacher, A.J. 1988. Resource development and environmental stress: environmental impact assessment and beyond in Australia and Canada. *Geoforum*, **19**, 339–352.

Conacher, A.J. and Conacher, J.L. 1995. *Rural Land Degradation in Australia.* Oxford University Press, Melbourne.

Conacher, A.J. and Conacher, J.L. 2000. *Environmental Planning and Management in Australia.* Oxford University Press, Melbourne, Chapter 11.

EPA SA, 1998. *State of the environment report for South Australia 1998.* Environment Protection Authority in co-operation with the Department for Environment, Heritage and Aboriginal Affairs, Natural Resources Council, Adelaide.

George, R.J. 1991. *Interactions between perched and deeper groundwater systems in relation to secondary, dryland salinity in the Western Australian wheatbelt: processes and management options.* PhD Thesis in Geography, University of Western Australia.

George, R.J. 1992a. Groundwater processes, sandplain seeps and interactions with perched aquifer systems. *Journal of Hydrology*, **134**, 247–271.

George, R.J. 1992b. Hydraulic properties of groundwater systems in the saprolite and sediments of the wheatbelt, Western Australia. *Journal of Hydrology*, **130**, 251–278.

George, R.J. and Conacher, A.J. 1992a. Interactions between perched and saprolite aquifers on a small, salt-affected and deeply weathered hillslope. *Earth Surface Processes & Landforms*, **18**, 91–108.

George, R.J. and Conacher, A.J. 1992b. Mechanisms responsible for streamflow generation on a small, salt-affected and deeply weathered hillslope. *Earth Surface Processes and Landforms*, **18**, 291–309.

George, R.J. and Conacher, A.J. 1993. The hydrology of shallow and deep aquifers in relation to secondary soil salinisation in southwestern Australia. *Geografia Fisica e Dinamica Quaternaria*, **16**, 47–64.

Harvey, N. 1998. *Environmental Impact Assessment: Procedures, Practice, and Prospects in Australia.* Oxford University Press, Melbourne.

Leopold, L.B., Clarke, F.E., Hanshaw, B.B. and Balsley, J.R. 1971. *A procedure for evaluating environmental impact.* Geological Survey Circular 645, Geological Survey, United States Department of the Interior, Washington DC.

Marshall, D., Sadler, B., Sector, J. and Wiebe, J. 1985. *Environmental management and impact assessment: some lessons and guidance from Canadian and international experience.* Federal Environmental Assessment Review Office Occasional Paper, Hull, Canada.

Munn, R.E. (ed.) 1979. *Environmental Impact Assessment* (2nd edition). SCOPE 5, Wiley, Chichester.

PADC Environmental Impact Assessment and Planning Unit, 1983. *Environmental Impact Assessment.* Martinus Nijhoff Publishers, The Hague.

Scheidegger, A.E. 1961. *Theoretical Geomorphology.* Springer-Verlag, Berlin.

Sonntag, N.C., Everitt, R.R., Rattie, L.P. and others 1987. *Cumulative Effects Assessment: a context for further research and development.* Background Paper prepared for the Canadian Environmental Assessment Research Council, Minister of Supply and Services Canada, Ottawa.

State Planning Commission, 1994. *Albany Regional Strategy*. Adopted by the State Planning Commission as a Basis for Coordination of Local Planning, Control of Subdivision and Advice to the Hon. Minister and other Government Agencies, Perth.

Thomas, I. 1996. *Environmental Impact Assessment in Australia: Theory and Practice*. The Federation Press, Sydney.

19 Contribution of Quantitative Geomorphic Analysis to the Evaluation of Geomorphological Hazards: Case Study in Italy

MAURIZIO DEL MONTE, PAOLA FREDI, ELVIDIO LUPIA PALMIERI AND ROBERTA MARINI
Dipartimento di Scienze della Terra, Università 'La Sapienza', Rome, Italy

ABSTRACT

The aim of this research is to establish a methodology which can aid the objective evaluation of geomorphological hazards. The study was carried out considering three Italian drainage basins which are representative of the different environmental conditions of central-southern Italy and are affected by geomorphological instability to different extents.

First, the main predicted *factors of instability* in the study areas were analysed through precise physiographic investigations (about lithological and tectonic characters, topography, drainage extent, etc.) supported by the examination of human impacts (quarrying, farming, urbanization, etc.). To make the analysis of the morphological factors of instability as unbiased as possible, some *morphometric parameters* were also calculated. They express the main geometric and morphodynamic characteristics of drainage basins that can contribute to the occurrence of potentially hazardous processes. Then, the most important morphological effects of hazardous *processes* acting in the sample basins were studied by means of both detailed geomorphological field survey and aerial photo interpretation as well as through the collection of indirect and direct data about denudation. The results of these investigations were shown through a series of thematic maps.

Successively, maps of hazardous processes were overlain on each thematic map of the predicted factors of instability. By evaluating the spatial join of hazardous processes and the attributes shown in each thematic map, it was pointed out that *lithologies, land use, slope gradient* and *drainage density* provide the best explanation of the occurrence of hazardous processes. Map overlay also allowed the weight of each category of the four factors to be determined. In order to evaluate the effects produced by the concomitant presence of the four selected factors of instability, the four thematic maps were overlain and the attributes of the different categories were combined. In this way *I* areas were obtained and some equations were derived which allowed the compilation of maps of hazard relevant to each process.

An attempt was made to obtain synthetic maps which would show for each basin the global hazard resulting from the destabilizing processes observed.

The proposed methodology is open to any improvement of the definition level of areas with different geomorphological hazards; therefore, it is intended to be a preliminary contribution of quantitative geomorphology to the unbiased evaluation of hazardous denudational processes, both present and potential, in a given territory.

Applied Geomorphology: Theory and Practice. Edited by R.J. Allison.
© 2002 John Wiley & Sons, Ltd.

19.1   INTRODUCTION

The last few decades have witnessed a growing interest among many authors in the study of geomorphological hazards, that is to say, the probability that certain geomorphological events of given intensities and capable of producing more or less appreciable damage will take place in given localities, even though the exact time of their occurrence is difficult to determine. These studies have a twofold aim: to identify areas in which hazardous geomorphological events are likely to occur; and to devise appropriate methods for risk reduction (Skempton and Delory 1957; Yong *et al.* 1977; Brunsden and Prior 1984; Einstein 1988; Embleton *et al.* 1989; Panizza and Piacente 1993; Reading 1993; Fell 1994; Allison 1996).

Considering the scientific and practical importance of such studies, the objective assessment of geomorphological hazards is desirable. Many preconditions of hazardous events can be translated parametrically through quantitative geomorphic analysis. The aim of this study is to examine the possible contribution of such analysis to the evaluation of geomorphological hazards in areas with a high susceptibility to destabilizing processes.

To this end three drainage basins in Italy were examined. They are representative of the various environmental conditions of central-southern Italy and are prone to marked and fast morphological changes.

The three sample basins are drained respectively by the rivers Orcia (Tuscany), Mignone (Latium) and Trionto (Calabria). The first two basins are in central Italy on the Tyrrhenian side of the Apennine chain, and extend respectively for 339.2 and 494.8 km^2. The third is located in southern Italy, extends for 292.5 km^2 and flows to the Ionian Sea (Figure 19.1).

The climatic conditions of the three study areas are quite different and reflect their different geographical locations.

In the Orcia basin the mean annual rainfall is 795.0 mm. The headwater receives most rainfall (1178 mm). Rainfall decreases towards the lower part of the basin where the lowest mean annual values are recorded (617 mm). Mean monthly values range from 106.0 mm in November to 31.0 mm in July. In this basin rainfall on consecutive days is frequent and sometimes exceeds 250 mm in five days. The mean annual temperature is 13.4°C; the maximum is registered in July (23.0°C) and the minimum in January (4.2°C).

Mean annual precipitation on the Mignone basin is higher (1035 mm). From the headwater to the lower basin, mean annual rainfall decreases progressively, so that close to the river mouth mean values (693 mm) are half that of the mountain zones (1405 mm). The wettest month is November (148 mm) and the minimum precipitation value is registered in July (17 mm). Consecutive days of rain are frequent and values higher than 250 mm are often recorded in five days. The mean annual temperature is 14.8°C. Monthly mean temperature reaches a maximum in July (23.5°C) and minimum in January (7.1°C).

The Trionto drainage basin shows the highest mean annual precipitation (1197.9 mm), due to its higher mean altitude. The headwater receives most rainfall (1609 mm); on the coastal plain mean annual rainfall is 915 mm. Two maxima are recorded in the mean monthly precipitation: the main one is in December (185.4 mm) and the secondary one in October (175.6 mm). The minimum precipitation value is registered in July (11 mm). Rainfall on consecutive days is very frequent and may exceed 450 mm in five days. Mean annual temperature is 13.4°C; the hottest month is August (22.9°C) and the coldest one is January (4.8°C).

For each of the chosen basins the main predicted factors of instability – both natural and human-induced – were examined through geological and land-use studies paralleled by morphometric analysis; as a result some thematic maps were obtained. Moreover, the most important destabilizing processes affecting the areas considered were identified and mapped

Figure 19.1 Geographical location of the study basins. 1, Fiume Orcia drainage basin; 2, Fiume Mignone drainage basin; 3, Fiume Trionto drainage basin

by studying their morphological effects, as revealed by aerial photo interpretation and field survey. In some cases, experimental measurements were also made in order to quantify the intensity of current morphogenetic processes.

Map overlay procedure allowed the selection of the most significant factors of instability in the study areas and consequently the drafting of maps of geomorphological hazards. The methodology followed is shown schematically in the flow diagram of Figure 19.2; the separate steps will be described in the following paragraphs.

19.2   ANALYSIS OF PREDICTED FACTORS OF INSTABILITY

19.2.1   Geological Conditions

The geological study, based on data available in existing literature supported by aerial photo interpretation and field survey, indicated different and complex geological arrangements in the three basins considered. The different categories of outcropping lithologies were examined in detail and then mapped on suitable maps (Figure 19.3).

The Orcia drainage basin is emplaced in a graben which originated during the Lower Pliocene as a result of an extensional tectonic phase following the Miocene compressive phase responsible for the emplacement of the Apennine folding (Lazzarotto 1993).

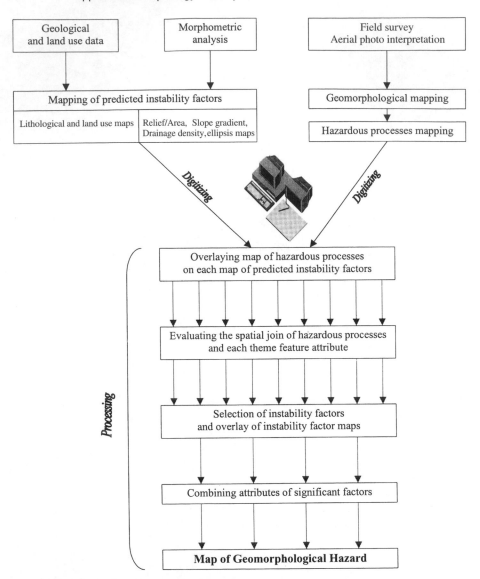

Figure 19.2 Flow diagram showing the various phases of the performed methodology

In structurally low areas the marine clays, sands and conglomerates of the neoauto-chthonous complex crop out (Pliocene). Structurally high positions are occupied by allo-chthonous mainly argillaceous and marly flysch (Lower Cretaceous–Paleocene) and by autochthonous carbonate rocks (Triassic–Eocene). In the southernmost part of the basin the trachybasaltic–andesitic neck of Radicofani is present. Finally, terraced and present alluvial deposits outcrop on the valley bottoms of Fiume Orcia and its tributaries (Figure 19.3a).

The geological arrangement of the Mignone basin is more complex. It is the result of different compressive and tensile tectonic phases of the general evolution of the

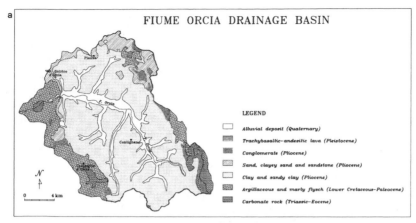

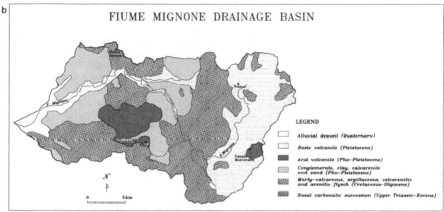

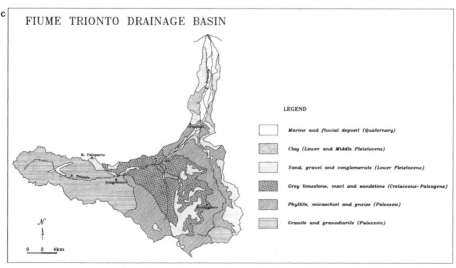

Figure 19.3 Lithological sketches of the study basins

central-northern Apennine belt and of its Tyrrhenian margin in particular (Conti and Corda 1985).

The most ancient outcropping rocks are the strongly tectonized limestones, dolomites and marls of the basal carbonatic succession belonging to Upper Triassic–Eocene. This succession is overlain, in tectonic contact, by the marly-calcareous, argillaceous, calcarenitic and arenitic flysch, Cretaceous–Oligocene in age, which crop out extensively from the coastal area as far as the upper part of the basin. After tensile tectonic events, the neoautochthonous complex was emplaced in the downthrown areas. This complex consists of conglomerates and marine clays, calcarenites and sands of Plio-Pleistocene age. In the upper and intermediate part of the basin these deposits are covered by the acid and basic products of the volcanoes of Latium (2–0.15 Ma). Finally, fluvial deposits – terraced in places – crop out along the valley bottom of the Fiume Mignone (Figure 19.3b).

The Trionto drainage basin is in an area with varying lithologies and a complex tectonic history (Gruppo Nazionale Geografia Fisica e Geomorfologia 1995).

In the upper and intermediate parts magmatic and low grade metamorphic rocks of Palaeozoic age crop out. They are overlain by Mesozoic transgressive sediments changing from fluvial deposits to pelagic turbidites in connection with local subsidence. These sediments are arranged in NE verging imbricated thrusts as a result of the compressive tectonic phase they underwent during Cretaceous–Palaeogene. Upward is an argillaceous-marly formation interbedded with thick clayey layers, Upper Miocene in age. It is slightly folded in consequence of a compressive tectonic phase, which took place after the Messinian. The following Quaternary tectonic phase was tensile and related to the uplift started at the end of the Pleistocene and still in progress. Pleistocene deposits are sands, gravels and conglomerates (Lower Pleistocene) overlain by marine clays (Lower and Middle Pleistocene) and marine and fluvial terraced deposits (Figure 19.3c).

19.2.2 Land Use

The important role of land use in geomorphological dynamics is very well known (De Graff and Canuti 1988; Nicolau *et al.* 1996; Parsons *et al.* 1996). Therefore this aspect was accurately examined and mapped (Figure 19.4). The results showed that the natural arrangement of the three basins was widely modified by human activity (CNR 1956; Regione Toscana 1985; Avena 1985; Dimase and Iovino 1988).

In the Orcia drainage basin human activity has involved the deforestation of wide areas and their conversion to farming and urbanization. At present most of the basin area is cropland and orchard while forests and reafforestation practice are confined to the most inaccessible slopes. Urbanized areas mainly consist of small rural centres, but mineral extraction sites are widespread along the river-beds of Orcia and of its major tributaries (Figure 19.4a).

The higher inclination of slopes has favoured the preservation of wider forest-lands in the Mignone basin. However, human impact on the natural environment is still evident and cropland and orchard are widespread. In particular, the need for a dense network of irrigation canals has markedly modified the surface drainage in the lower portion of the basin. Urbanized areas are small and mineral extraction sites are found mainly on the outcrops of volcanites, exploited for paving and building stone (Figure 19.4b).

In the Trionto drainage basin cropland and orchard are present in the headwater and in the northern-central area. Abandoned fields represent about one-third of the basin surface, which has strong effects on the intensity of denudation. In this basin, forest lands and reafforestation are also confined to the most inaccessible slopes. Urbanized areas are scant and mineral extraction sites are restricted to river-bed digging. The basin area close to the river

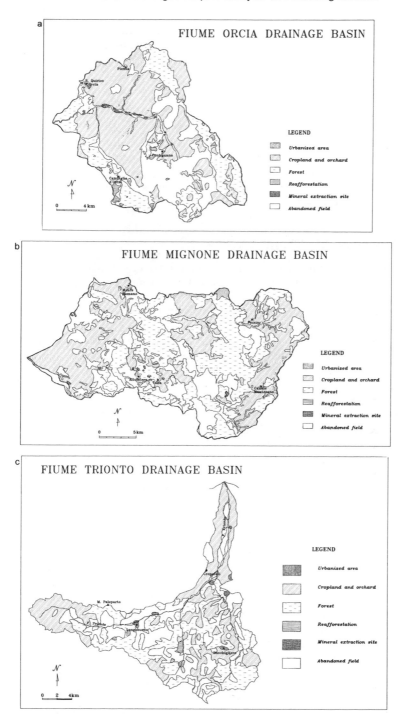

Figure 19.4 Land-use maps of the study basins

Table 19.1 Morphometric parameters of the study basins

	Fuime Orcia	Fiume Mignone	Fiume Trionto
Area (km^2)	339.82	494.78	292.50
Relief/area (m km^{-2})	2.69	1.28	5.64
Slope gradient (%)	14.60	18.48	37.57
Drainage density (km km^{-2})	4.83	3.49	5.65
Density of hierarchical anomaly (ga)	13.62	15.58	32.64
Index of hierarchical anomaly (Δa)	1.07	2.14	2.06
Hypsometric integral ($\int v$)	0.283	0.406	0.468

mouth has suffered from the most intense urbanization, mainly related to tourist facilities at seaside resorts (Figure 19.4c).

19.2.3 Morphometric Characteristics

The main features, which may contribute to the occurrence of potentially hazardous processes in the study basins, were examined and translated into synthetic parameters. The differences in elevation and the degree of surface inclination – expressed by the *relief/area ratio* and *slope gradient* – were accurately calculated. The extent and the degree of organization of drainage networks were also considered and quantified by *drainage density* (*D*: Horton 1945) and some *hierarchical parameters* (*Rb*: Horton 1945; *Rbd, Ga, Δa and ga*: Avena *et al.* 1967; Ciccacci *et al.* 1987). Previous studies have indicated that these parameters are significantly related to the intensity of denudational processes in Italian drainage basins (Ciccacci *et al.* 1981, 1992; Lupia Palmieri *et al.* 1995). Finally, hypsometric analysis was performed after Strahler (1952). It showed that hypsometric integrals are usually low, which testifies to plano-altimetric configurations of the drainage basins strongly dependent on the effectiveness of denudational processes (Ciccacci *et al.* 1992, 1995; Ohmori 1993).

The above parameters were calculated for partial basins or for units of area, depending on their typology. Moreover, the distribution of their different categories throughout the study areas was shown on specific thematic maps.

The mean values of the morphometric variables of the three drainage basins considered are given in Table 19.1.

19.3 ANALYSIS OF HAZARDOUS PROCESSES

This analysis was carried out by surveying the main morphological effects of destabilizing slope processes through aerial photo interpretation and geomorphological field survey. In some cases, direct measurements of denudation were carried out.

All landforms surveyed were classified on the basis of their genesis and mapped at the scale 1:25 000. The interpretation of these maps and the multi-year field controls allowed the identification of those processes which are responsible for the morphological evolution of different areas. Although they differ in intensity and distribution, the potentially hazardous processes which mark the study basins are essentially due to the same exogenous agents, gravity and surface running water being the most important.

On the basis of the prevailing processes which affect their evolution, three kinds of areas were singled out and mapped on simplified sketches (Figure 19.5):

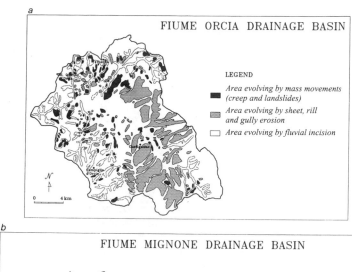

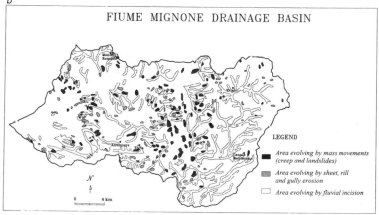

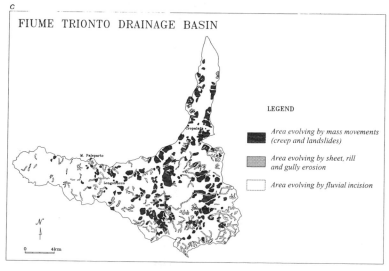

Figure 19.5   Distribution of main hazardous processes in the study basins

- areas evolving by mass movements (*Am*);
- areas evolving by sheet, rill and gully erosion (*Ad*);
- areas evolving by fluvial deepening (*Al*).

The Orcia drainage basin undergoes, on the whole, severe denudational processes, which the lack of continuous vegetation cover does not mitigate.

Most of the basin is affected by widespread erosion due to surface running waters. Rounded and knife-edged badlands are the most dramatic landforms, which characterize the Pliocene argillaceous outcrops. Rounded badlands are predominant in the northern-central part of the basin, with lower angles of slope. In spite of the human attempts to reduce its intensity, badland erosion is particularly severe and morphological changes were observed in the 24–48 hours following intense rainfalls. Gravity effects are evident where higher slope gradients are found; slumps occur on the flysch outcrops in the southwestern part of the basin, whereas creep typically acts, in the middle Orcia valley, on the low-gradient slopes of the clayey outcrops. At places, creep evolves into mudflows, which have sometimes damaged buildings of artistic interest. Erosion by channelled waters is marked as well. Downcutting is particularly evident on flysch lithologies but it is also brought about by the main streams draining areas affected by badland erosion.

The Mignone drainage basin is widely affected by denudational processes, which can cause rapid and hazardous morphological changes. The intensity of such processes is strongly variable, depending on the varying lithologies and slope gradients.

Areal erosion effects are recognizable on flysch and clayey outcrops, but are less frequent on the more resistant volcanites. Mass movements are perhaps the most widespread processes. Slumps and rock falls are typical of the highly tectonized flysch, whereas creep prevails on the clayey Plio-Pleistocene lithologies on which rare earthflows and mudflows have also been observed. Stream deepening is distributed all over the basin. This process is often an important external cause of landslides, but its effect is particularly evident on the volcanites, which are cut by deep and steep-sided valleys.

The Trionto drainage basin is characterized by denudational processes which are particularly intense where metamorphic and sedimentary lithologies crop out. Less intense processes are observed at the basin headwater where igneous intrusive rocks are present.

Surface running waters are important morphogenetic agents all over the basin. However, their action has a particularly marked effect on the metamorphic and pelitic sedimentary outcrops, which are cut by deep and steep-sided valleys, and on the clayey lithologies, often affected by badlands. Mass movements are the most widespread hazardous processes; rotational slides, rock falls and flows are present over most of the basin area. In particular, the northernmost part is widely affected by large flows. Furthermore, creep sometimes affects entire valley slopes.

19.3.1 Denudation Rate

The erosion rate in the study areas was evaluated indirectly through multiple regressions relating mean annual suspended load (t km^{-2} a^{-1}) to the morphometric variables D and Δa, which resulted from the statistical analysis of a large number of Italian drainage basins (Ciccacci *et al.* 1987). The 'erosion index' thus obtained (in t km^{-2} a^{-1}) was then translated into denudation rate (cm a^{-1}).

Values obtained for the whole basins and for their partial basins are representative of average conditions (Table 19.2). To quantify experimentally the maximum intensity of present slope erosion, short-term topographic modifications were also measured. To this end, some

Table 19.2 Mean denudation rates measured in the field

Station	Lithology	Slope gradient (°)	Exposure	Mean denudation rate (cm a^{-1})
Fiume Orcia drainage basin*				
A	Clay	35	South	4.8
B	Clay	40	East	1.3
C	Clay	22	East	3.0
D	Clay with sand and conglomerate interbedding	40	East	3.0
E	Clayey sand	30	East	1.6
F	Clay	55	West	1.0
G	Clay	32	South	1.2
Fuime Mignone drainage basin**				
A	Clay	35	South	3.7
B	Clay	42	South	6.0
C	Clay	40	South	6.0
Fiume Trionto drainage basin***				
A	Clay	25	South	0.3
B	Clay	30	South	6.4
C	Clay	30	South	2.5
D	Clay	40	West	5.4
E	Clay	25	North	2.8
F	Clay with gypsum beds	40	Northwest	4.3
G	Clay with gypsum beds	38	Northeast	3.3

* Denudation index = 1358 t km^{-2} a^{-1}; denudation rate = 0.0543 cm a^{-1}
** Denudation index = 639 t km^{-2} a^{-1}; denudation rate = 0.0256 cm a^{-1}
*** Denudation index = 3176 t km^{-2} a^{-1}; denudation rate = 0.1270 cm a^{-1}

measuring stations were located in areas of the study basins affected by accelerated erosion and having different conditions in terms of slope gradient and exposure. Topographic measures were carried out every two months and for a three-year period, using an electronic theodolite.

In Table 19.2 the mean denudation rates measured at the experimental stations are shown; lithology, gradient and exposure of the considered slopes are also indicated. The measured values of denudation rate must be considered the highest in the study basins as they refer to slopes which suffer from marked and continuous morphological changes due to sheet, rill and gully erosion. The lowest value of mean erosion rate was calculated for the Fiume Mignone basin. In this same basin the highest values of denudation rate were measured at the experimental stations. However, it should be noted that these measurements relate to sites characterized by accelerated erosion, which in the Mignone basin consists of isolated phenomena affecting small areas with very high slope gradients.

19.4 IDENTIFICATION OF THE MAIN FACTORS OF INSTABILITY

In order to identify which of the factors considered (lithology, land use, morphometric variables) plays the most important role in the geomorphological instability of the study area, the map of hazardous processes of each basin was overlain in turn on each of the other

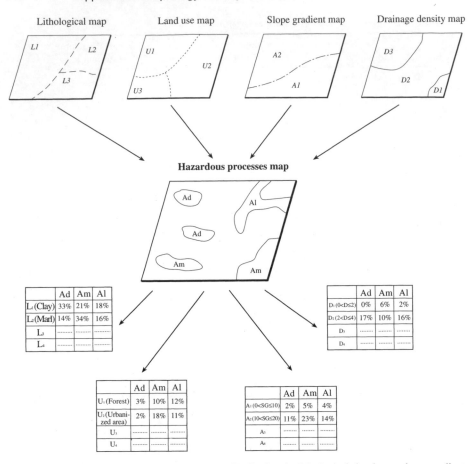

Figure 19.6 Evaluation of hazardous process distribution in lithological, land use, slope gradient and drainage density categories. Spatial joins of the input themes are expressed numerically as percentage ratios between the area of a given category affected by a given type of hazardous processes and the total area of the same category of instability factors

thematic maps. This comparative analysis of cause/effect relationships showed that lithology, land use, slope gradient and drainage density provide the best explanation of the occurrence of hazardous processes (Figure 19.2). Moreover, map overlay allowed the role of each category of the four factors to be weighed in terms of percentage area affected by a given type of process (Figure 19.6, Table 19.3).

As would be expected, the influence of lithology in determining fast morphological changes is particularly strong (cf. Figures 19.3 and 19.5).

Land use has an important role too. In all three study basins, polygon-on-polygon overlay showed that abandoned fields are extensively affected by the action of surface running waters and in particular by sheet, rill and gully erosion. Mass movements generally prevail in urbanized areas and where mineral extraction sites are present. Forest lands are usually the most stable areas (cf. Figures 19.4 and 19.5).

Slope gradient and lithology are linked in determining the predominance of a given destabilizing process (cf. Figures 19.7 and 19.5).

Table 19.3 Percentage area of each category of instability factor affected by the hazardous processes surveyed

Fiume Orcia drainage basin	Ad	Am	Al	Fiume Mignone drainage basin	Ad	Am	Al	Fiume Trionto drainage basin	Ad	Am	Al
Lithology				**Lithology**				**Lithology**			
Alluvial deposits	2	1	5	Alluvial deposit	1	1	2	Alluvial deposit	0	1	0
Trachybasaltic-andesitic lava	0	0	0	Basic volcanite	1	1	23	Clay	28	37	10
Conglomerate	2	11	17	Acid volcanite	2	3	11	Sand, gravel and conglomerate	11	8	13
Sand, clayey sand and sandstone	4	13	18	Conglomerate, clay, calcarenite	4	11	14	Grey limestone, marl and sandstone	0	15	11
Clay and sandy clay	58	14	9	Marly-calcareous, arenitic, flysch	3	15	19	Phyllite, micaschist and gneiss	0	16	11
Argillaceous and marly flysch	5	12	34	Basal carbonatic succession	0	0	16	Granite and granodiorite	1	3	8
Carbonate rock	0	0	5								
Land use				**Land use**				**Land use**			
Urbanized area	0	10	0	Urbanized area	0	4	15	Urbanized area	0	12	0
Cropland and orchard	17	18	15	Cropland and orchard	2	7	10	Cropland and orchard	9	13	6
Forest	1	2	26	Forest	1	2	18	Forest	0	11	8
Reafforestation	3	4	3	Reafforestation	0	0	0	Reafforestation	0	0	2
Mineral extraction site	0	2	0	Mineral extraction site	2	0	0	Mineral extraction site	0	1	0
Abandoned field	67	7	18	Abandoned field	4	11	27	Abandoned field	14	11	8
Slope gradient (%)				**Slope gradient**				**Slope gradient**			
0 < SG ≤ 10	18	16	9	0 < SG ≤ 10	0	1	4	0 < SG ≤ 10	0	5	1
10 < SG ≤ 20	35	16	23	10 < SG ≤ 20	2	5	8	10 < SG ≤ 20	1	5	6
20 < SG ≤ 30	38	4	29	20 < SG ≤ 30	4	12	19	20 < SG ≤ 30	11	12	9
30 < SG ≤ 45	44	4	26	30 < SG ≤ 45	2	8	21	30 < SG ≤ 45	3	12	11
SG > 45	12	4	28	SG > 45	0	0	26	SG > 45	4	14	10
Drainage density (%)				**Drainage density**				**Drainage density**			
0 < D ≤ 2	4	2	6	0 < D ≤ 2	0	2	4	0 < D ≤ 2	0	0	0
2 < D ≤ 4	6	5	15	2 < D ≤ 4	2	2	15	2 < D ≤ 4	3	26	3
4 < D ≤ 6	17	5	9	4 < D ≤ 6	4	5	22	4 < D ≤ 6	2	17	8
6 < D ≤ 8	33	7	5	6 < D ≤ 8	1	14	35	6 < D ≤ 8	13	7	11
D > 8	64	3	1					D > 8	21	7	13

Ad, area affected by sheet, rill and gully erosion
Am, area affected by mass movements
Al, area affected by stream incision

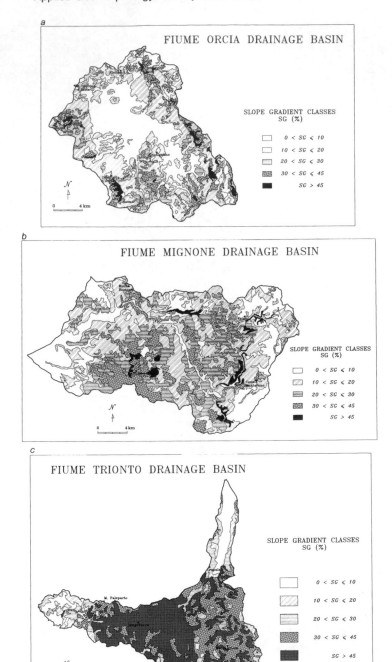

Figure 19.7   Slope gradient (SG) maps of the sample basins

In the Fiume Orcia drainage basin the action of surface running waters becomes progressively stronger up to gradient values of about 30%. Higher slope gradients seem to enhance the effectiveness of sheet, rill and gully erosion rather than favour stream deepening. In addition, where the value exceeds 45%, surface waters have little or no effect, especially where calcareous and quartzarenitic lithologies are present. Creep is widespread where clayey outcrops are associated with low slope gradients (up to 20%), but badlands and slides occur on the same lithologies if slope gradients exceed 20%.

Fluvial incision also increases with slope gradient in the drainage basin of Fiume Mignone and its effects are still evident when the gradient is very high (>45%). Sheet, rill and gully erosion is not widespread. It usually occurs on slopes with gradients ranging from 20% to 30%, which are the values most commonly attained by lithologies particularly prone to this kind of process, such as Plio-Pleistocenic clays and, subordinately, argillaceous flysch. Mass movements are mainly found on the same lithologies but they are connected with the higher slope gradients (up to 45%). In particular, rotational slides occur on the flysch outcrops. Creep and flows, by contrast, are common on the clays. In addition, rock falls are observed on the steep slopes of the volcanite outcrops.

Valley slopes of the Fiume Trionto drainage basin are characterized by greater gradient mean values. The most frequent values actually exceed 45%. Mass movements and fluvial incision are much more marked than sheet, rill and gully erosion. Their intensity increases with the increasing values of slope gradients. In particular, mass movements prevail on very steep slopes of the lower Trionto valley. In this same part of the drainage basin accelerated erosion phenomena are also frequent where clayey lithologies crop out. By contrast, fluvial incision dominates on very steep slopes where metamorphic and marly lithologies crop out.

Drainage density depends on lithology and the variability of its values reflects the variability of process typology (cf. Figures 19.8 and 19.5).

In the Fiume Orcia basin the polygon of lower values of drainage density ($2 < D \leq 4$) overlap on areas where linear incision prevails rather than areal erosion and mass movements. With increasing values of D, together with Pliocene clay outcrops, sheet, rill and gully erosion gradually become predominant: it affects 64% of the area with $D > 8$.

The Mignone basin shows, on the whole, lower mean values of drainage density. Generally speaking, the increase in D values corresponds to a wider spread of destabilizing processes. The lowest D values (<4) are found where intense denudational processes are lacking or at most limited to fluvial incision. Higher values ($4 < D \leq 6$) occur on pelitic lithologies affected mainly by stream deepening and on flysch chiefly dominated by mass movements.

In the Fiume Trionto basin drainage density attains higher mean values. As in the previous cases, low values (D < 4) are scarcely represented and characterize areas affected at most by moderate fluvial incision. Higher values go together with increasing intensity of stream channel deepening, mass movements and accelerated erosion.

19.5 EVALUATION OF GEOMORPHOLOGICAL HAZARDS

The map overlay results allowed us to assess which of the different categories of the selected factors of instability (lithology, land use, slope gradient and drainage density) were more or less likely to bring about a given instability process, depending on the actual occurrence of the same process. It must be underlined that some of the basin characteristics considered as factors of instability, such as the slope gradient, may depend, in turn, on analogous geomorphic processes of the past. However, it is also evident that these same characteristics are potential causes of new destabilizing processes.

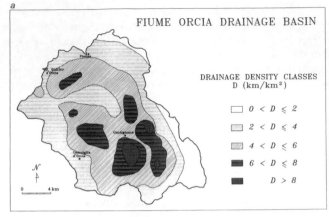

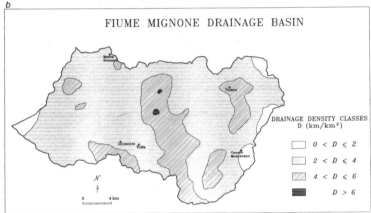

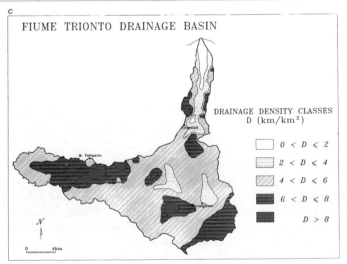

Figure 19.8   Drainage density (D) maps of the sample basins

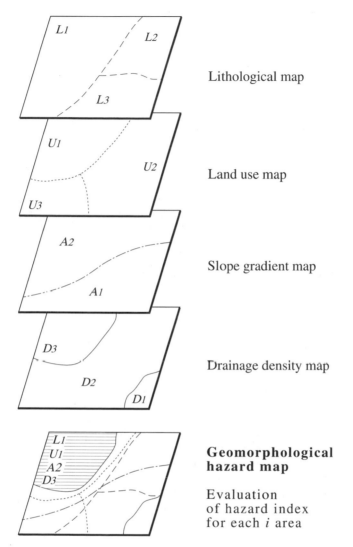

Figure 19.9 Overlay of maps of selected instability factors. The striped area is a hypothetical representation of the concomitant presence of the four given categories of instability factors. The different weight of the four factors is evaluated following the scheme of Figure 19.6

In order to evaluate the effects produced by the concomitant presence of the four factors of instability, the four thematic maps were overlain and the attributes of the different categories were combined (Figure 19.9). In this way I areas were obtained which represent all the possible intersections among the different categories of the factors of instability:

$$a_i = a_{j,k,n,l} = \left[(L_j \bigcap_{j=1}^{J} \bigcap_{k=1}^{K} U_k) \bigcap_{n=1}^{N} A_n \right] \bigcap_{l=1}^{L} D_l$$

where $I = JKNL$ = n. of max possible areas.

On the basis of the present morphodynamic conditions, the following products can be considered expressive of the probability that a given destabilizing process continues or starts in each of the I areas:

$$Pm(a_i) = Am_j/L_j \times Am_k/U_k \times Am_n/A_n \times Am_l/D_l \qquad (19.1)$$
$$Pd(a_i) = Ad_j/L_j \times Ad_k/U_k \times Ad_n/A_n \times Ad_l/D_l \qquad (19.2)$$
$$Pl(a_i) = Al_j/L_j \times Al_k/U_k \times Al_n/A_n \times Al_l/D_l \qquad (19.3)$$

where $Pm(a_i)$ = hazard index for landslide occurrence; $Pd(a_i)$ = hazard index for sheet, rill and gully erosion occurrence; $Pl(a_i)$ = hazard index for stream incision occurrence; Am_j, Am_k, . . . = area affected by mass movements in the classes j, k,. . .; Ad_j, Ad_k, . . . = area affected by sheet, rill and gully erosion in the classes j, k,. . .; Al_j, Al_k, . . . = area affected by stream incision in the classes j, k,. . .; L_j = area covered by the lithology j; U_k = area covered by land use k; A_n = area covered by slope gradient n; D_l = area covered by drainage density l.

Obviously, Pm, Pd and Pl values range between 0 and 1. The lower bound indicates that the area considered has no hazard related to a given typology of processes. The upper bound would indicate, on the contrary, that the same typology of processes is already spread all over the area.

On the basis of Equations 19.1, 19.2 and 19.3, the maps of hazard – in the sense specified above – relevant to each category of processes were obtained for the three study basins. In Figure 19.10, the map of mass movement hazard in the Trionto basin is shown as an example.

19.6   ELABORATION OF MAPS OF SYNTHESIS

Analysis of hazards often has practical implications. Therefore, it can be useful to have a general picture of areas which suffer from higher or lower geomorphological hazards within the same basin. For this reason an attempt was made to produce schematic maps of synthesis, which show the 'global hazard' resulting from the total destabilizing processes observed. To this end, the calculated values of Pm, Pd and Pl were all considered together and subdivided into eight classes. It is worth emphasizing that these classes were not defined a priori but derived from the comparative analysis between causes and effects of instability processes. On this basis, areas with values of Pm, Pd and Pl belonging to the same class interval were graphically unified, since they have the same tendency to undergo one of the processes considered. To these unified areas ($\sum a_i$) was attributed a 'global hazard', which ranged between the minimum and the maximum values of Pm, Pd and Pl relevant to the sample areas a_i.

It should be emphasized that the unification of areas a_i in the synthetic maps of Figures 19.11 to 19.13 was necessary to reduce the number of cartographic representations; but this simplification should be avoided if a more detailed picture is needed.

Apart from the general considerations about variability of hazards in the study basins, the analysis of these maps showed that the four factors of instability considered play roles of varying importance in determining the hazardous processes in the different basins. Therefore, the weight of each factor is more properly determined through field survey and morphometric analysis, while a priori attributions must be avoided.

19.7   CONCLUSION

The comparative analysis of the factors and effects of destabilizing processes considered led to the development of some equations which express the relationship between the occurrence

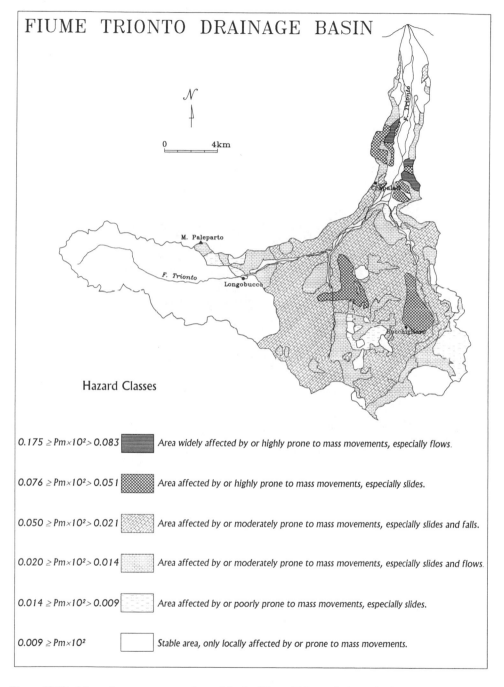

Figure 19.10 Map of mass movement hazard in the Fiume Trionto drainage basin

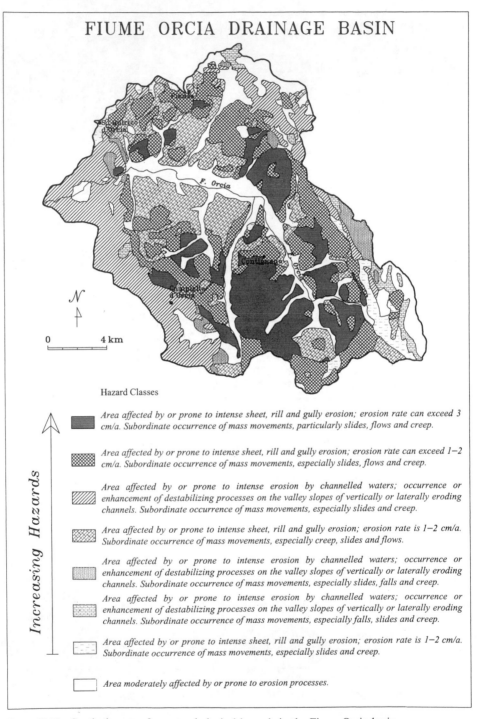

FIUME ORCIA DRAINAGE BASIN

Hazard Classes

Area affected by or prone to intense sheet, rill and gully erosion; erosion rate can exceed 3 cm/a. Subordinate occurrence of mass movements, particularly slides, flows and creep.

Area affected by or prone to intense sheet, rill and gully erosion; erosion rate can exceed 1–2 cm/a. Subordinate occurrence of mass movements, especially slides, flows and creep.

Area affected by or prone to intense erosion by channelled waters; occurrence or enhancement of destabilizing processes on the valley slopes of vertically or laterally eroding channels. Subordinate occurrence of mass movements, especially slides and creep.

Area affected by or prone to intense sheet, rill and gully erosion; erosion rate is 1–2 cm/a. Subordinate occurrence of mass movements, especially creep, slides and flows.

Area affected by or prone to intense erosion by channelled waters; occurrence or enhancement of destabilizing processes on the valley slopes of vertically or laterally eroding channels. Subordinate occurrence of mass movements, especially slides, falls and creep.

Area affected by or prone to intense erosion by channelled waters; occurrence or enhancement of destabilizing processes on the valley slopes of vertically or laterally eroding channels. Subordinate occurrence of mass movements, especially falls, slides and creep.

Area affected by or prone to intense sheet, rill and gully erosion; erosion rate is 1–2 cm/a. Subordinate occurrence of mass movements, especially slides and creep.

Area moderately affected by or prone to erosion processes.

Increasing Hazards

Figure 19.11 Synthetic map of geomorphological hazards in the Fiume Orcia basin

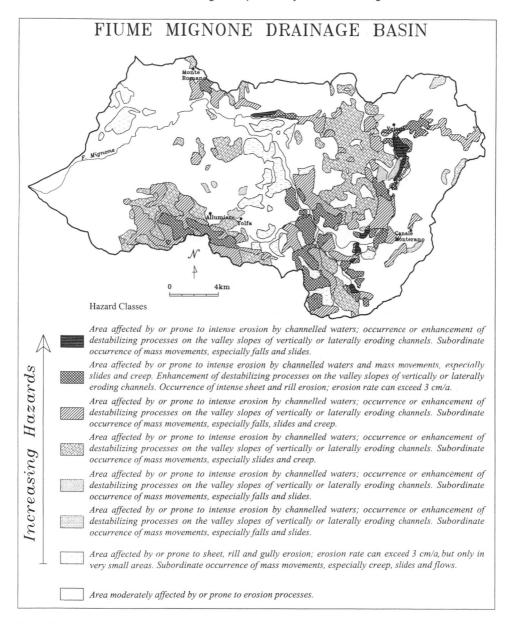

Figure 19.12 Synthetic map of geomorphological hazards in the Fiume Mignone basin

of a process and some physical and anthropic characteristics of a given area. The equations obtained were used to generate thematic maps, based on quantitative data, which allowed the identification of areas at present affected by and more or less prone to undergo a given instability process.

The analysis confirmed that the problem of evaluating geomorphological hazards is a very complex one and further and more detailed studies are necessary which follow both the

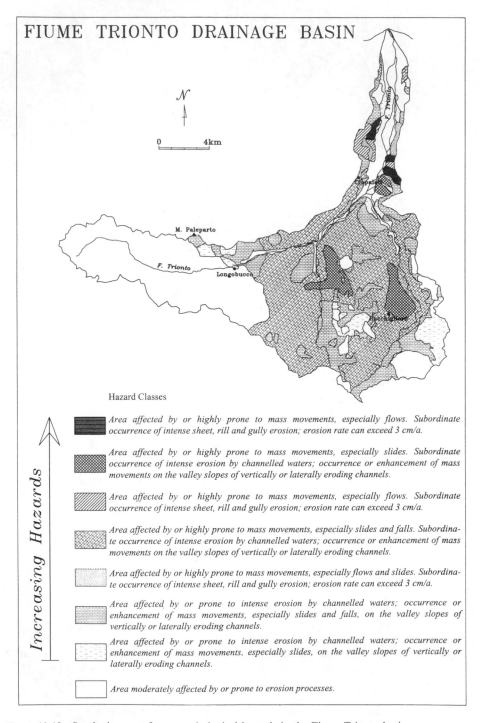

Figure 19.13   Synthetic map of geomorphological hazards in the Fiume Trionto basin

statistical and the deterministic approach. This research was carried out following this twofold approach and with the aim of achieving an objective evaluation of geomorphological instability, which is essential when geomorphological events capable of producing enormous damage may occur.

Results so far obtained show that quantitative geomorphic analysis can provide a useful – although partial – contribution to the precise evaluation of geomorphological hazards. At the same time they suggest the possibility of pursuing this line of research further in order to obtain a more in-depth analysis.

ACKNOWLEDGEMENTS

This work was financially supported by the National Council for Research (CNR) and by the Ministry for the University and Scientific and Technological Research (MURST); Director of research is Professor Elvidio Lupia Palmieri.

We are grateful to Dr Peter Clarke of the British School at Rome for improving the English in this chapter.

REFERENCES

Allison, R.J. 1996. Slopes and slopes processes. *Progress in Physical Geography*, **20**(4), 453–465.

Avena, G.C. 1985. Carta del paesaggio vegetale del bacino del Mignone. In *Valutazione sullo stato dell'ambiente nel bacino idrografico del Fiume Mignone*. Provincia di Roma – Assessorato Ambiente.

Avena, G.C., Giuliano, G. and Lupia Palmieri, E. 1967. Sulla valutazione quantitativa della gerarchizzazione ed evoluzione dei reticoli fluviali. *Bollettino della Società Geologica Italiana*, **86**, 781–796.

Brunsden, D. and Prior, D.B. (eds) 1984. *Slope Instability*. Wiley, New York.

Ciccacci, S., Fredi, P., Lupia Palmieri, E. and Pugliese, F. 1981. Contributo della analisi geomorfica quantitativa alla valutazione dell'entità dell'erosione nei bacini fluviali. *Bollettino della Società Geologica Italiana*, **99**, 455–516.

Ciccacci, S., Fredi, P., Lupia Palmieri, E. and Pugliese, F. 1987. Indirect evaluation of erosion entity in drainage basins through geomorphic, climatic and hydrological parameters. In Gardiner, V. (ed.), *International Geomorphology*, **II**. Wiley, Chichester, 33–48.

Ciccacci, S., D'Alessandro, L., Fredi, P. and Lupia Palmieri, E. 1992. Relations between morphometric characteristics and denudational processes in some drainage basins of Italy. *Zeitschrift für Geomorphologie N.F.*, **36**, 53–67.

Ciccacci, S., Del Monte, M, Fredi, P. and Lupia Palmieri, E. 1995. Plano-altimetric configuration, denudational processes and morphodynamics of drainage basins. *Geologica Romana*, **XXXI**, 1–13.

CNR, 1956. *Carta della utilizzazione del suolo d'Italia (scala 1:200.000)*. Foglio 20, Touring Club Italiano, Milan.

Conti, M.A. and Corda, L. 1985. Caratteristiche geolitologiche del bacino idrografico del Fiume Mignone. In *Valutazione sullo stato dell'ambiente nel bacino idrografico del Fiume Mignone*. Provincia di Roma – Assessorato Ambiente, 9–19.

De Graff, J. and Canuti, P. 1988. Using isopleth mapping to evaluate landslide activity in relation to agricultural practices. *IAEG Bulletin*, **38**, 61–71.

Dimase, A.C. and Iovino, F. 1988. *Carta dei suoli dei bacini idrografici del Trionto, Nicà e torrenti limitrofi (Calabria)*. CNR, Progetto Finalizzato IPRA – Sottoprogetto 2 – Area Problema 2.2.

Einstein, H.H. 1988. Special lecture: Landslide risk assessment procedure. *Proceedings of the 5th International Symposium on Landslides*, Lausanne, **2**, 1075–1090.

Embleton, C., Federici, P.R. and Rodolfi, G. (eds) 1989. Geomorphological hazards. *Geografia Fisica e Dinamica Quaternaria*, **Suppl. II**.

Fell, R. 1994. Landslide risk assessment and acceptable risk. *Canadian Geotechnical Journal*, **31**(2), 261–272.

Gruppo Nazionale Geografia Fisica e Geomorfologia, 1995. *Carta geomorfologica del bacino del Trionto*. Selca, Firenze.

Horton, R.E. 1945. Erosional development of streams and their drainage basins; hydrophysical approach to quantitative morphology. *Bulletin of the Geological Society of America*, **56**, 275–370.

Lazzarotto, A. 1993. Elementi di geologia. In *La storia naturale della Toscana meridionale*. Monte dei Paschi di Siena, 20–87.

Lupia Palmieri, E., Ciccacci S., Civitelli, G., Corda L., D'Alessandro L., Del Monte M., Fredi P. and Pugliese F. 1995. Geomorfologia quantitativa e morfodinamica del territorio abruzzese: I – Il bacino idrografico del Fiume Sinello. *Geografia Fisica e Dinamica Quaternaria*, **18**, 31–46.

Nicolau, J.M., Solé-Benet, A., Puigdefábregas, J. and Gutiérrez, L. 1996. Effects of soil and vegetation on run-off along a catena in semi-arid Spain. *Geomorphology*, **14**, 297–309.

Ohmori, H. 1993. Changes in the hypsometric curve through mountain building resulting from concurrent tectonics and denudation. *Geomorphology*, **8**, 263–277.

Panizza, M. and Piacente, S. 1993. Geomorphological assets evaluation. *Zeitschrift für Geomorphologie N.F., Suppl. Bd.*, **87**, 13–18.

Parsons, A.J., Abrahams, A.D. and Wainwright, J. 1996. Responses of interrill run-off and erosion rates to vegetation change in southern Arizona. *Geomorphology*, **14**, 311–317.

Reading, A.J. 1993. A Geographical Information System (GIS) for the prediction of landsliding potential and hazard in south-west Greece. *Zeitschrift für Geomorphologie N.F., Suppl. Bd.*, **87**, 141–149.

Regione Toscana, 1985. *Carta dell'uso del suolo*. Regione Toscana.

Skempton, A.W. and Delory, F.A. 1957. Stability of natural slopes in London Clay. *Proceedings of 4th International Conference on Soil Mechanics*, London, **2**, 378–381.

Strahler, A.N. 1952. Hypsometric (area-altitude) analysis of erosional topography. *Geological Society of America Bulletin*, **63**, 1117–1142.

Yong, R.N., Alonso, E., Tabba, M.M. and Fransham, P.B. 1977. Application of risk analysis to the prediction of slope instability. *Canadian Geotechnical Journal*, **14**, 540–543.

20 Geomorphological Changes and Hazard Potential by Eruption and Debris Discharge, Unzen Volcano, Japan

MASARU KEN IWAMOTO
Department of Civil Engineering, Nishi-nippon Institute of Technology, Fukuoka, Japan

ABSTRACT

The volcanic disaster at Unzen is the worst eruption in the modern history of Japan because it is located near the city and activity still continues. In this disaster, pyroclastic flows were frequently generated by the fall of lava domes. On rainy days, debris flows occurred and caused much damage. However, these potential risks could be gradually predicted by research regarding the characteristics of geomorphological changes. For instance, pyroclastic flow was predicted by the relationships between the earthquake, magma supply, lava dome growth and shape changes. On the pyroclastic plateau, erosion mechanisms and river struggle were investigated by the decrease of permeability due to the volcanic ash. The characteristics of debris flow were also simulated under heavy rainfall conditions over a short period of time, and clarified the mechanical difference from pyroclasic flow. Since then, the evacuation system and countermeasures have been conducted using these geomorphological changes and hydrological conditions: the evacuation system for pyroclastic flow was set up in the midstream and the warning system for debris flow was carried out on rainy days.

20.1 INTRODUCTION

It is reported that 40–50 volcanoes erupt in the world every year. Figure 20.1 shows the distribution of tectonic plates. Especially in Asian countries, there are many active volcanoes and their activities are very varied, including earthquake, eruption, pyroclastic flow, landslide and debris flow. They cause serious disasters, after which there remain problems of recovery and restoration of the affected area. In these disasters, the potential risk of pyroclastic and debris flows depends mainly on geomorphological changes, because the occurrence condition and moving energy of these flows are basically characterized by the altitude, slope angle and materials at the resource.

This chapter reports on the process of geomorphological changes and hazard potential caused by volcanic activity at Unzen volcano in Japan during 1989–1996.

Applied Geomorphology: Theory and Practice. Edited by R.J. Allison.
© 2002 John Wiley & Sons, Ltd.

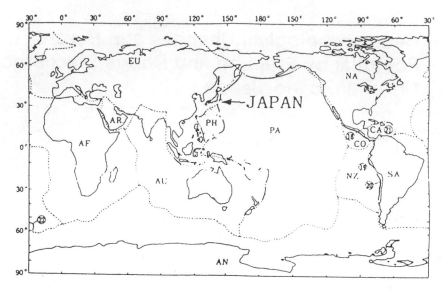

Figure 20.1   The distribution of tectonic plates worldwide

20.2   CHARACTERISTICS OF VOLCANIC HAZARD IN JAPAN

20.2.1   Japanese Volcanoes

Japan is an arc of islands located at the boundary zone of the Pacific, Philippine and Eurasian plates. There are about 250 volcanoes around the islands, and 77 of them are still active. Japan is thus a volcanic island, as shown in Figure 20.2. Kyushu island, southern Japan, has large calderas (Aso, Aira and Chijiwa, etc.) and various types of pyroclastic flow deposits are widely distributed around there. In addition, these areas are active crust zones with many small volcanoes (Japan Construction Agency 1966).

20.2.2   Types of Volcanic Hazards

Volcanic activity varies depending on the magma structure, components and eruption types. Table 20.1 summarizes the various types of volcanic hazards. In general, most Japanese volcanoes effuse viscous, slow-moving, andesite–dacite lavas which cannot reach further than a few kilometres. However, in the cases of Mt Asama (1783) and Mt Unzen (1991), many people living near the volcano were killed by pyroclastic flow. In contrast, some volcanoes (Mt Sakurashima, Mt Miyakeshima) effused slow, fluid basaltic lavas which spread more than 10 km and victims were fewer. The largest and worst hazard happened at Unzen volcano in 1792 when about 15 000 people were killed by a landslide and tidal wave caused by the earthquake (Ohta 1969).

20.2.3   Unzen Volcano

Unzen volcano is located at the Unzen graben in western Kyushu island, which is spreading in the N–S direction at the rate of 1.4 cm a^{-1} and subsiding at the rate of 2–3 mm a^{-1}

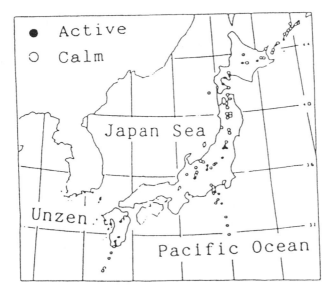

Figure 20.2 The distribution of volcanoes around Japan

Table 20.1 A history of volcanic disasters in Japan

Disaster	Volcano	Year	No. of deaths
Lava flow	Sakurashima	1914	58
	Miyakeshima	1983	(340 houses burned)
Pyroclastic flow	Asamayama	1783	1151
	Ususan	1822	51
	Unzen	1991	44
Debris flow	Tokachidake	1926	144
	Sakurashima	1974	8
Tephras fall	Asamayama	1947	11
	Asosan	1958	12
Volcanic gas	Shiranesan	1976	3
Landslide	Bandaisan	1888	481
	Ontakesan	1984	15
Tidal wave	Komagatake	1640	700
	Ohshima	1741	1475
	Unzen	1792	15000

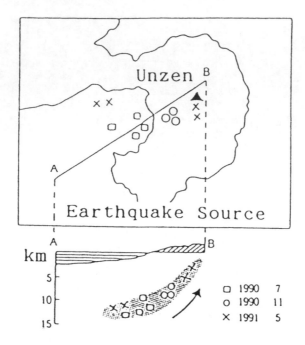

Figure 20.3   Activity of Unzen volcano

(Shimizu *et al.* 1988). It is a composite volcano characterized by many dacitic domes. Mt Mayu-yama is one of these, and collapsed as a result of the earthquake of 1792 (Table 20.1).

20.3   ACTIVITY OF UNZEN VOLCANO SINCE 1990

20.3.1   Volcanic Activity

Unzen volcano began regular activity in November 1989, when a series of earthquakes (main shock, magnitude 5.7) occurred after a dormant period of 197 years. The focal mechanisms of these earthquakes changed with each event (Figure 20.3). A new crater appeared at the summit, and caused ash falls around the summit area (deposit depth, $D = 1.5$ m). Subsequently, the first lava dome was gradually formed at the rate of $0.3 \times 10^6 \, \text{m}^3$ per day in May 1991, and the 13 lava domes and pyroclastic deposits had a volume of about $200 \times 10^6 \, \text{m}^3$ by the end of 1995. Figure 20.4 shows a history of volcanic activities, such as earthquake, pyroclastic flow, magma supply and amount of deposit. The various types of volcanic disasters were caused by fallen lava domes.

20.3.2   Pyroclastic Flow

Pyroclastic flow was originally generated by the collapse and fall of lava domes. It was a highly mobilized turbulent flow consisting of hot, vesiculated juvenile fragments derived from the basement formation. The flows spread laterally from the source, and the larger ones rapidly crossed topographic ridges more than 500–1000 m in height, with a maximum velocity

Earthquake   Pyroclastic Flow

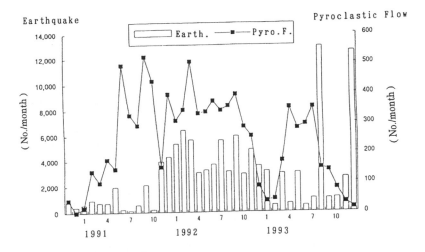

Relationships between Frequencies of Earthquake and Pyroclastic Flow

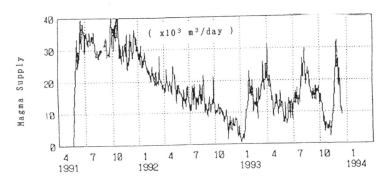

Changes of Daily Ratio of Magma Supply

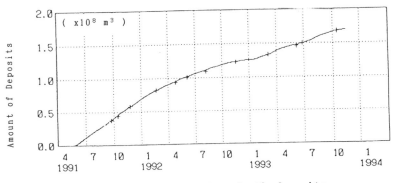

Changes of Amount of Pyroclastic Deposits

Figure 20.4   Volcanic activity, including earthquakes, pyroclastic flow, magma supply and deposits

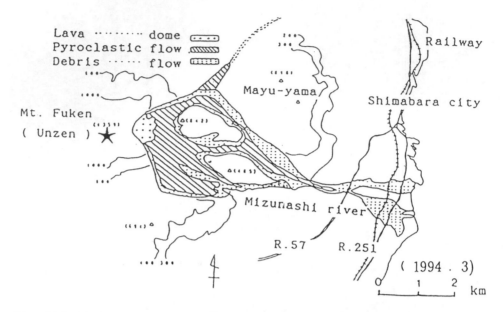

Figure 20.5 Distribution of deposits caused by debris disasters

of $100\,\mathrm{m\ s^{-1}}$. Consequently, 43 people were killed by pyroclastic flow on 3 June 1991. The distribution of pyroclastic flow deposits is shown in Figure 20.5.

20.3.3 Debris Flow

Subsequently, debris flows began to occur around the lower pyroclastic deposit area under heavy rainfall conditions. Especially in a volcanic area, the potential risk of debris flow becomes extremely high because of the decrease of permeability due to fine ash and the rapid occurrence of surface flow (Iwamoto 1996).

20.3.4 Damages

Figure 20.5 shows the changes in distribution of disasters due to pyroclastic and debris flows. It is clear that debris flows reach further than pyroclastic flows, and they severely damaged the downstream areas. Table 20.2 summarizes the volcanic activity and total damage during 1990–1995. The damage increased with each new stage of volcanic activity, namely by the appearance of new domes. More than five years after the eruption, 912 persons (289 families) were still evacuated (Takahashi 1995).

20.4 GEOMORPHOLOGICAL CHANGES BY VOLCANIC ACTIVITY

After the eruption, the mountain profile gradually changed. The characteristics and scale of pyroclastic and debris flows were controlled mainly by topographical conditions, such as

Table 20.2 Volcanic activity and total damage at Unzen

	Scale
Activity	
Lava discharge	200×10^6 m^3
Lava dome	13
Altitude	1359–1494 m
Earthquakes	163 955
Pyroclastic flows	9425
Debris flows	38
Damage	
Deaths	44
Burned houses	819
Inundated area	1.95 km^2
Damaged houses	1695
Refugees (1991)	11 012
Refugees (1996)	912

altitude, slope gradient, sediment material and devastated areas. In this section, the process of geomorphological changes by volcanic activity are summarized for the assessment of volcanic hazard.

20.4.1 Changes by Lava Domes

New craters constantly appeared near the summit with the frequent earthquakes and they effused viscous, slow-moving andesite–dacite lavas at a rate of 0.3×10^6 m^3 per day. The total volume of the 13 lava domes is about 200×10^6 m^3. The summit rose 130 m (from 1359 m to 1494 m), and the domes overhung the summit forming an arch.

20.4.2 Changes by Pyroclastic Flows

Pyroclastic flows were constantly generated by the collapse and fall of lava domes. They flowed down the valley first, with some overflowing the low ridge of the valley and spreading laterally out of the valley. The valley was covered by pyroclastic deposits and in vertical cross-section, the upper valley (30–60 m deep) was gradually buried to form gentle slopes ($4° < \theta < 30°$; Figure 20.6). The process and speed of slope change were dependent on the scale of pyroclastic materials, such as diameter, and slope gradient (Takahashi *et al.* 1995).

20.4.3 Changes in River Course

After filling up the valley (upstream of Mizunashi River), pyroclastic flow began to overflow to the neighbouring river (Nakao River). Their upstream watersheds were almost unified, and finally the river began to change its course. Since then, pyroclastic flow has easily overflowed to other river areas. As a result, the volcanic risk began to spread to other rivers as shown in Figures 20.5 and 20.7.

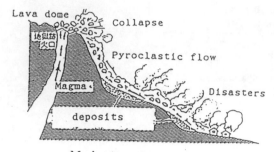

Mechanism of pyroclastic flow

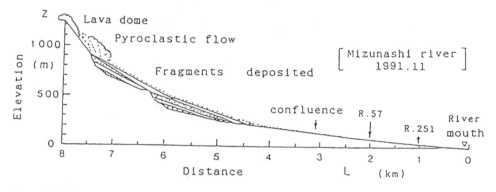

Figure 20.6 Topographic changes caused by pyroclastic flow and debris flow

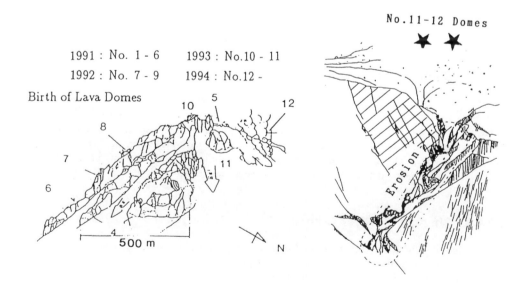

Figure 20.7 Changes in river courses caused by pyroclastic flows

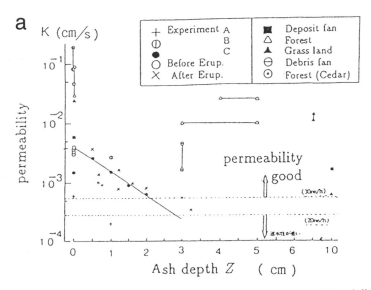

Figure 20.8   (a) Change of permeability. (b) Relationships between total rainfall and discharge

20.4.4   Changes by Erosion and Debris Flow

Permeability was decreasing in the pyroclastic plateau because of the sealing effect of fine ash (Chikushi and Iwamoto 1995; Figure 20.8a). In such a case, the surface can be easily eroded by rainfall. Kyushu island is the most rainy in Japan. In the rainy and typhoon seasons, rainfall of more than 30 mm h $^{-1}$ (200 mm per day) occurs every two years. The topographic relations between gully erosion and channel deformation are illustrated in Figure 20.7. These deposits were transported and deposited at the point of slope change and at the fan. Subsequently, debris flows occurred frequently even in heavy rainfall of more than 7 mm h^{-1}. Figure 20.8b shows the annual relationships between total rainfall and debris discharge during 1991–1994. The equation of the guide line is the average discharge ratio in Japan. When comparing these equations, debris discharges in Unzen were clearly increasing year after year. These differences depend on the annual increment of devastated area, especially due to gully erosion (Iwamoto 1995).

20.4.5   River Morphology

The motion energies of pyroclastic and debris flows were both very large and they flowed straight down at high speed. The average characteristics of river morphology can be summarized as follows: pyroclastic flow has a high speed ($v_p > 100$ m s^{-1}) and flows down to the mouth of the valley (flow length, $L_p < 3$ km); debris flow has a moderate speed ($v_d > 10$ m s^{-1}) and flows down to the river mouth ($L_d > 5$–7 km). When stopping and depositing near the mouths of valley and river, the larger materials are deposited first, then smaller ones spread laterally. In the case of the pyroclastic fan, the slope gradient ($\theta_p < 10°$) was a little steeper than that of the debris flow ($\theta_d < 4°$), because these differences were dependent on the existence of dry sand or water in these flows (interlocking effect or flowability). These

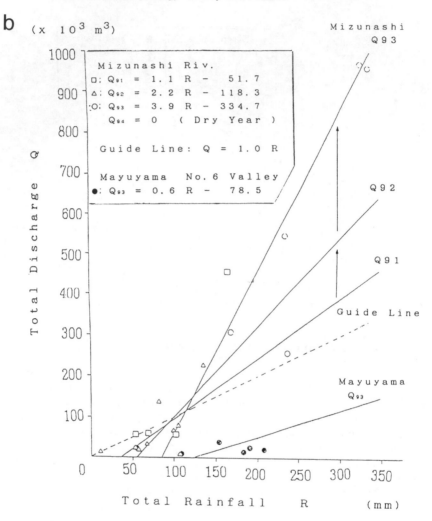

b (x 10³ m³)

Mizunashi Riv.

□: Q₉₁ = 1.1 R - 51.7
△: Q₉₂ = 2.2 R - 118.3
:O: Q₉₃ = 3.9 R - 334.7
      Q₉₄ = 0 (Dry Year)

Guide Line: Q = 1.0 R

Mayuyama No. 6 Valley
●: Q₉₃ = 0.6 R - 78.5

Figure 20.8 (*cont.*)

patterns were repeated again and again in proportion to the discharge scales. Finally, typical alluvial fans were formed in midstream and downstream (Figures 20.5 and 20.6).

20.5 DISASTER PREVENTION PLANS

20.5.1 Hazard Risk Map

In the volcanic area, a hazard risk map should be made to save human lives. In general, this map can be made in two ways; statistical data analysis and theoretical simulation (Shimabara City 1995).

Pyroclastic flow

At the beginning of this event, there were few data concerning hot gas in Japan and the prediction accuracy of the dynamic mechanisms, such as velocity and distribution scale, was not sufficient. This accuracy was gradually improved to be reliable in proportion to advanced data analysis supported by infrared photography and various field observations. Figure 20.9 shows the typical patterns of seismic shock waves of the earthquakes due to magma supply, pyroclastic flow and debris flow. The high frequency of pyroclastic flow (2–7 Hz) was much lower than that of debris flow (5–20 Hz) (Suwa 1992). By using these data, the potential risk of pyroclastic flow can be predicted just before its generation. In addition to this, in the dynamic analysis, the main geomorphological factors are the slope angle, valley depth, material size and density. From these statistical relationships and simulation method, the hazard risk map (evacuation area) was designed at the upstream and midstream areas (A, B) as shown in Figure 20.10.

Debris flow

Debris flows frequently occurred due to heavy rainfall every year in Japan. Especially for Unzen volcano, a rather small rainfall intensity ($r > 7\,\mathrm{mm\ h^{-1}}$) was the occurrence condition of debris flow because of the low permeability caused by fine ash (Figure 20.8a). Furthermore, the course, scale and distribution area of debris flows were also controlled by the topographical conditions as well as pyroclastic flow. Therefore, the hazard risk map (warning and evacuation areas) was designed at the middle and downstream areas (B, C) as shown in Figure 20.10.

20.5.2 Debris Control Works

The typical inundation fans of pyroclastic and debris flows were clearly divided and illustrated by aerial photographs taken after the disasters. By using the geomorphological information concerning the course and scale of the flows, a temporary levee and sand pockets were constructed in midstream before 1995. After regular volcanic activity ended in March 1995, full-scale works were under construction as shown in Figure 20.11. The ground height and site position of the works were designed in a safety zone away from the disasters.

20.6 CONCLUSION

The eruption of Unzen volcano over five years provided many significant lessons to be learned and problems to be solved in preventing a disaster. In order to save human lives, it is important to realize that the process and patterns of geomorphological changes by volcanic activity are the best information for the next stage and action. This information is summarized as follows.

(1) The first change caused by volcanic activity was the earthquake and appearance of the lava dome. Then, the altitude and volume of the mountain slope gradually changed, becoming higher and smaller than before the eruption.
(2) The lava dome continued to grow and it began to collapse due to the effect of gravity. Subsequently, pyroclastic flow occurred; its scale was dependent on the relationships between viscosity and volume of the fallen lava dome.

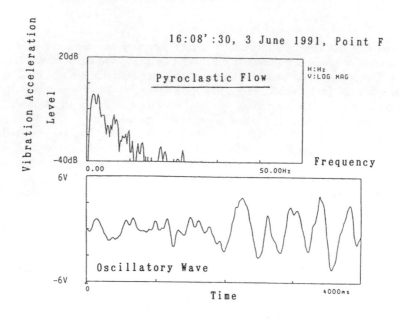

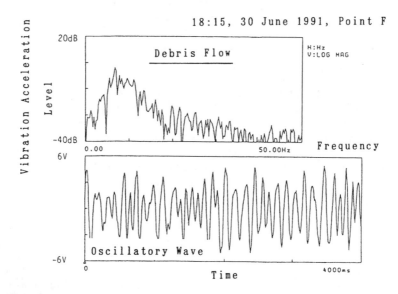

Figure 20.9 Typical pattern of seismic shock waves (Suwa 1992)

(3) Pyroclastic flow occurred first down the valley. Large flows ran straight over the small hills and deposited laterally. These flows were successively generated and filled the undulating topography resulting in the pyroclastic plateau ($\theta < 10°$).

(4) After filling up the valley, the conical mountain slope was gradually formed and the river's course began to change.

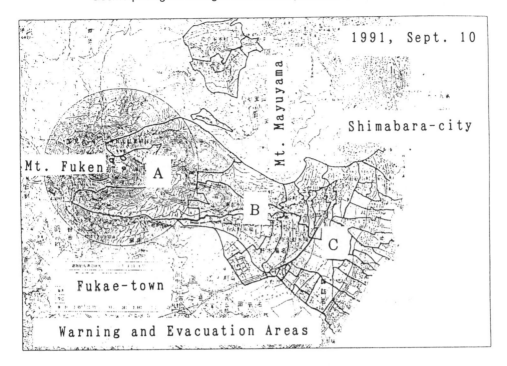

1991, Sept. 10

Mt. Mayuyama

Shimabara-city

Mt. Fuken

A

B

C

Fukae-town

Warning and Evacuation Areas

Figure 20.10 Hazard risk map (pyroclastic and debris flows)

(5) In the pyroclastic plateau, erosion began to occur because of the decrease in permeability and the debris was transported and deposited at the point of slope change and at the fan.

(6) Subsequently, debris flows occurred frequently even in heavy rainfall of more than 7 mm h^{-1}. The rate of debris discharge due to debris flow was much larger than that of flood water, for instance more than ten times larger.

(7) Debris flow ran further than pyroclastic flow because of the presence of water and the low viscosity. Debris discharges were deposited on a gentler slope ($\theta < 4°$) and formed the alluvial debris fan.

(8) In the assessment of volcanic hazard, the fundamental research is observation and analysis based on the characteristic changes of mountain shapes. From these results, the prediction of and countermeasures against pyroclastic and debris flows are possible, and can be improved by combining knowledge among the various relevant agencies.

ACKNOWLEDGEMENTS

I wish to thank Professors Muneo Hirano and Kazuya Ohta (Kyushu University) and Professor Kazuo Takahashi (Nagasaki University), and many staff at Shimabara City for their cooperation and advice. Finally, I express my deep regret over the deaths of citizens, volcanic researchers, police and fire fighters.

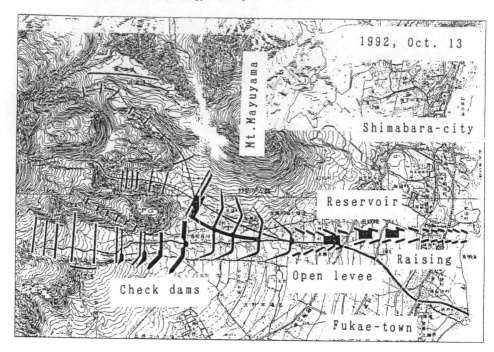

Figure 20.11   Master plan for soil conservation works

REFERENCES

Chikushi J. and Iwamoto, M. 1995. Surface flow simulation considering soil sealing effect. *Proceedings of International Sabo Symposium*, Japan, 369–376.

Iwamoto, M. 1995. An estimation for debris discharge in active volcanic area. *Proceedings of International Sabo Symposium*, Japan, 23–30.

Iwamoto, M. 1996. Prevention of disasters caused by debris flows at Unzen Volcano. In Slaymaker, O. (ed.), *Geomorphic Hazards*. Wiley, Chichester, 95–110.

Japan Construction Agency, 1996. Volcanic hazard prevention project at the Unzen. *Journal of Sabo*, **48**(5), 39–44 (in Japanese).

Ohta, K. 1969. *Study on the collapses in the Mayu-yama*. Report of Shimabara Volcanic Institute, Kyushu University, No. 5, 7–35 (in Japanese).

Shimabara City, 1995. *New Master Plans for Post-disaster Restoration* (in Japanese).

Shimizu, H. *et al.* 1988. Seismic activity and tectonic stress in the Unzen Volcanic Region. *International Volcanic Symposium*, Japan, 337–338.

Suwa, H. *et al.* 1992. *Seismic analysis among volcanic activity and flows*. Report of Unzen Disaster, Ministry of Education, No. B-3-1 (in Japanese).

Takahashi, K. 1995. *Research on Volcanic Disaster and Restoration Plans at Unzen*. Nagasaki University, Vol. 1 (in Japanese).

Takahashi, T. *et al.* 1995. Fluid dynamics of pyroclastic flow. *Proceedings of International Sabo Symposium*, Japan, 1–8.

PART 4 Coasts

Overview: Using Geomorphology at the Coast

DENISE J. REED
Department of Geology and Geophysics, University of New Orleans, Louisiana, USA

INTRODUCTION

Nowhere are the pressures of population growth and resource exploitation felt more keenly than at the coast. Already the majority of the US population lives within 130 km of the coast with 53% of the population living in approximately 17% of the land area that is considered coastal (Culliton 1998). Increased population density brings pollution and habitat degradation – decreasing the value of many of the resources that initially attract coastal development. More people means more infrastructure on or near the coast. Tourist developments, port facilities and fishery processing facilities are normal components of coastal population growth, but as the development increases so does the vulnerability of developed coastal areas to natural hazards. Our ability to predict the paths and intensity of hurricanes and typhoons continually improves and this should be reflected in reduced risks from storm impacts through better forecasting and preparedness. However, growing population and development at the coast ensures that damage caused by such hazards will continue to be catastrophic.

The resources on which coastal populations depend are threatened by overexploitation and natural system degradation. Mangrove forests that provide essential nursery habitat for fishery species are harvested for firewood. Aquaculture ponds, producing high value shrimp, pollute adjacent estuarine waters with their discharges. Rivers that enter the coastal ocean are dammed for power or irrigation, or leveed to prevent flooding of coastal communities, fundamentally changing the nature of estuarine and coastal waters, sediments and natural resources. But in addition to these human pressures, which once recognized could at least theoretically be mitigated, coastal systems face the threats associated with climate change. There is debate over trends in the historical frequency and intensity of tropical cyclones (Henderson-Sellers *et al.* 1998), as the global pattern is complicated by regional variability and decadal-scale changes. However, Timmerman *et al.* (1999) suggest that increasing greenhouse gas concentrations may change the tropical Pacific to a state similar to present-day El Niño conditions. Such conditions would suggest that the USA would experience fewer Atlantic hurricanes but more extratropical storms on the west coast. More certain perhaps is the trend in sea level where debate centres on the magnitude and rate of the rise, and coastal island states around the globe are clearly vulnerable (Pernetta and Hughes 1990). Changes will occur in coastal areas in the future, whether from storm impacts, rising sea levels, or changes in the magnitude and frequency of freshwater inflows (NAST 2000). Applying our knowledge of these systems will not alter the changes but could be fundamental to our ability to adapt and cope with what lies ahead.

Applied Geomorphology: Theory and Practice. Edited by R.J. Allison.
© 2002 John Wiley & Sons, Ltd.

THE ROLE OF GEOMORPHOLOGY

Geomorphology has a critical role to play in coastal issues. The physical landscape is an underlying control on the location and function of many water-dependent contemporary human activities near the coast, such as ports and resort hotels, as well as the historic communities and cultural resources that have developed in areas both accessible and strategically important to earlier societies. The physiography of the land–ocean margin determines the suitability of specific locations for certain human activities. Many of the great coastal cities of the world like Hong Kong, Singapore and Rio de Janeiro are centred on either naturally sheltered deep-water harbours, or on navigable routes between land-based commercial centres. Even more fundamentally, the geomorphologic character of the coast both in terms of morphology and process controls the nature of coastal ecosystems from coral reefs to sandy barrier islands to rocky intertidal communities.

While the physical landscape exerts great influence on human activities at the coast, geomorphology as a science is essentially equipped to work at the interface between the understanding of coastal landscapes and the pressures on these systems brought about by massive coastal development. The process–response approach to understanding how and why coasts work allows ready application of geomorphological understanding to coastal planning, risk assessment and restoration. In Chapter 24, Claudino-Sales and Peulvast explore the origin and development of coastal dune features in Brazil. Such understanding of the auto-genic sediment supply within the dune systems and the role of open water bodies in dune dynamics is essential to both effective restoration of natural dunes following disturbance or development, and projecting the environmental impacts of adjacent developments or tourist pressure.

While geomorphology as a science in its own right may be little known by coastal planners and developers, the increasing use of environmental regulations that require Environmental Impact Statements or similar products to be developed will lead to a greater appreciation for the understanding of landscape dynamics. A brief evaluation of the role of applied geomorphology in understanding the effect of human activities at the coast and the risks faced by coastal communities will be used to highlight some of the issues addressed by the chapters in Part 4 of this volume.

UNDERSTANDING THE EFFECT OF HUMAN ACTIVITIES

In many cases the effects of human structures and developments at the coast have been historically integrated into our understanding of coastal geomorphology. The extensive reclamation of coastal lowlands and marshes for agriculture in the Netherlands and in southeast England since the 13th century limits our geomorphological understanding of these systems to the modified form–process relationship. For instance, in southeast Essex in England considerable attention is being paid to the role of coastal marshes in protecting seawalls from storm and wave erosion, and the potential effects of sea-level rise (Pethick 1993). The sea-walls that back the salt marshes must alter the tidal flooding and sedimentation regime relative to a marsh with a gradual landward upland transition. They must function as marshes fringing sea-walls although it is currently unclear how much difference this makes to their response to sea-level rise or marginal wave erosion. French and Reed (2001) describe efforts at managed realignment in southeast Essex where sea-walls are breached with the expectation of recreating marshes in formerly drained areas, thus providing a more extensive 'buffer' for coastal communities. These 'restoration' sites are essentially marshes almost entirely surrounded by sea-walls with one or more breaches that provide the tidal connection. Cahoon *et al.* (2000) have documented early sedimentation patterns within one of these breached sites, and show increased rates in lower elevation areas as might be expected based on flooding frequency–sedimentation relationships for coastal marshes (e.g. French 1993). Studies of sedimentation in UK coastal marshes have frequently pointed to the role of major creeks in supplying sediment (e.g. Stoddart *et al.* 1999). However, marshes which are limited by sea-walls at their landward margin, and/or those which have limited tidal connections such as in these levee breach restoration sites, are unlikely to develop extensive creek networks. Reed (1988) showed how sedimentation patterns in a marsh on the Dengie Peninsula in Essex, backed by a sea-wall, were dominated by sediment delivery across the extensive boundary between the marsh and the intertidal flats. The levee–breach sites do not have such an extensive open boundary – most of the perimeter apart from the breach locations consists of pre-existing sea-walls. Thus the sedimentary dynamics of these essentially man-made landforms is not readily understood. Such features provide opportunities for geomorphologists to develop knowledge of fundamental marsh processes, but also challenges, as human performance measures for such projects are considered in terms of the needs of coastal communities. Geomorphologists know that coastal marshes, in their natural form, can be important, sustainable buffers from storm surges and waves – we are currently less sure of our ability to recreate these functions in the human landscape.

While the influence of sea-walls and human structures in the landscape have a clear influence on geomorphological processes, many human activities have more subtle effects. In Chapter 21, Bondesan *et al.* document vertical ground movements in the Po Plain. The history of human settlement in this area is well known and internationally recognized through the cultural and artistic legacy of the Venetians. Many know of the increasing threat of flooding in Venice, and the ambitious engineering structures proposed as remedies. However, less widely recognized is the regional change in land elevation throughout the Po Plain associated with extraction of hydrocarbons and groundwater. Bondesan *et al.* identify recent rates of subsidence of up to 10 cm per year near Bologna – itself a town of great historic and cultural importance. Groundwater withdrawal is identified here as the main causative factor of the contemporary problem in the southern Po Plain. The consequences of such high rates of subsidence for both the human and natural landscapes are phenomenal. Subsidence associated with groundwater extraction is well documented around the world (e.g. Bangkok; Arunin 1999). Here geomorphology provides the tools for problem identification and geomorphologists must carry the message of their science into the public arena.

The influence of human activities on the coastal landscape is frequently difficult to specifically identify relative to natural variability in geomorphic processes. For instance in the coastal plain of east Texas, White and Tremblay (1998) show how movement on geologic faults is associated with submergence of wetlands. Faulting itself is often considered a process beyond human control. The earthquake hazard associated with fault movement is something we mitigate the effects of rather than attempt to directly reduce. However, White and Tremblay (1998) show how fault activity in east Texas is triggered by the extraction of oil and gas from subsurface deposits. Teasing apart the role of natural subsidence in deltas, like the Po, or geologically young coastal plains, such as east Texas, from surface lowering associated with fluid extraction is frequently difficult. While planners and policy-makers may prefer to attribute landscape problems to natural, and therefore uncontrollable, factors, geomorphologists can help to identify the specific roles of various processes. Such an example is the study of D'Alessandro *et al.* (Chapter 22) who identify the role of changing fluvial inputs of sediment to the coastal system as a critical factor in controlling erosion rates of Calabrian beaches. While some of these changes in the catchments draining to this coast may be due to climate change, the effect of human activities in modifying sediment delivery to the coast can be seen throughout the system. This information is crucial to those responsible for managing coastal resources, who otherwise might direct their efforts at combating erosion solely to storm protection measures rather than seeing the beach in a much broader sedimentary context.

UNDERSTANDING FUTURE RISKS

Increasing understanding of climate change and the improved ability of models to project future changes presents geomorphologists with an unparalleled opportunity to contribute their skills to planning coping and adaptive strategies. For coastal systems, while changes in the supply of fresh water and fluvial sediments to the coast my be substantial in some systems, the most prominent long-term risk for coastal communities and the geomorphological systems on which they depend is sea-level rise. For the extensive human infrastructure at the coast, such as docks, sea-walls and hotels, sea-level rise projections themselves provide a direct indication of the response required. The situation is more complex for coastal landforms.

Arens *et al.* (2001) see coastal dunes as the result of both marine and terrestrial steering processes, with geomorphological processes more important in the seaward zone and bio-logical processes becoming dominant inland. Adjustments in the coastal dune system, both geomorphological and biological, take place as a result of long-term shoreline erosion or sea-level rise. The problem for the future is whether continually adjusting systems such as dunes and barrier islands can cope with changes in sea-level rise rates and/or the frequency of storm events. In Chapter 23, Palacio-Prieto and Otriz-Pérez consider the effect of Hurricane Roxanne on the coastal geomorphology of southeastern Mexico. Their findings demonstrate the variation in impact associated with a single storm, and the role of geomorphology in controlling the degree of damage to human infrastructure. If storm surges are found to 'reoccupy' overwash areas, cuts and inlets from previous storms, then geomorphological studies can determine the likelihood of recovery of the island system between storms, and assist in assessing the risks associated with storm impacts to human structures placed in or near these areas.

Our coast is continually changing – because of human action and natural forces. In many areas, catastrophic storms occur so infrequently that coastal residents have no personal experience of storm impacts. Structures, from houses to roads to industrial facilities, are built in vulnerable areas. The geomorphologists' understanding that future climate change may cause greater impacts than those we have recently experienced must be translated into the public context and forthrightly presented to coastal planners. In the ecological sciences, protocols have been developed to link human activities with environmental stressors that produce ecological responses and such approaches are being applied to large-scale ecosystem management (Harwell 1998). The undoubted effects of future climate change of coastal systems, and the critical role of geomorphology in providing services for the human popu-lation and essential structure for coastal ecosystems, point to the clear need for development of similar approaches in geomorphology.

APPLYING GEOMORPHOLOGY AT THE COAST

This short overview and the more detailed studies presented in the following chapters illustrate the many current opportunities for geomorphologists to apply their skills in the fields of coastal planning and management. To take advantage of this, and to meet the challenges involved in working with other applied sciences, it is essential that courses and curricula are developed to provide society with geomorphologists who can work effectively in this context. Some established professionals have developed approaches to communicating and using their findings in an applied context through years of trial and error. We must do better for young geomorphologists and provide them with the tools necessary to ensure their contributions to solving coastal problems are heard. Some strides are already being made in this arena, with efforts to develop postgraduate courses in applied geomorphology. It is also important to integrate applied approaches into our routine professional lives. Several developments are still needed in this area.

- As a discipline we must give recognition and status to those who work successfully in the applied field. All too often, promotion, tenure or other critical career decisions are made solely on the basis of scholarly contributions measured using traditional measures such as citation indices or journal hierarchy. The type of documentation and synthesis needed to effectively communicate geomorphological understanding of coastal issues to planners and resource managers rarely ranks highly in such evaluations. Professionally we must find a way to reward those who excel in this area.
- Our professional societies can work to develop outlets for geomorphological studies that have broad application, that are accessible and actually used by resource managers. Products might include syntheses of existing knowledge in topical areas, or systemic evaluations of common problem areas from a geomorphological perspective, and should be both peer-reviewed and widely publicized. Most important here is that product is geared towards the user in the applied context. Frequently, and for good reason, we orient our products, from conferences to journals, to inform other geomorphologists. To do justice to our science and our skills, our audience must be broader.
- At the coast, as in many other environments, the application of geomorphological skills requires some familiarity with other disciplines, such as ecology and engineering. While geography or Earth science departments can provide training opportunities for students in this realm, integration with other fields, especially those with direct links to resource exploitation or commercial interests, can provide young geomorphologists with a breadth of expertise that will serve them well, both in solving real-world problems and also in their professional careers.

Many coastal geomorphologists have made great strides in furthering the effective management of coastal resources through applied science. The immediate effect of climate change on the coast, in addition to the pressures presented by rising coastal populations, means that here more than in any other environment our professional community must move forward to develop and apply the science necessary to sustain coastal communities.

REFERENCES

Arens, S.M., Jungerius, P.D. and van der Meullen, F. 2001. Coastal dunes. In Warren A. and French, J.R. (eds), *Habitat Conservation: Managing the Physical Environment*. Wiley, Chichester, Ch. 9.

Arunin, S. 1999. Impacts of sea-level rise on coastal areas of Thailand – highlighting on agriculture. *Current Topics in Wetland Biogeochemistry*, **3**, 142–151.

Cahoon, D.R., French, J.R., Thomas, S. Reed, D.J. and Moller, I. 2000. Vertical accretion versus elevational adjustment in UK salt-marshes: an evaluation of alternative methodologies. In Pye, K. and

Allen, J.R.L. (eds), *Coastal and Estuarine Environments: sedimentology, geomorphology and geoarchaeology*. Geological Society, London, Special Publications 175, 223–238.

Culliton, T.J. 1998. *Population: distribution, density, and growth*. NOAA State of the Coast Report, NOAA, Silver Spring, MD. Available online: http://state-of-coast.noaa.gov/topics/html/pressure.html.

French, J.R. 1993. Numerical simulation of vertical marsh growth and adjustment to accelerated sea-level rise, North Norfolk, UK. *Earth Surface Processes and Landforms*, **18**, 63–81.

French, J.R. and Reed D.J. 2001. Physical contexts for saltmarsh conservation. In Warren, A. and French, J.R. (eds), *Habitat Conservation: Managing the Physical Environment*. Wiley, Chichester, 179–227.

Harwell, M.A. 1998. Science and environmental decision making in South Florida. *Ecological Applications*, **8**(3), 580–590.

Henderson-Sellers, A., Zhang, H., Bertz, G., Emanuel, K., Gray, W., Landsea, C., Holland, G., Lighthill, J., Shieh, S-L., Webster, P. and McGuffie, K. 1998. Tropical cyclones and global climate change: A post-IPCC assessment. *Bulletin of the American Meteorological Society*, **79**(1), 19–38.

National Assessments Synthesis Team, 2000. *Climate change impacts on the United States: the potential consequences of climate variability and change*. US Global Change Research Program, Washington, DC.

Pernetta, J.C. and Hughes, P.J. 1990. Implications of expected climate change in the South Pacific region: an overview. UNEP Regional Seas Reports and Studies 128.

Pethick, J. 1993. Shoreline adjustments and coastal management: physical and biological processes under accelerated sea-level rise. *The Geographical Journal*, **159**, 162–168.

Reed, D.J. 1988. Tidal currents and glacial discharge, Laguna San Rafael, Southern Chile. *Journal of Coastal Research*, **4**(1), 93–102.

Stoddart, D.R., Reed, D.J. and French, J.R. 1999. Understanding salt-marsh accretion, Scolt Head Island, Norfolk, England. *Estuaries*, **12**(4), 228–236.

Timmermann, A., Oberhuber, J., Bacher, A., Esch, M., Latif, M. and Roeckner, E. 1999. Increased El Niño frequency in a climate model forced by future greenhouse warming. *Nature*, **398**; 694–697.

White, W.A. and Tremblay, T.A. 1995. Submergence of wetlands as a result of human-introduced subsidence and faulting along the upper Texas Gulf Coast. *Journal of Coastal Research*, **11**(3), 788–807.

21 Vertical Ground Movements in the Eastern Po Plain

MARCO BONDESAN

Dipartimento di Scienze Geologiche e Paleontologiche, Università degli Studi di Ferrara, Ferrara, Italy

AND

MARCO GATTI AND PAOLO RUSSO

Dipartimento di Ingegneria, Università degli Studi di Ferrara, Ferrara, Italy

ABSTRACT

The main results of a study of vertical ground movements carried out for the publication of the *Map of relief and vertical movements of the Po Plain*, an appendix to the *Geomorphological Map of the Po Plain* at a scale of 1:250 000, are presented.

The study was carried out in the eastern part of the plain, which is the area most affected by both natural and artificial subsidence. Here subsidence is the main cause of the presence of vast depressions in the coastal plain; in addition it interferes with the run-off of surface water, raises the height of groundwater in the fields, provokes the infiltration of sea water into groundwater and causes beaches to retreat.

The data derived from various high-precision levellings carried out by the Istituto Geografico Militare and the Italian Cadastre were processed in an Information System for Levelling Data; the results are shown using graphs which represent the vertical recorded movements.

Until the 1970s subsidence was very marked especially to the north of the river Po (and in particular towards the delta) and in the coastal plain to the south of the river Reno; it was mainly caused by the extraction of water containing methane and by hydraulic reclamation, as well as the pumping of groundwater for agricultural, industrial and civil purposes.

Analysing the most recent levellings we have observed that to the north of the river Po and in the delta the phenomenon has been drastically reduced, while it is increasing in Modena, around Forlì and towards Ferrara, in Ravenna and, above all, towards Bologna (where the rate is up to 10 cm/year). In general the greatest subsidence is registered in areas which have been recently industrialized, and it can be interpreted as the effect of the excessive exploitation of groundwater.

21.1 INTRODUCTION

The Po Plain has for a long time been an important area for the study of vertical ground movements. Most of these studies have been carried out in the eastern Po Plain (Figure 21.1),

Applied Geomorphology: Theory and Practice. Edited by R.J. Allison.
© 2002 John Wiley & Sons, Ltd.

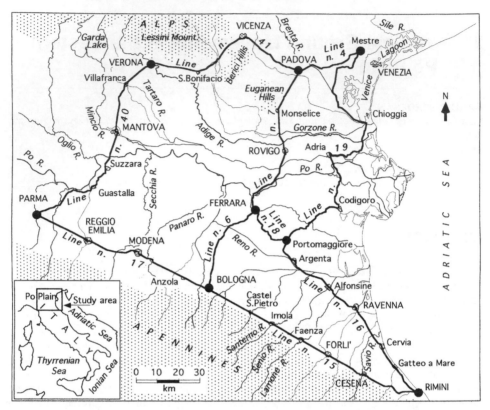

Figure 21.1 Position of the area studied and of the ten levelling lines analysed, in the IGM first-order levelling network. In Figures 21.4, 21.5 and 21.6, the graphs showing the subsidence that has occurred are drawn from left to right, following the south–north component of the orientation of the single levelling lines. From Bondesan *et al.* (2000c). Reproduced with permission of CNR – SISOLS 2000

since it is characterized by marked subsidence, as demonstrated by historical and geological data (Braglia 1955; Ciabatti 1959, 1967, 1968). During the 20th century these movements have been recorded by direct measurement of the elevation of the ground.

Geological studies have demonstrated that natural subsidence in the Po Plain is caused by the tectonic activity that resulted in the orogenesis of the Alps and the Apennines, and by the compaction of sediments in the Po–Adriatic basin which were produced by the erosion of these two mountain chains (Sacco 1900; TCI 1957).

The orogenesis of the Apennines is still continuing beneath the Po Plain to the south of the present position of the Po river in the folds of the buried Apennines (Pieri and Groppi 1981). To the north there is a homocline which reaches the Alpine piedmont and which is characterized by the presence of various faults, as well as by the presence of the so-called 'Lessino-Berico-Euganea dorsal' which is formed by the Lessini mountains and the Berici and Euganean hills (Figures 21.2 and 21.3).

Clear evidence for the existence of natural subsidence is provided by the Pliocene and Quaternary sediments. Although they have always been deposited at depths close to sea-level (Ciabatti 1967), they are in fact of notable thickness, especially in the area of buried Apennine orogenic belt where they exceed 8000 m (Figure 21.2). In the areas away from the Apennine

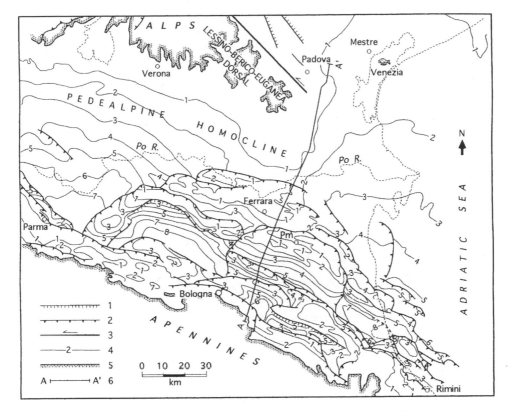

Figure 21.2   Geological map of the eastern Po Plain: 1, normal faults; 2, overthrusts and reverse faults; 3, strike-slip faults; 4, isobaths of the base of Pliocene (in km); 5, margin of relief rising above plain; 6, geological cross-section shown in Figure 21.3. Source: Pieri and Groppi (1981)

orogenesis the thickness increases from the foot of the Alps to 1500 m at Venice and to the north of Ferrara, and to 3500 m on the southern side of the modern Po delta (see also CNR 1992).

For these sediments, which are still undergoing lithification, the coefficients of compaction are higher in the upper levels (middle and late Quaternary) than in the lower ones, where the compaction process has largely been completed.

In this context it is reasonable to think that even today the natural subsidence varies in different areas, and that this spatial variation is related not only to the different lithology and thickness of the loose sediments but also to the tensions and movements which are still present, or in other words the 'neotectonic movements' (Zanferrari *et al.* 1982; Bartolini *et al* 1983; Arca and Beretta 1985). In fact we have reached the conclusion that in the study area the subsidence rates which can be attributed to natural phenomena have been generally quite low north of the Po where they are estimated to be about 0.4 mm/year, while south of the Po they are about 1 mm/year (Elmi 1984; Groppi and Veggiani 1984), with local maxima of up to 2–3 mm/year (Bosellini 1971).

In the study area there are also phenomena of artificial subsidence, which are rather greater, and it is this fact that makes it difficult to make direct measurements of natural subsidence.

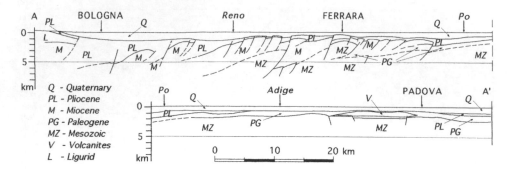

Figure 21.3 Cross-section of the eastern Po Plain (the position is shown in Figure 21.2). One may observe the difference between the part south of the Po, characterized by the buried Apennines, and the part north of the Po, characterized by the Pedealpine Homocline. Source: Pieri and Groppi (1981)

Furthermore, the current altimetric situation in the eastern Po Plain is primarily attributable to artificial subsidence.

Human activity to reduce river flooding and to avoid sedimentation outside the river channels by controlling the hydrographic network and constructing embankments along the rivers has impeded compensation of the natural subsidence. However, this is not sufficient to explain the notably low elevations which characterize a large part of the eastern Po Plain and, in particular, the Po delta, where they reach −5 m.

In recent centuries, and in particular in the last hundred years, other human activity has accelerated the rate of subsidence (Carbognin 1986); repeated measurements of the ground elevation have shown that in the study area this phenomenon has mainly been caused by:

- land reclamation, which started on a large scale at the end of the 16th century and intensified between 1870 and 1970: most of the marshes and salt marshes have been hydraulically reclaimed and transformed into agricultural land;
- withdrawal of water and hydrocarbons from the subsoil, and in particular the extraction of methane between 1938 and 1964 from the Quaternary strata down to a depth of 200 m.

In the eastern Po Plain this subsidence was, and continues to be, a very important problem. For example, it is clear that artificial subsidence has been the main cause of the low elevations of the eastern part of the Po Plain, where an area of 2400 km² lies below the mean sea level.

The danger of this situation, and that of further possible subsidence, is above all clear in light of global sea-level rise (Pirazzoli 1993). In the last 30 years there has also been a crisis in the sediment transport of rivers, which led to a reduction in the sediment supply of beaches, greater erosive activity of rivers, and a general increase in the depth of river channels. These phenomena, coupled with subsidence, are producing notable damage, especially because of reductions in the run-off of surface water, modifications of river channels, alterations to the slopes of drainage and irrigation channels, modifications of the height differences which the pumping stations have to overcome in order to drain water towards the sea, the rising water table, increasing salinity of rivers, increasing infiltration of sea water into the groundwater, increasing beach erosion and increased danger of flooding (ERSA 1978; Montori 1983; Bondesan *et al.* 1995). Uplifting ground movements can clearly also give rise to problems, but they are generally much rarer and on a smaller scale.

21.2   AIMS AND METHODOLOGIES OF THE STUDY

The research presented here was carried out for the production of the *Map of relief and vertical movements of the Po Plain* (MURST 1997b), an appendix to the *Geomorphological Map of the Po Plain* at a scale of 1:250000 (MURST 1997a).

During this research, vertical ground movements which have taken place in the eastern Po Plain during the 20th century were reconstructed on the basis of measurements of ground elevation by levelling.

Levelling has been undertaken periodically in the eastern Po Plain since the end of the 19th century. The Italian first-order levelling network (Nuova Rete Altimetrica Fondamentale) carried out by the Istituto Geografico Militare (IGM) is particularly important for two reasons. The first is the updating and standardization of the heights of the existing benchmarks. The network in fact constitutes the vertical datum according to which all the other lines and networks in the plain are regulated. Periodic surveys not only make it possible to update the reference benchmarks of the individual local networks, but they also provide important indications for the comparison of benchmarks in different networks or those measured in different periods. The second important result is that it makes possible the quantification of ground movement over a much larger area. Since the survey, made between 1943 and 1956 in the eastern Po Plain, the national levelling network has furnished data of great interest on the movements which took place in the first half of the century (Salvioni 1957; Arca and Beretta 1985). Subsequent levelling carried out between 1968 and 1973 not only confirmed some previously known artificial subsidence, but also registered new examples. The surveys carried out by the Italian Cadastre between 1974 and 1977 also produced important information.

It is therefore clear that the recent series of measurements carried out by the IGM in the eastern Po Plain between 1986 and 1992 was destined to arouse the interest of scientists and workers in the area. The results of these measurements were used by the IGM to calculate the new heights of the benchmarks. These data have been acquired, stored and processed in an Information System for Levelling Data, by the University of Ferrara (Bitelli *et al* 1993; Bondesan *et al* 1997). In this chapter the results of the study of the ground movements which have taken place in the second half of the 20th century, together with some interpretative notes on the observed phenomena, are presented.

21.3   THE LEVELLING INVESTIGATIONS IN THE EASTERN PO PLAIN

Table 21.1 shows the levellings carried out by the IGM and the Italian Cadastre in the eastern Po Plain since the end of the 19th century. These measurements refer to the levelling lines illustrated in Figure 21.1; the most recent measurement campaigns, carried out since 1985, are marked in bold (Table 21.1).

The graphs in Figures 21.4, 21.5 and 21.6 show the vertical ground movements which have taken place since the IGM measurements for the 'Nuova Rete Altimetrica Fondamentale' (late 1940s and 1950s) and the subsequent levellings up to and including 1990.

For each line only the levellings for the largest number of benchmarks have been plotted. Only the levellings of the IGM are of this type, therefore the levellings of the Cadastre were not shown in the graphs. Each graph includes only the benchmarks which were measured in all of the different IGM levellings presented. The first levelling is shown conventionally by a

Table 21.1   IGM levelling lines and measurement dates

Levelling line	Measurement dates
4 Padova–Mestre	1883; 1950/53; 1967/70; 1974*; **1986**
6 Bologna–Ferrara	1886/1900; 1943/49; 1973; **1990**
7 Ferrara–Padova	1884; 1942/47; 1970; 1978; 1984; **1986**
15 Rimini–Bologna	1889/1900; 1949/53; 1970/72; 1977*; **1990**
16 Rimini–Portomaggiore	1885/1902; 1950/53; 1970; 1977*; **1990**
17 Bologna–Parma	1887/1900; 1949/52; 1972/73; 1974*; **1980**
18 Portomaggiore–Ferrara	1885/1902; 1950/53; 1970; 1986; **1990**
19 Portomaggiore–Mestre	1884/1909; 1925; 1950/51; 1956/57; 1958/59; 1970; 1977*; **1988**
40 Parma–Verona	1952/53; 1974*; **1985**
41 Padova–Verona	1883; 1950/53; 1974*; **1986**

* Levelling carried out by Italian Cadastre.
From Bondesan *et al.* (1997) with permission of Istituto Geografico Militare Italiano

straight line, and for each benchmark the height change with respect to the preceding levelling is shown.

In the graphs of lines 4, 40 and 41, along which the movements have only been very slight, the vertical scale has been expanded (Figure 21.6).

21.4   ANALYSIS OF THE DATA

21.4.1   Error Section

With regard to the significance of the movements of levelling benchmarks, it must be remarked that the height variations resulting from the comparison of levellings indicate ground movements when they exceed measurement errors. Unfortunately neither the IGM nor the Cadastre published the values of the mean square errors (m.s.e.) of the adjusted heights of the benchmarks.

Nevertheless, an estimate of the uncertainty affecting the computed ground movements can be attempted.

The levellings performed by the IGM and the Cadastre are 'high-precision levellings' as regards instruments, observing procedures and tolerances; for this class of measurement the error $\sigma_{\Delta H}$ of the measured height difference ΔH between two benchmarks is less then $\sqrt{L}$ mm, where L is the length of the section expressed in km. The vertical datum for the Italian first-order levelling is the mean sea level recorded at Genova Tide Gauge in 1942, while levellings carried out later in Po Plain by the IGM and the Cadastre were locally adjusted assuming benchmarks of the first-order levelling network as constraints. Under the hypothesis that such benchmarks, placed in the Apennines or in the Alps, are not affected by vertical movements, the height variation δH of any benchmark recorded between two levellings presents an m.s.e. $\sigma_{\delta H} \leq \sqrt{2L}$ mm, where L is the length of the levelling line connecting the 'local reference benchmark' and the benchmark itself. In the Po Plain generally $L \leq 100$ km, so the m.s.e. is $\sigma_{\delta H} \leq 1.4$ mm. Assuming $T = 2\,\sigma_{\delta H}$ as tolerance, we find that the uncertainity of the ground movements should not exceed 3 cm. However, smaller values for a benchmark can be considered significant if they present the same sign in successive levellings and agree with the average movements of adjacent benchmarks.

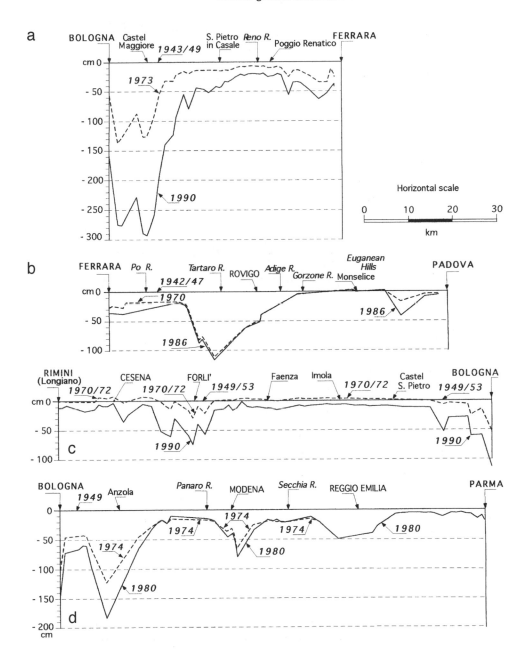

Figure 21.4 Ground movements in the second half of the 20th century along the IGM levelling lines (see Figure 21.1): (a) line 6 Bologna–Ferrara; (b) line 7 Ferrara–Padova; (c) line 15 Rimini–Bologna; (d) line 17 Bologna–Parma. Based on figures in Bondesan *et al.* (1997, 2000). Reproduced with permission of Istituto Geografico Militare Italiano and CNR – SISOLS 2000

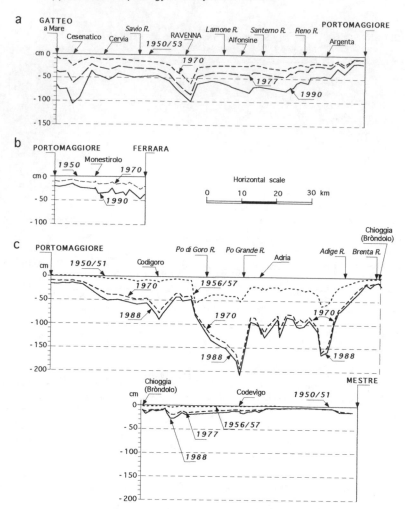

Figure 21.5 Ground movements in the second half of the 20th century along the IGM levelling lines (see Figure 21.1): (a) line 16 Rimini–Portomaggiore; (b) line 18 Portomaggiore–Ferrara; (c) line 19 Portomaggiore–Mestre. Based on figures in Bondesan *et al.* (1997, 2000). Reproduced with permission of Istituto Geografico Militare Italiano and CNR – SISOLS 2000

21.4.2   Pre-1970 Levellings

As regards the IGM levellings carried out in the 1940s and 1950s and those of the 1970s (see also IGM 1971), the most marked subsidence has been registered in the following zones.

- At Bologna and in the area immediately to the north: more than 4.5 cm/year in the period between 1943 and 1973 (Figure 21.4a, line 6).
- To the north of the Po river in the Rovigo area: values up to 4 cm/year in the period between 1942 and 1970 (Figure 21.4b, line 7). Between Rimini and Bologna (Figure 21.4c, line 15) the maximum annual rate of subsidence between 1942 and 1972 was registered at

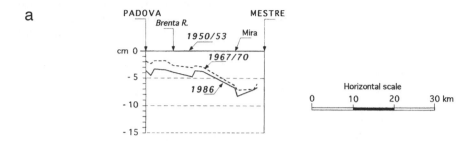

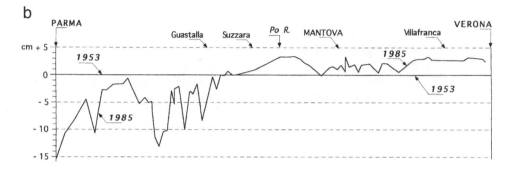

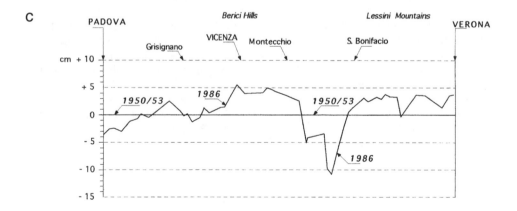

Figure 21.6 Ground movements in the second half of the 20th century along the IGM levelling lines (see Figure 21.1): (a) line 4 Padova–Mestre; (b) line 40 Parma–Verona; (c) line 41 Padova–Verona. The vertical scale has been expanded with respect to the other graphs. Based on figures in Bondesan *et al.* (1997, 2000). Reproduced with permission of Istituto Geografico Militare Italiano and CNR – SISOLS 2000

Bologna, Forlì and Cesena, although their quantification is uncertain (Pieri and Russo 1978). Between Bologna and Parma the maximum rate of subsidence was registered near Anzola (more than 4.5 cm/year) and Modena (1.6 cm/year) between 1940 and 1974 (Figure 21.4d, line 17).

- In the coastal plain to the south with maximum rates at Ravenna: from 3 cm/year in the period from 1950 to 1970 and 6–7 cm/year between 1970 and 1977 (Figure 21.5a, line 16).
- To the south of the Po delta: rates of more than 6 cm/year between 1950 and 1956 and more than 4 cm/year between 1956 and 1970 (Figure 21.5c, line 19).
- Still to the north of the Po but further east, near Adria (Figure 21.5c, line 19): maximum rate of 10 cm/year in the period between 1956 and 1970 (here the rates were also very high between 1950 and 1956).

Around the Venice lagoon (Figure 21.5c, line 19) the most marked subsidence in the 1956–1970 period was registered at Chioggia (between 7 and 9 mm/year), on the western edge of the lagoon (more than 9 mm/year) and near Mestre (1 cm/year). Notable but not exceptional rates have also been registered between Padova and Mestre (Figure 21.6a, line 4) between 1956 and 1970, with maximum rates of about 5 mm/year.

21.4.3 Latest Levellings

A comparison of pre-1970 and recent levellings shows that the rates of subsidence have diminished to the north of the Po and towards the Po delta; to the north of the Po some upward movement of the ground has also been noted.

The rates have also diminished, though less drastically, in the coastal plain between Rimini and Ravenna. On the other hand there has been an increase between Ravenna and Porto-maggiore, except in the tract immediately to the northeast of Ravenna (Figure 21.5a, line 16).

The greatest tendency towards a marked increase in the rate of subsidence has been registered in Bologna and to the north of the city (Figure 21.4a, line 6), with maximum values of 10 cm/year; in this area artificial subsidence has now reached comparable rates with those recorded in the Po delta in the 1950s and 1970s. Marked subsidence has also taken place between Bologna and Anzola, and in Modena (Figure 21.4d, line 17).

Other increases in the rate of subsidence have taken place near Ferrara (Figure 21.4a, line 6) and between Rimini and Bologna (Figure 21.4c, line 15).

The last IGM levelling campaign also made it possible to analyse line 40 (Parma–Verona) and line 41 (Padova–Verona); for the first time in the second half of the 20th century, it is possible to make a comparison with the IGM data on these two lines (Figures 21.6a and c). Along both lines the movements have been modest. Along line 40 to the south of the Po there has been a prevalence of subsidence, while to the north of the Po the ground has only moved upwards. Along line 41 there is an alternation between tracts of subsidence and tracts of upward movement, with a prevalence of the latter near Vicenza (the tract at the foot of the Berici Hills) and towards Verona (the tract at the foot of the Lessini Mountains).

Factors of artificial subsidence should not affect these two lines, except for brief tracts. It is, however, necessary to be very prudent in interpreting any upward movement as true ground movement, as many of the values are in fact too small to be significant. However, a general tendency towards upward movement in the northwest sector of the area studied may be plausible. This will have to be verified by further IGM levellings in the Verona node before it can be accepted as a trend.

21.5 INTERPRETATION OF THE RESULTS

The analysis of these data makes it possible to evaluate the evolution of subsidence in the study area, and to identify its main causes. In order to do this it is obviously necessary to incorporate information on past and present land use.

With regard to the subsidence registered between the 1950s and the 1970s, interpretations can be made that incorporate data from levellings carried out by other agencies and on specific local networks.

For the area between the rivers Po and Adige and the area to the southwest of the Po (Figure 21.1), it has been possible to demonstrate that the subsidence is mainly the result of the extraction of water containing methane from the Quaternary strata between 1938 and 1964, which resulted in notable subsidence of the piezometric surfaces (Caputo *et al.* 1970; Barbujani 1973; Mazzalai *et al.* 1978; Borgia *et al.* 1982); the marked subsidence which took place here in the 1970s also seems to have been caused by these activities (Elmi 1984; Bondesan *et al.* 1986, 1990).

The rates recorded in the most recent surveys are decidedly lower in this area, although in general they are still higher than those attributable to natural subsidence.

In the Ravenna area and the coastal plain as far as Rimini, subsidence is mainly due to the intensive extraction of underground water for uses connected with industry, tourism and agriculture, especially in the 1960s and 1970s, and to a lesser extent to the extraction of methane from deep formations (Carbognin *et al.* 1978a; Bertoni *et al.* 1987, 1995); in these areas marked subsidence of the piezometric surfaces has also been recorded. Subsequently the area has been served with surface water brought by new canals and aqueducts. The latest IGM levellings again registered marked subsidence in various zones, once again probably due to the effects of water extraction and other causes related to recent urban and industrial developments, although in general the phenomenon is diminishing; this reduction is also demonstrated by the recent surveys carried out by another society, Idroser (Regione Emilia Romagna 1994a).

In the areas near Bologna and Modena too, subsidence observed from previous levelling has been primarily attributed to excessive exploitation of groundwater (Arca and Cardini 1977; Pieri and Russo 1977, 1978, 1980, 1984, 1985; Barbarella *et al.* 1990). Lowering of the piezometric surfaces was found here as well, and analogous phenomena were recorded in the 1970s near Ferrara and in various tracts along line 15. The most recent levellings show that the rate of subsidence has since then increased in all these areas, especially near Bologna, in Modena and to the east of Forlì. In general these are areas where there has been, in recent times, an expansion of urban and industrial areas accompanied by the extraction of water from the subsoil (Regione Emilia Romagna 1994b).

The artificial subsidence of the Venice hinterland has also been attributed to exploitation of deep groundwater primarily for industrial use (Caputo *et al.* 1972; Mozzi *et al.* 1973; Carbognin *et al.* 1978b). Extraction here has now been strictly limited, and the most recent levellings show a considerable reduction in subsidence. Although subsidence has never been very great in this area, it has nevertheless been important, given the vulnerability of the area.

The causes of uplift movements may be varied. For example, those indicated by the Rimini–Bologna levelling in 1972 may be due to problems of stability of one of the benchmarks used for the height reference. However, one cannot exclude the possibility of the presence of neotectonic phenomena in some tracts characterized by particular movements, as noted by Arca and Beretta (1985) for the movements between 1897 and 1957. In the present study it seems reasonable to use this explanation for the majority of the uplift observed to the north of the Po along the Parma–Verona line, and in correspondence with the 'Lessino-Berico-Euganea dorsal' along the Padova–Verona line.

21.6 CONCLUSIONS

The method adopted in the present study, using an Information System of Levelling Data for numerous levellings all carried out with the same degree of precision, has made it possible to compare homogeneous data over a long period and over a very large area.

This has enabled us to update the picture which emerged from the levellings carried out up to the 1970s; at the same time it has been possible to investigate the causes underlying the observed movements.

Comparing the data for the 1970s with the most recent data, it has been shown that the rates of subsidence recorded between the rivers Po and Adige, and in particular in the area around the Po delta, have markedly diminished: here the positive consequences of halting (since 1964) the extraction of water containing methane from shallow strata have been clearly demonstrated.

Our comparisons have also made it possible to ascertain a reduction in the high rate of subsidence which had been recorded in the Venice hinterland and in the coastal plain between Rimini and Ravenna, where the main cause was in fact the excessive extraction of groundwater for industrial and agricultural purposes. These needs have clearly been satisfied by the construction of new aqueducts and new irrigation canals.

Nonetheless, groundwater withdrawal still seems to be the main cause behind the artificial subsidence in the eastern Po Plain. It is this factor which is responsible for the high and steadily increasing rates of subsidence recorded at Bologna and in the surrounding area, at Modena, between Forlì and Cesena and in other areas, especially to the south of the Po: in general in these areas there has recently been substantial urban development with, in particular, the arrival of new small and medium-sized industrial plants.

It is still necessary to explain various cases of subsidence, and especially those recorded during the latest levellings, in areas where there has been no exploitation of groundwater. This is especially the case for the agricultural land situated in the coastal depressions or located nearby; despite a reduction in the rate of subsidence, rates have been recorded which are higher than the natural rates in areas where water containing methane was extracted; it is, however, unlikely that this is only the residual effect of such activities, since they ceased 30 years ago.

It therefore seems likely that the main causes, which are well known, probably act alongside other causes: one example could be the artificial control of groundwater in fields. In many agricultural areas there is excessive irrigation, which leads to excessive drainage by pumping stations, and when the water in the surrounding canals is lowered too much, to excessive lowering of the water table.

The possibility of recording ground movements over long periods of time and over large areas has enabled general observations to be made about the influence of geology on subsidence. For example, this research has made it possible to identify a generalized tendency for uplift towards Verona. Here, as on the Berici Hills, it seems reasonable to suspect the role of neotectonic phenomena.

ACKNOWLEDGEMENTS

The authors would like to thank the engineer Luciano Surace and the cartographer Alfonso Marchioni for information about the surveys carried out by the IGM, and Professor Giovanni Battista Castiglioni of the Department of Geography at the University of Padua for documentation and advice. This research was carried out using MURST 40% funds: Eastern

Po Plain, regional and applied geomorphological studies (Professor M. Bondesan); Po Plain, new regional geomorphological themes (Professor G.B. Castiglioni); CNR funds: Venetian Lagoon System Project (Professor M. Bondesan).

REFERENCES

Arca, S. and Beretta, G.P. 1985. Prima sintesi geodetico-geologica sui movimenti verticali del suolo nell'Italia Settentrionale (1897–1957). *Bollettino di Geodesia e Scienze Affini*, **44**(2), 125–156.

Arca, S. and Cardini, A. 1977. Analisi degli spostamenti verticali del suolo nella città di Bologna. *Bollettino di Geodesia e Scienze Affini*, **36**(4), 439–450.

Barbarella, M., Pieri, L. and Russo, P. 1990. Studio dell' abbassamento del suolo nel territorio bolognese mediante livellazioni ripetute: analisi dei movimenti e considerazioni statistiche. *Inarcos*, **506**, Bologna, 1–19.

Barbujani, E. 1973. Evoluzione recente del delta del Po. *Annali Università di Ferrara*, suppl. **1**, 29–54.

Bartolini, C., Bernini, M., Carloni, G.C., Costantini, A., Federici, P.R., Gasperi, G., Lazzarotto, A., Marchetti, G., Mazzanti, R., Papani, G., Pranzini, G., Rau, A., Sandrelli, F., Vercesi, P.L., Castaldini, D. and Francavilla, F. 1983. Carta neotettonica dell'Appennino Settentrionale. Note illustrative. *Bollettino Società Geologica Italiana*, **101**(4), 523–549.

Bertoni, W., Brighenti, G., Gambolati, G., Gatto, P., Ricceri, G. and Vuillermin F. 1987. *Risultati degli Studi e delle Ricerche sulla subsidenza a Ravenna*. Comune di Ravenna.

Bertoni, W., Brighenti, G., Gambolati, G., Ricceri, G. and Vuillermin, F. 1995. Land subsidence due to gas production in the on- and off-shore natural gas fields of the Ravenna area. *Proceedings of the 5th International Symposium on Land Subsidence*. The Hague, The Netherlands, IAHS, **234**, 13–20.

Bitelli, G., Gatti, M. and Russo, P. 1993. An information system for levelling data in land deformation surveys. *Proceedings of the 7th International Symposium on Deformation Measurements*, 3–5 May 1993, Banff, Alberta, Canada, 398–409.

Bondesan, M., Minarelli, A. and Russo, P. 1986. Studio dei movimenti verticali del suolo nella provincia di Ferrara. *Studi idrogeologici sulla Pianura Padana*, **2**, Clup, Milano, 1–31.

Bondesan, M., Minarelli, A. and Russo, P. 1990. Analisi dei movimenti verticali del suolo avvenuti nel periodo 1970–1978 lungo l'asta del Po a est di Polesella e nel delta. In *PO AcquAgricolturAmbiente, 2, L'alveo e il Delta*. Il Mulino, Bologna, 385–404.

Bondesan, M., Castiglioni, G.B., Elmi, C., Gabbianelli, G., Marocco, R., Pirazzoli, P.A. and Tomasin, A. 1995. Coastal areas at risk from surges and sea-level rise in northeastern Italy. *Journal of Coastal Research*, **11**(4), 1355–1379.

Bondesan, M., Gatti, M. and Russo, P. 1997. Movimenti verticali del suolo nella Pianura Padana orientale desumibili dai dati I.G.M. fino a tutto il 1990. *Bollettino di Geodesia e Scienze Affini*, **56**(2), 141–172.

Bondesan, M., Gatti, M. and Russo, P. 2000. Subsidence in the eastern Po Plain (Italy). In Carbognin, L. (ed.), *Proceedings of the 6th International Symposium on Land Subsidence*. Ravenna, Italy, 24–29 September 2000, **2**, 193–204.

Borgia, G., Brighenti, G. and Vitali, D. 1982. La coltivazione dei pozzi metaniferi del bacino polesano e ferrarese. Esame critico della vicenda. *Inarcos*, **425**, Bologna, 13–23.

Bosellini, A. 1971. *Atti tav. Rot. I movimenti del suolo nel Ravennate*. Lions Club, Rotary Club, Ravenna, 11–21.

Braglia, A. 1955. *La Grande Bonificazione Ferrarese. Sec. XV–Sec. XX*. Ferrara, 1–59.

Caputo, M., Pieri, L. and Unguendoli, M. 1970. Geometric investigation of the subsidence in the Po delta. *Bollettino di Geofisica Teorica e Applicata*, **47**, 187–207.

Caputo, M., Folloni, G., Gubellini, A., Pieri, L. and Unguendoli, M. 1972. Survey and geometric analysis of the phenomena of subsidence in the region of Venice and its hinterland. *Rivista Italiana di Geofisica*, **21**(1–2), 19–26.

Carbognin, L. 1986. La subsidenza indotta dall'uomo nel mondo. I casi più significativi. *Bollettino Associazione Mineraria Subalpina*, **23**(4), 433–468.

Carbognin, L., Gatto, P., Mozzi, G. and Gambolati, G. 1978a. Land subsidence of Ravenna and its similarities with the Venice case. In Saxena, J.K. (ed.), *Evaluation and Prediction of Subsidence*. ASCE, New York, 254–266.

Carbognin, L., Gatto, P., Mozzi, G., Gambolati, G. and Ricceri, G. 1978b. Lo sfruttamento delle risorse idriche sotterranee e l'abbassamento del suolo a Venezia. *Atti Convegno Problemi della subsidenza nella politica e difesa del Territorio*. Comune di Pisa, Comune di Ravenna, Regione Emilia Romagna, Regione Toscana, Pisa, **4**, 39–49.

Ciabatti, M. 1959. Ricerche sul costipamento dei terreni quaternari polesani. *Giornale di Geologia*, **27**, 31–101.

Ciabatti, M. 1967. Ricerche sull'evoluzione del Delta Padano. *Giornale di Geologia*, **34**, 1–26.

Ciabatti, M. 1968. Gli antichi delta del Po anteriori al 1600. *Atti Convegno Internazionale di Studi sulle Studi Antichità di Classe*, Ravenna 1967, 23–33.

CNR, 1992. *Structural Model of Italy*. Progetto Finalizzato Geodinamica, Sottoprogetto Modello strutturale tridimensionale. Editor Selca, Firenze.

Elmi, C. 1984. Subsidenza regionale e locale nel Delta del Po. *Atti Tavola Rotonda Metano e Polesine* (Ordine degli Ingegneri di Rovigo e Associazione Nazionale Ingegneri Minerari), 8–11.

ERSA Servizio Bonifiche, 1978. Subsidenza nel territorio del basso ferrarese. *Atti Convegno Problemi della subsidenza nella politica e difesa del Territorio*. Comune di Pisa, Comune di Ravenna, Regione Emilia Romagna, Regione Toscana, Pisa, **7**, 58–60.

Groppi, G. and Veggiani, A. 1984. Geotectonic stucture of the Po Plain from Venice to Ravenna. *Guidebook of the 4th International Symposium on Land Subsidence*, Venice, 1–13.

IGM, 1971. *Relazione sui lavori di livellazione geometrica di alta precisione eseguiti nel 1970 dall'I.G.M. nelle zone della Laguna Veneta e del Delta Padano*. Comitato per lo studio dei provvedimenti a difesa della città di Venezia, 1–170.

Mazzalai, P., Ricceri, G. and Schrefler, B. 1978. Studio della subsidenza nel delta padano. *Atti Convegno Problemi della subsidenza nella politica e difesa del Territorio*. Comune di Pisa, Comune di Ravenna, Regione Emilia Romagna, Regione Toscana, Pisa, **4**, 50–57.

Montori, S. 1983. Effetti della subsidenza sui territori di bonifica. *L'Agricoltore Ferrarese*, **3**, 87–99.

Mozzi, G., Masutti, M., Gatto, P. and Benini, G. 1973. Relazione tra abbattimento della pressione di strato negli acquiferi artesiani ed abbassamento del suolo a Venezia. *Technical Report* 66, Istituto Studio Dinamica Grandi Masse, CNR, Venezia.

MURST, 1997a. *Geomorphological Map of Po Plain*. Selca, Firenze.

MURST, 1997b. *Map on relief and vertical movements of Po Plain*. Selca, Firenze.

Pieri, M. and Groppi, G. 1981. *Subsurface geological structure of the Po Plain, Italy*. Prog. Fin. Geodinamica, sottoprog. Modello Strutturale, **13**(7), Rome, CNR Publication 414, 1–11.

Pieri, L. and Russo, P. 1977. Studio del fenomeno di abbassamento del suolo in atto nella zona di Bologna. *Bollettino di Geodesia e Scienze Affini*, **36**(3), 365–388.

Pieri, L. and Russo, P. 1978. Nuovi contributi allo studio di abbassamento del suolo in atto nella zona di Bologna. *Atti Convegno Problemi della subsidenza nella politica e difesa del Territorio*, Comune di Pisa, Comune di Ravenna, Regione Emilia Romagna, Regione Toscana, Pisa, **5**, 3–24.

Pieri, L. and Russo, P. 1980. Abbassamento del suolo nella zona di Bologna: considerazioni sulle probabili cause e sulla metodologia per lo studio del fenomeno. *Collana di orientamenti geomorfologici ed agronomico-forestali, MTB 6*, Regione Emilia Romagna, Pitagora, Bologna, 1–43.

Pieri, L. and Russo, P. 1984. The survey of soil vertical movements in the region of Bologna. *Proceedings of 3rd International Symposium on Land Subsidence*, Venice, IAHS, **151**, 1–11.

Pieri, L. and Russo, P. 1985. Situazione attuale delle ricerche sull'abbassamento del suolo nel territorio bolognese. *Inarcos*, **456**, Bologna, 57–61.

Pirazzoli, P.A. 1993. Global sea-level changes and their measurement. *Global and Planetary Changes*, **8**, 34–56.

Regione Emilia Romagna, 1994a. *Piano progettuale per la difesa della costa adriatica; aggiornamento e integrazione*. Idroser, Bologna.

Regione Emilia Romagna, 1994b. *Sistematizzazione dei dati ambientali relativi al territorio regionale soggetto a subsidenza (Legge 845/80)*. Idroser, Bologna, 1–111.

Sacco, F. 1900. La Valle Padana. *Atti Regia Accademia Agricola Torino*, **18**, 1–27.

Salvioni, G. 1957. I movimenti del suolo nell'Italia centro-settentrionale. Dati preliminari dedotti dalla comparazione di livellazioni. *Bollettino di Geodesia e Scienze Affini*, **16**(3), 325–366.

TCI, 1957. L'Italia fisica. *Conosci l'Italia*, **1**, Milano, 1–320.

Zanferrari, A., Bollettinari, G., Carobene, L., Carton, L., Carulli, G.B., Castaldini, D., Cavallin, A., Panizza, M., Pellegrini, G.B., Pianetti, F. and Sauro, U. 1982. Evoluzione neotettonica dell'Italia nord-orientale. *Memorie di Scienze Geologiche*, Padova, 44–121.

22 Natural and Anthropogenic Factors Affecting the Recent Evolution of Beaches in Calabria (Italy)

LEANDRO D'ALESSANDRO, LINA DAVOLI, ELVIDIO LUPIA PALMIERI AND
ROSSANA RAFFI
Dipartimento di Scienze della Terra, Università degli Studi di Roma 'La Sapienza', Rome, Italy

ABSTRACT

Calabria has one of the longest coasts of all the Italian regions. This region extends between the Tyrrhenian Sea (west) and the Ionian Sea (east) and its coastline stretches for 736 km. Long-term studies provide a complete and unified picture of the recent variations which have affected its shoreline since the second half of the 19th century and particularly in the last few decades.

This study, which attempts to define both natural and anthropogenic causes of these variations, relies on analysis of documentary evidence as well as on studies of geomorphology, anemometry and pluviometry.

Shoreline dynamics has been investigated using available maps and aerial photographs, which has made it possible to reconstruct its history starting from 1870. On this basis it has been shown that Tyrrhenian beaches were the first to undergo a strong erosional crisis, with the loss of about 3 300 000 m^2 between the mid-1950s and the end of the 1970s; in the following decade this coastline reached relative stability. During the first time interval Ionian beaches were stable, except for some deltaic cusps affected by marked erosion processes; however, in the following decade widespread and significant erosion took place, decreasing the surface of these beaches by about 4 000 000 m^2.

The different trends in evolution of the Tyrrhenian and Ionian beaches may be ascribed to the differences in morphological setting of the catchment basins supplying solid load, as well as to the differences in exposure of the shores to the directions of wind and wave approach.

The erosional crisis may have been triggered, at least in part, by short-term climatic changes. Anemometry studies on the Tyrrhenian coastal zone indicate a sharp increase in wind speed and frequency and a drop in calms between the mid-1950s and the end of the 1970s; such variations are likely to have played a role in the erosional crisis of the Tyrrhenian beaches. In parallel, pluviometry studies carried out throughout Calabria have shown a conspicuous decrease in precipitation in the time interval 1978–1987 compared to 1954–1977 (100 mm a^{-1} less, equal to 1.55×10^9 m^3 a^{-1}). It may be assumed that such a decrease has contributed considerably to the reduction of the solid load supplied to the beaches, as well as the anthropogenic interventions carried out on the drainage basins (including landslide-retaining devices, dams and dredgings).

Applied Geomorphology: Theory and Practice. Edited by R.J. Allison.
© 2002 John Wiley & Sons, Ltd.

22.1   INTRODUCTION

This work describes the geomorphological dynamics observed over more than one century, for the Tyrrhenian and Ionian beaches of Calabria, which extend for more than 690 km of the total shore length of about 736 km. It expands previous research (Lupia Palmieri *et al.* 1981; D'Alessandro *et al.* 1982, 1983, 1991, 1992) which focused mainly on the historical and recent evolution of part of the Tyrrhenian beaches.

The coastal evolution has been investigated by means of maps, aerial photographs and direct field surveys, using a very detailed scale (1:10 000); the results are summarized in a set of maps showing the linear changes of the beaches.

In addition, on the basis of geomorphological studies, anemometry and pluviometry, an attempt has been made to define thoroughly both natural and anthropogenic causes of these variations.

22.2   PHYSIOGRAPHIC CHARACTERS OF THE SHORE

Calabrian coasts extend for about 736 km between the mouth of the Fiume Noce on the Tyrrhenian Sea and Capo Spulico on the Ionian Sea (Figure 22.1). The shore and shelf shapes reflect the complex structural arrangement of the Calabrian arc, which definitively assumed its present features as a result of the different tectonic phases acting from middle Pliocene to Pleistocene. Such movements caused differential uplift and fragmentation of the Calabrian arc into structures having longitudinal and transverse trends, bounded by tectonic lines that match the whole geometric structure of the shore and the shelf (Ghisetti 1980).

The Tyrrhenian shore of Calabria stretches for 257 km from the mouth of the Fiume Noce, marking the boundary with the Basilicata region, to Villa S. Giovanni where the Stretto di Messina begins. The coast is slightly sinuous in its northern part and is shaped into two wide gulfs in the southern part. It can be divided from north to south into three main sectors differing in physiographic and morphodynamic characteristics: F. Noce–Capo Suvero, Golfo di S. Eufemia and Golfo di Gioia.

The first sector extends from the mouth of the F. Noce to Capo Suvero. Its northern portion (F. Noce–Capo Bonifati) is slightly irregular and broken by a few rocky spurs. The southern portion (Capo Bonifati–Capo Suvero) is bordered to the east by the Catena Costiera and its shoreline is often the edge of narrow coastal plains.

The shore extending from Capo Suvero to Villa S. Giovanni is characterized by two wide neighbouring gulfs: S. Eufemia and Gioia. The Golfo di S. Eufemia stretches from Capo Suvero to Capo Vaticano; the northern and central portions have the largest beaches of the whole sector, while in the southern part indented rocky cliffs prevail, in places associated with narrow pebbly or sandy belts and pocket beaches. The Golfo di Gioia stretches from Capo Vaticano to Villa S. Giovanni; the coast of the northern and southern portions has irregular and rocky cliffs with pocket beaches, whereas wide beaches can be found in the centre of the Gulf.

The studied Ionian shore stretches for 435 km from Capo Spulico to Villa S. Giovanni and can be subdivided into five sectors. The first, between Capo Spulico and Punta Alice, is located in the southwestern part of the Golfo di Taranto; it has a rather sinuous trend due to the presence of several deltas. The second sector extends from Punta Alice to Capo Rizzuto; its most important morphological feature is the cusped delta of the Fiume Neto that breaks off the almost meridian trend of the shoreline. This sector ends with the peninsula of Crotone, characterized by a rocky promontory with pocket beaches.

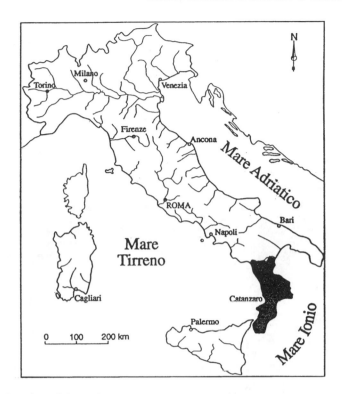

Figure 22.1 Location of the study area

To the southwest of Capo Rizzuto the third sector begins: it extends as far as Punta Stilo and corresponds to the Golfo di Squillace. Its shore, at places dotted with rocky cliffs, consists mainly of beaches which are rectilinear, because of the lack of well-developed emerged deltas, in spite of the very large number of streams.

The sector between Punta Stilo and Capo Spartivento forms a slightly curved inlet oriented NE–SW; the beaches are fed by solid load from numerous short-lived streams draining rather steep slopes.

The southernmost sector of the Ionian coast extends from Capo Spartivento to Villa S. Giovanni; its shore changes orientation several times, reaches the Stretto di Messina and ties up with the Tyrrhenian coasts. Numerous small deltas make this shoreline sinuous.

22.3 RECENT CHANGES TO BEACHES

The shoreline variations recorded from 1870 to 1990 have been analysed. The cartographic documents relevant to the earlier period (1870–1954) are often lacking in reference points, which has prevented precise comparison between the shorelines of different years; however, it has been possible to single out evolutionary trends over the whole period (Figure 22.2). In contrast, the more detailed documents of the most recent period (1954–1990) allowed the shoreline variations to be quantitatively analysed (Figure 22.3).

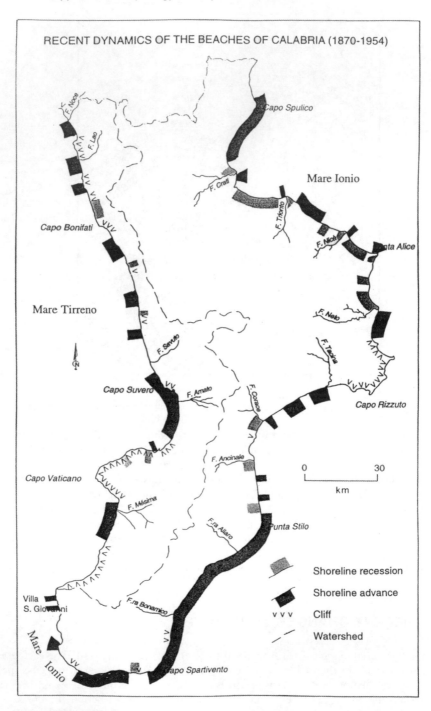

Figure 22.2 Evolutionary trends of the beaches of Calabria in the oldest of the considered periods (1870–1954)

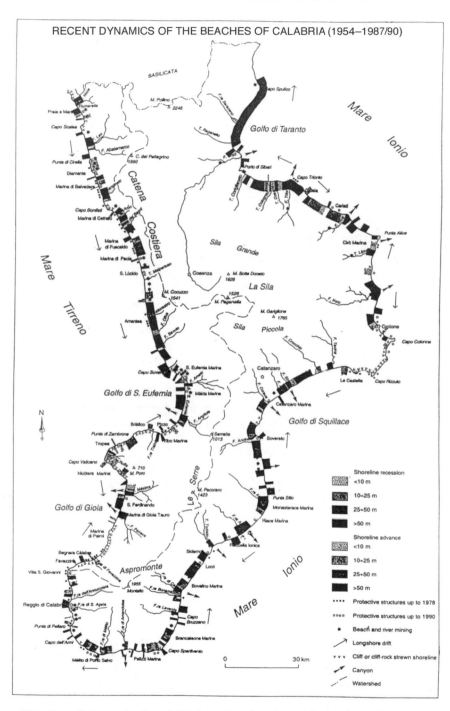

Figure 22.3 Quantitative evaluation of Calabrian shoreline changes during the period 1954–1987/90

As already evidenced in other Italian areas, the Calabrian shores show a particular evolutionary trend characterized by a widespread widening of the beaches during the early period, and by their strong recent reduction (Caputo *et al.* 1991).

In the case of Calabria the trend towards beach widening or reduction has been diachronous on the two sides. In particular, a general progression of the shoreline took place along the Tyrrhenian side in the period 1870–1954 (Figure 22.2) and was especially marked in correspondence to the main deltas. In the same time interval a general widening characterized the southern beaches of the Ionian side, while the northern ones were affected by discontinuously spread phenomena of feeding and erosion. In the most recent period (1954–1987 for the Tyrrhenian beaches, 1954–1990 for the Ionian ones) a generalized erosion prevails throughout (Figure 22.3); it occurred on the Tyrrhenian shores starting in the second half of the 1960s and successively, starting in the 1980s, on the Ionian shores (Figures 22.4, 22.5).

22.3.1   The Tyrrhenian Beaches

Widespread instability conditions were recorded between 1954 and 1978 along the whole sector from F. Noce to Capo Suvero and particularly in its southern portion.

The northern portion (F. Noce–Capo Bonifati) was characterized by erosion, especially along the main deltas; the process affected in particular the alluvial plain of the Fiume Lao-Abatemarco, which lost about $200\,000\,m^2$ of its delta lobes. This conspicuous erosional process was probably linked to the spreading of river and beach quarrying. Moreover, a marked shoreline regression occurred along the beaches of many coastal towns, for instance Diamante and Alba Calabra; consequently numerous longitudinal and transverse protecting structures were built.

In summary, during the time interval 1954–1978 the beaches between F. Noce and Capo Bonifati underwent erosion for 60% of their length, losing more than $640\,000\,m^2$. During the following interval (1978–1987) erosion was less marked and much less widespread; in fact, the beaches lost only about $35\,000\,m^2$. This testifies to a general equilibrium, although shoreline recession still continued along the main deltas.

In the Capo Bonifati–Capo Suvero portion the beach sedimentary budget relevant to the last decade strongly differed from that of the previous ones.

Between 1954 and 1978 the beaches lost more than $2\,000\,000\,m^2$, the maximum recorded erosion rate in Calabria (Table 22.1). This erosional crisis occurred along more than 90% of beach extent; it affected many coastal towns and caused considerable damage to the coast communication lines (railways and roads), making it necessary to build sea defences.

In contrast, between 1978 and 1987 erosion affected only 15% of the total beach length; at the same time accretion of more than $560\,000\,m^2$ was recorded which, however, failed to restore the 1954 situation (Table 22.1). This type of dynamics characterized mainly the beaches protected by sea defences built from the 1970s until more recent times (D'Alessandro *et al.* 1991).

The study of the beach dynamics at Golfo di S. Eufemia has evidenced a prevailing erosional trend between 1954 and 1978, which affected above all the beaches in the central-southern portion of the gulf; in contrast, along the northern part there was an accretion of beaches. The reducing beaches cover 56% of the overall length of the gulf coastline and are located essentially in the Fiume Angitola–Capo Vaticano stretch. As a whole, the beaches lost about $337\,000\,m^2$ (Table 22.1).

Analogously to the previously examined sectors, the dynamics of the Golfo di S. Eufemia beaches changed in the time interval 1978–1987. In fact, these years were characterized by a

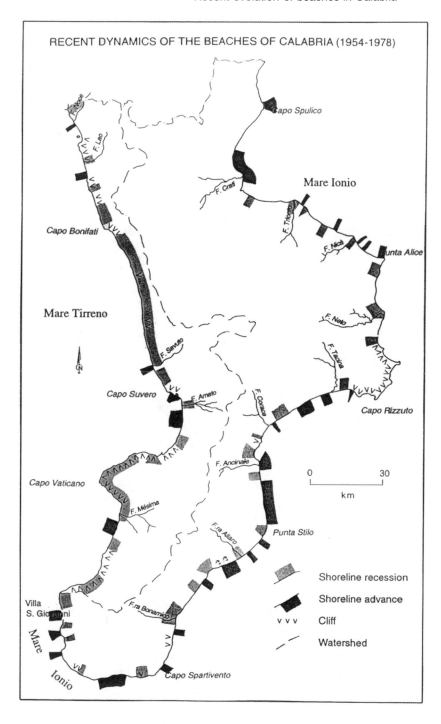

Figure 22.4 Evolutionary trends of the beaches of Calabria during the period 1954–1978

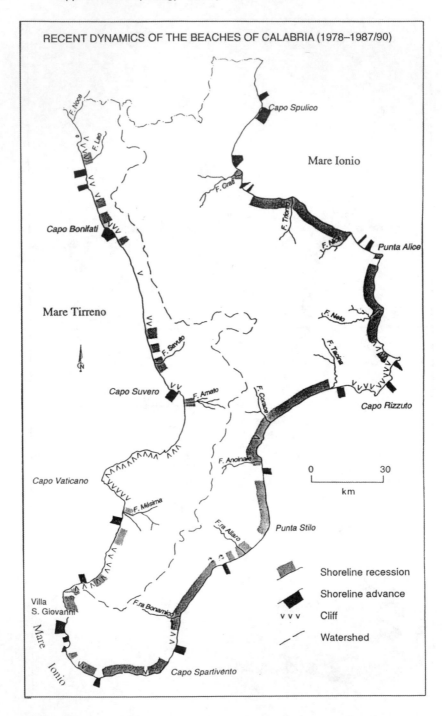

Figure 22.5 Evolutionary trends of the beaches of Calabria in the period 1978–1987/90

Table 22.1 Erosion data for Tyrrhenian beaches

Period	Areal change (m^2)	Annual mean of areal change (m^2)	Linear mean change (m)	Annual mean of linear change (m)
F. Noce–Capo Bonifati (46 km)				
1954–1978	−641 075	−26 711	−13.94	−0.58
1978–1987	−35 435	−3937	−0.77	−0.09
1954–1987	−676 510	−20 500	−14.71	−0.45
Capo Bonifati–Capo Suvero (75 km)				
1954–1978	−2 170 929	−90 455	−28.95	−1.21
1978–1987	586 991	63 221	7.59	0.84
1954–1987	−1 601 938	−48 544	−21.36	−0.65
Capo Suvero–Capo Vaticano (68 km)				
1954–1978	−337 700	−14 071	−4.97	−0.21
1978–1987	270 237	30 026	3.97	0.44
1954–1987	−67 463	−2044	−0.99	−0.03
Capo Vaticano–Villa S. Giovanni (68 km)				
1954–1978	−153 528	−6397	−2.26	−0.09
(1977–1978)	511 014*			
1978–1987	−103 916	−11 546	−1.53	−0.17
	115 763*			
1954–1987	−257 444	−7801	−3.79	−0.11
	626 777*			
Total: F. Noce–Villa S. Giovanni (257 km)				
1954–1978	−3 303 232	−137 635	−12.85	−0.54
1978–1987	699 877	77 764	2.72	0.30
1954–1987	−2 603 355	−78 890	−10.13	−0.31

* Artificial accretion due to the building of the Giola Tauro Harbour

prevailing accretion of beaches along the northern and central portions of the gulf and by an equilibrium along the southern one, with the exception of the beach at the mouth of Fiume Amato.

In the Golfo di Gioia (Capo Vaticano–Villa S. Giovanni) the analysis carried out for the 1954–1978 period has shown a prevailing erosional trend, variable in intensity from place to place. Erosion affected 49% of the gulf beaches and the maximum shoreline regression was recorded at the apex of the Mesima and Petrace river deltas (Table 22.1). The other beaches of this sector showed a poorly marked erosional trend due to the effects of protective structures or remarkable widening tied to the building of the Gioia Tauro harbour.

Excluding the artificial beaches connected to the harbour, erosional processes continued during the 1978–1987 period, and a severe and persistent beach line retreat was recorded at the Mesima and Petrace river deltas.

The morphological evolution of the Tyrrhenian beaches was thus characterized by different trends in the two periods 1954–1978 and 1978–1987 (Table 22.1). During the first period erosion phenomena prevailed and were particularly marked in the northern part; the beach total loss attained 3 000 000 m^2 (Table 22.1), making the building of the sea defences necessary. During the second period other protective structures were added which significantly contributed to reducing erosion, although it persisted along the deltaic apexes (Figure 22.5).

The dynamics of the central portion of the Golfo di Gioia were the most peculiar and complex; in fact the presence of a submarine canyon cutting the continental platform, controlled the sedimentary budget in the littoral zone favouring the persistent shifting of sediments basinwards, and thus their definitive loss (Colantoni *et al.* 1992).

22.3.2 The Ionian Beaches

In the period 1954–1978, the sector between Capo Spulico and Punta Alice was characterized by relative stability, particularly evident in the northernmost portion (delta of Fiume Crati); the southern part, including the deltaic flats of the rivers Trionto and Nicà, was an exception as it underwent remarkable erosion which involved about 15% of the beaches. It should be underlined that this relative general stability of the northern Ionian beaches is in contrast with the significant reduction which affected the northern Tyrrhenian beaches during the same period (Figure 22.4).

During the following period (1978–1990), in contrast, a widespread regression trend prevailed; more than 70% of the beaches of this Ionian sector underwent a remarkable reduction and area losses came to over $1\,400\,000\,m^2$ (Figure 22.5; Table 22.2). The already critical situation of numerous deltas became worse because of erosion increasing in the later years.

The second sector, which stretches from Punta Alice to Capo Rizzuto, showed a general instability trend in the period 1954–1978. The most conspicuous erosional phenomena occurred along the northern area of the Fiume Neto delta and along the beaches north of Crotone. Along the southernmost portion the small pocket beaches in the rocky overhangs retained their overall balance (Figure 22.4; Table 22.2).

In 1978–1990 a more widespread and remarkable erosion occurred; it involved up to 60% of the beaches of the whole sector, which had already been suffering erosive crisis. The most noticeable reductions were recorded along the whole delta of the F. Neto and came to a total loss of more than $510\,000\,m^2$, with a yearly mean rate of 2.5 m (Figure 22.5). For this sector as well, therefore, the last time period witnessed severe erosional processes, which showed themselves with a delay of about ten years in comparison to the Tyrrhenian side.

The Golfo di Squillace, between Capo Rizzuto and Punta Stilo, recorded a general progression along the northern and southern beaches from 1954 to 1978; only the mouth areas of most streams flowing into the central portion of the gulf showed a severe erosional crisis (Figure 22.4).

During 1978–1990 widespread erosion affected the beaches of the gulf. This phenomenon took place in stretches of shore that had previously been characterized by wide beaches; in the areas where narrow beaches existed, the shoreline regression involved urban facilities. On the whole, between 1978 and 1990, $862\,000\,m^2$ of the Golfo di Squillace beaches disappeared, equal to a mean linear reduction of 8.3 m a^{-1} (Figure 22.5; Table 22.2). The erosive phenomenon, which involved 67% of beaches, was remarkably intense along the central and northern portions of the gulf.

In 1954–1978 the dynamics of Punta Stilo–Capo Spartivento beaches generally showed negative area variations of limited extent, erosion being restricted mainly to the deltaic lobes (Figure 22.4).

During the following time period (1978–1990), in contrast, a widespread and remarkable trend to regression prevailed (Figure 22.5); 81% of the beaches of the whole sector were subject to erosion with area losses exceeding $1\,400\,000\,m^2$, which implied the highest linear

Table 22.2 Erosion data for Ionian beaches

Periods	Areal change (m²)	Annual mean of areal change (m²)	Linear mean change (m)	Annual mean of linear change (m)
Roseto Capo Spulico–Punta Alice (108 km)				
1954–1978	530 374	22 099	4.91	0.20
1978–1990	−1 433 401	−119 450	−13.27	−1.11
1954–1990	−903 027	−25 084	−8.36	−0.23
Punta Alice–Capo Rizzuto (63 km)				
1954–1978	−157 678	−6750	−2.50	−0.10
1978–1990	−658 545	−54 879	−10.45	−0.87
1954–1990	−816 223	−22 673	−12.96	−0.36
Capo Rizzuto–Punta Stilo (104 km)				
1954–1978	233 890	9745	2.25	0.09
1978–1990	−861 941	−71 828	−8.29	−0.69
1954–1990	−628 051	−17 446	−6.04	−0.17
Punta Stilo–Capo Spartivento (80 km)				
1954–1978	−71 342	−2973	−0.89	−0.04
1978–1990	−1 431 880	−119 323	−17.90	−1.49
1954–1990	−1 503 222	41 756	−18.79	−0.52
Capo Spartivento–Villa S. Giovanni (80 km)				
1954 1978	26 212	1092	0.33	0.01
1978–1990	−309 745	−25 812	−3.87	−0.32
1954–1990	−283 533	−7876	−3.54	−0.10
Total: Roseta Capo Spulico–Villa S. Giovanni (435 km)				
1954–1978	561 456	23 394	1.29	0.05
1978–1990	−4 695 512	−391 293	−10.79	−0.90
1954–1990	−4 134 056	−114 835	−9.50	−0.26

recession among those recorded on the whole Ionian shoreline (Table 22.2). The intensity of this phenomenon can be compared with that affecting the Capo Spulico–Punta Alice sector, but its effects were much more disastrous because it involved beaches that were narrower than the northern ones by nature.

The Capo Spartivento–Villa S. Giovanni sector was marked by a substantial stability of the beaches between 1954 and 1978 (Figure 22.4; Table 22.2). However, it is noteworthy that along the westernmost portion of this sector, facing the Stretto di Messina, the beaches downcurrent of the S. Elia harbour were subject to strong erosion; actually this coastal stretch has always suffered the highest anthropogenic impact of the whole of Ionian Calabria and therefore its natural dynamics have been deeply disturbed by beach defences and harbour structures.

During the following period (1978–1990), erosion became intense and widespread, involving up to 56% of beaches; a positive trend mainly produced by artificial nourishment was recorded at places along the westernmost portion of this sector.

Summing up, the beaches between Capo Spulico and Villa S. Giovanni have shown a moderate instability during the 1954–1978 period, which has mainly affected the mouths of the numerous streams; in the last time period (1978–1990), a more widespread and intense erosion involved 70% of the whole shoreline, with a total area loss of about $4\,700\,000\,m^2$ (Table 22.2).

22.3.3 Remarks

On the Ionian beaches the erosional processes have taken place with a delay of over a decade in comparison to the Tyrrhenian ones. The different evolution of the Tyrrhenian and Ionian beaches can be ascribed to the different configuration of the catchment basins and to the exposure of the coasts to different directions of wind and wave approach. The Tyrrhenian beaches are thin strips extending at the foot of steep slopes, whose watershed is displaced towards the Tyrrhenian Sea, involving considerable dissymmetry between the Tyrrhenian and the Ionian side. Consequently, Tyrrhenian streams are short and their catchment basins are small. Conversely, wider beaches are found on the Ionian side, with gentler and longer slopes, especially in its northern portion. These beaches are supplied by streams with larger catchment basins, characterized by frequent outcrops of highly erodible lithologies (Ibbeken and Scheleyer 1991). Furthermore, the two coastal areas also have differently shaped seabeds: along the Tyrrhenian Sea, the bottom is steeper and the abrasion platform is narrower than along the Ionian Sea (Rossi and Gabbianelli 1978; Colella and Normark 1986; Pennetta 1992). These differences imply that Tyrrhenian coasts are more vulnerable and thus they were the first to undergo erosion.

22.4 CAUSES OF BEACH EROSION

Investigations of the dynamics of Calabrian beaches have concerned both the anthropogenic and the physical factors.

The factors directly or indirectly contributing to shore erosion are many and interacting, and it is very difficult to single out the role that each one plays in erosional processes. However, information drawn from previous research and from this study on Calabrian beaches allows the main causes of shoreline regression to be identified.

Among the anthropogenic factors, the building of settlement facilities has been of particular significance. Towards the end of the 19th century the whole coastal belt showed extremely different environmental conditions in comparison with the present day. The only noteworthy facility was the railway, which was discontinuous; a coastal road ran only along some stretches of the southern Ionian shore linking a few settlements that constituted the littoral municipalities of some inland villages. Towards the end of the 1960s the growing recreational interest in the still uncontaminated beaches produced widespread urbanization that often involved the levelling of the backshore and coastal dunes and the modification of riverbeds. The Tyrrhenian coast was the first to suffer from the enlargement of urban resorts; later, between the end of the 1970s and the beginning of the 1980s, the Ionian coast was also affected in a remarkable way by this phenomenon. Small settlements near railway stations spread towards the seaside increasing in area by up to ten times; urbanization extended up to a few metres from the seashore, deeply altering the natural arrangement and size of the beaches.

As already mentioned, the erosional crisis affected the Tyrrhenian beaches first (Figures 22.6, 22.7 and 22.8). Aiming to reverse this crisis, various perpendicular and longitudinal beach defences were erected (Table 22.3); these unquestionably contributed to slowing the erosive crisis, but could not restore the previous width of the beaches and gave rise to further strong erosion on the downcurrent beaches.

Figure 22.6 Along the Tyrrhenian side the beach stretching from Diamante to Alba Calabria has sustained severe erosion threatening both the road and the railway (Aerial view 1990 – Authorization SMA no. 524, 4/11/1997)

Figure 22.7   Along the Tyrrhenian side the beach between Capo Tirone and the Fiume Sangineto mouth has been severely eroded and beach nourishment and sea defences have been necessary (Aerial view 1990 – Authorization SMA no. 524, 4/11/1997)

Figure 22.8   Marina di Cetraro: to the south of the harbour, shoreline recession affects wide urbanized areas and the beach of Marina di Cetraro in particular. The sea defences in this area have been responsible for sand accumulation updrift and beach erosion downdrift (Aerial view 1990 – Authorization SMA no. 524, 4/11/1997)

Table 22.3   Beach defences on Tyrrhenian beaches

Period	Length (km)	%
Fiume Noce–Capo Bonifati (46 km)		
1954–1978	5.375	11.68
1978–1987	2.025	4.40
1954–1987	7.400	16.09
Capo Bonifati–Capo Suvero (75 km)		
1954–1978	11.875	15.83
1978–1987	4.250	5.67
1954–1987	16.125	21.50
Capo Suvero–Capo Vaticano (68 km)		
1954–1978	2.850	4.19
1978–1987	2.890	4.25
1954–1987	5.740	8.44
Capo Vaticano–Villa S. Giovanni (68 km)		
1954–1978	6.980	10.26
1978–1987	0.730	1.07
1954–1987	7.710	11.34
Total: Fuime Noce–Villa S. Giovanni (257 km)		
1954–1978	27.080	10.54
1978–1987	9.920	3.86
1954–1987	37.000	14.40

The Ionian beaches reached a crisis later on; however, the sparser urbanization as well as the more inland positions of the railway and road did not require such widespread protection as the ones on the Tyrrhenian side, where the communication lines are very close to the beaches due to orographic conditions. In fact sea defences erected along the Ionian side were less extensive and mainly located near urban settlements (Table 22.4).

The quarrying activity connected with the building boom that took place starting in the 1960s was another important cause of beach erosion. This activity affected river beds and beaches as well.

Official data relevant to the Tyrrhenian territory show that in the 1970s and 1980s licences were issued for quarrying more than $1\,300\,000\,m^3$ (D'Alessandro et al. 1983); an amount of illegal quarrying that is impossible to establish must be added to this figure. There is no doubt that quarrying contributed to the reduction of the solid load supplied to the beaches, together with the many dykes built across most of the 'fiumare' flowing to the Tyrrhenian shore.

Along the Ionian side too, intense quarrying took place between Capo Rizzuto and Villa S. Giovanni; licensed quarrying from 1970 to 1986 along the watercourses and beaches amounts to $1\,200\,000\,m^3$.

Another factor which dramatically contributed to the reduction of the solid load to the sea was the complex of landslide-retaining devices and channellization (Figure 22.9) carried out to a massive extent in Calabria since the beginning of the 1950s, to contain the extended and remarkable mass movements troubling the region. For example, in the drainage basins discharging into the sea between Capo Rizzuto and Villa S. Giovanni, 370 km of embankments and 1915 dykes along stream channels were built between 1955 and 1985, in addition to a large number of reservoirs; moreover, landslide consolidation and reafforesting have been executed,

Table 22.4 Beach defences on Ionian beaches

Period	Length (km)	%
Roseto Capo Spulico–Punta Alice (108 km)		
1954–1978	1.050	0.97
1978–1990	0.825	0.76
1954–1990	1.875	1.74
Punta Alice–Capo Rizzuto (63 km)		
1954–1978	1.250	1.98
1978–1990	3.250	5.16
1954–1990	4.500	7.14
Capo Rizzuto–Punta Stilo (104 km)		
1954–1978	2.800	2.69
1978–1990	0.000	0.00
1954–1990	2.800	2.69
Punta Stilo–Capo Spartivento (80 km)		
1954–1978	3.400	4.25
1978 1990	0.000	0.00
1954–1990	3.400	4.25
Capo Spartivento–Villa S. Giovanni (80 km)		
1954–1978	7.100	8.88
1978–1990	2.000	2.50
1954–1990	9.100	11.38
Total: Roseto Capo Spulico–Villa S. Giovanni (435 km)		
1954–1978	15.600	3.59
1978–1990	6.100	1.40
1954–1990	21.700	4.99

amounting to a total of 47 885 hectares, equivalent to approximately 20% of the territory above. These interventions, although helping to lessen the geomorphological instability of the catchment basins, were responsible for a conspicuous decrease in the solid load allotted to the sedimentary balance of the beaches.

Among the natural phenomena affecting shore dynamics, eustatic oscillations of sea level should be mentioned. The present rate of sea-level rise is too slow (1.1–1.6 mm a^{-1}) to have much influence on short-term coastline changes; furthermore, it is balanced by the continuous uplift which has been been affecting this area as testified by Quaternary marine terraces (Cosentino and Gliozzi 1992). Likewise, a close relationship between shoreline regression and coastal plain subsidence has not been proved. On the contrary, short-term climatic changes had a role in triggering the erosive crisis of Calabrian beaches, as proved by a study of both the anemometry and pluviometry.

22.4.1 Anemometry

The Capo Palinuro station (184 m a.s.l.), located on a rocky head of the Tyrrhenian shore, is the only one that has continuously recorded anemometry for more than 50 years (Figure 22.10). Observations have evidenced a decrease in calms and a parallel marked increase in

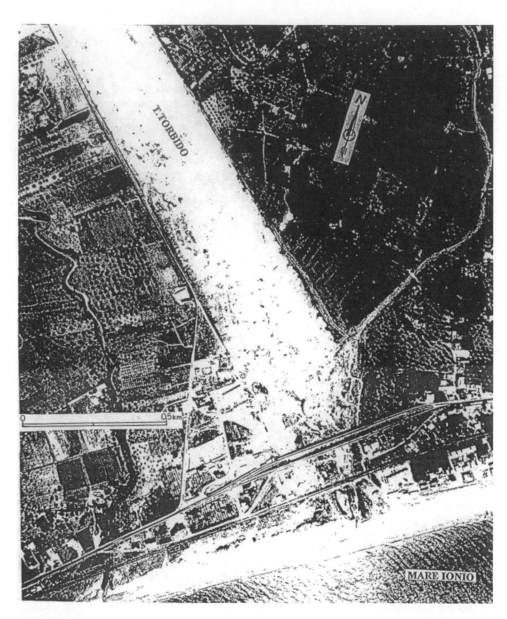

Figure 22.9 Old channellization of the Torrente Torbido close to the mouth; more recent quarrying sites are evident as white spots along the whole river bed (Aerial view 1978 – Authorization SMA no. 524, 4/11/1997)

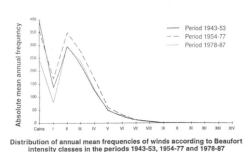

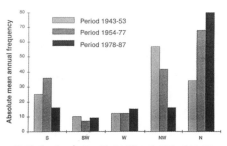

Distribution of annual mean frequencies of winds according to Beaufort intensity classes in the periods 1943-53, 1954-77 and 1978-87

Distribution, for octants and in the different periods, of annual mean frequencies of onshore winds having speed higher than 10 knots

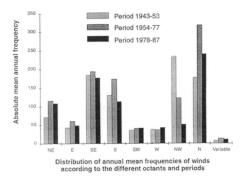

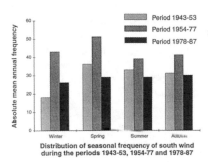

Distribution of annual mean frequencies of winds according to the different octants and periods

Distribution of seasonal frequency of south wind during the periods 1943-53, 1954-77 and 1978-87

Figure 22.10   Station of Capo Palinuro (Tyrrhenian side): wind climate

strong winds recorded in the 1954–1977 time period, in contrast to the previous period from 1943 to 1953 (D'Alessandro *et al.* 1992).

Changes in the frequency of onshore winds have also been observed (Figure 22.10); in particular, during the years 1954–1977 the mean annual frequency of northerly and southerly winds strongly increased, and doubled for northerlies with speeds exceeding 10 knots. Furthermore, the seasonal distribution of northerlies, responsible for the most effective waves, changed: minimum and maximum frequencies reversed, so that in the 1954–1977 period the maximum value occurred in summer and the minimum in winter. As a consequence, it is likely that the severe erosion of the Tyrrhenian beaches over the years 1954–1977 was also caused by changes in the meteorological marine climate.

Wind conditions in the periods 1954–1977 and 1978–1987 were significantly different. During the latter, a strong decrease in the wind mean frequency was recorded as well as a 24% decrease of winds up to 10 knots and a 21% decrease of winds between 11 and 27 knots (Figure 22.10).

Concerning the onshore winds, from 1978–1987, mean annual frequency of the mistral and southern winds decreased respectively by 60% and 24%, while the mean annual frequency of other onshore winds remained almost unchanged. For wind speed lower than 10 knots, the mean annual frequency of northerly winds decreased by 36%. Taking into account only winds with speeds exceeding 10 knots, southerly winds decreased by 53% over the years 1978–

1987; the mistral also underwent a strong reduction (62%). In contrast, northerly winds recorded an increase of 19%; such an increase, however, was not great enough to balance the decrease recorded by the whole of the onshore winds from northern octants.

In conclusion, the analysis of anemometry for the Capo Palinuro station has shown that between 1978 and 1987 winds were less frequent and less intense. Although it is valid only for the Tyrrhenian beaches of Calabria, this favourable change in the wind may have helped to reduce the erosional trend.

22.4.2 Precipitation Changes

As stated above, the massive erosive crisis that involved the beaches of Calabria can be partly related to the sediment yield deficit due to quarrying and landslide-retaining devices. To single out other possible natural causes of sediment yield reduction, an analysis of pluviometric flow has been carried out, lacking suitable data on surface run-off and solid load. By analysing precipitation data recorded at 123 stations, the pluviometric regime of Calabria has been examined. The data set refers to the period between 1954 and 1987; where possible, precipitation analysis has been extended back to 1921 for the most significant stations.

In a first phase the annual, monthly and seasonal mean values of precipitation and rainy days have been calculated both for the whole period 1954–1987 and for the two subsets 1954–1977 and 1978–1987. Moreover, the mean annual precipitation has been elaborated for the Tyrrhenian and Ionian slopes separately according to the Thiessen polygonal mapping method, in order to compare precipitation in the different periods with changes in the beach dynamics.

To provide evidence of the possible existence of a pluviometric trend, the five-point moving average has been calculated for annual precipitation. On the basis of the percentage differences between moving and general averages the linear regression lines (linear model $y = bx + a$), whose parameters have been estimated by quadratic regression methods, have been obtained (Tables 22.5 and 22.6). In Figures 22.11 and 22.12 the results obtained for two sample stations are shown. To the same end the statistics tests of Mann–Kendall and of Spearman (Giuffrida and Conte 1991) have also been applied both to annual precipitations and rainy days (Tables 22.5, 22.6 and 22.7). Finally the data sets have been submitted to the homogeneity test of Thom, according to which the series is assumed as homogeneous with an 80% confidence interval (Tables 22.5, 22.6 and 22.7).

As far as the seasonal distribution of precipitation is concerned, the autumn and winter months are the most rainy ones, supplying more than 70% of the annual total; they are followed by the spring and the summer months. Comparing the percentage differences of the annual and seasonal averages of precipitation and rainy days of 1954–1977 with 1978–1987, it has been noticed that the latter period was less rainy than the former and that the months mostly suffering from this decrease are the autumn and the summer ones, followed by the winter ones. As to the annual averages for the whole of Calabria, 80% of 123 stations taken into account recorded a decrease in precipitation and 93% a decrease in the number of rainy days between 1978 and 1987.

A difference between the Tyrrhenian and Ionian slopes was evident: for the latter, the precipitation decrease was recorded for 93% of the stations, in comparison with 78% of the Tyrrhenian side. Such a difference between the two slopes is remarkable in autumn, winter and summer. During the spring some increase in precipitation in 62% of stations on both slopes was noticed. The Ionian slope, however, has the highest number of stations with negative variations (38% against 33% for the Tyrrhenian slope). Actually, although the two

Table 22.5 Statistical analysis of precipitation data

Stat.	b	a	r	r²	Spear.	M-K	Thom
1	-0.50	+8.04	-0.59	0.35	s	s	OM
2	-0.33	+6.72	-0.32	0.10	s	s	/
3	-0.21	+3.77	-0.89	0.80	s	s	OM
4	-0.12	+1.07	-0.62	0.39	s	s	OM
5	-1.45	+22.90	-0.88	0.77	–	–	/
6	-1.19	+18.63	-0.65	0.43	–	–	/
7	-0.11	+1.17	-0.73	0.53	s	s	OM
8	-1.40	+22.81	-0.76	0.58	–	–	/
9	-3.90	+61.19	-0.96	0.91	–	–	/
10	+0.02	+0.89	+0.10	0.01	s	s	/
11	-1.24	+18.59	-0.84	0.71	–	–	/
12	-0.72	+12.06	-0.67	0.44	s	s	OM
13	-0.38	+7.92	-0.35	0.12	s	s	OM
14	-0.99	+14.51	-0.65	0.47	–	–	OM
15	-0.18	+1.76	-0.69	0.48	s	s	OM
16	-1.00	+14.22	-0.74	0.54	–	–	OM
17	-0.65	+8.98	-0.52	0.27	s	s	OM
18	-0.47	+8.21	-0.39	0.15	s	s	OM
19	+0.84	-13.70	-0.61	0.37	s	s	OM
20	+0.16	-2.15	+0.41	0.17	s	s	OM
21	+0.16	-2.55	+0.92	0.85	s	s	OM
22	-0.44	+7.61	-0.53	0.28	s	s	OM
23	+0.05	+0.63	+0.49	0.24	s	s	OM
24	-1.78	+28.39	-0.83	0.69	–	–	/
25	-0.21	+3.32	-0.43	0.19	s	s	OM
26	-0.73	+10.17	-0.40	0.16	s	s	OM
27	-0.41	+6.38	-0.38	0.14	s	s	/
28	-0.45	+6.21	-0.90	0.81	s	s	/
29	+0.61	-8.84	+0.40	0.16	s	s	OM
30	+1.89	-36.63	+0.53	0.28	s	s	/
31	-0.74	+11.73	-0.47	0.22	s	s	OM

Stat.	b	a	r	r²	Spear.	M-K	Thom
63	+0.01	+0.75	+0.01	0.00	s	s	OM
64	+0.39	-7.33	+0.39	0.15	s	s	OM
65	-0.52	+8.14	-0.53	0.28	s	s	OM
66	+0.09	-0.38	+0.14	0.02	s	s	OM
67	+0.17	-1.98	+0.10	0.01	s	s	OM
68	+0.04	-0.66	+0.06	0.00	s	s	/
69	-0.30	+4.79	-0.33	0.11	s	s	OM
70	-0.19	+3.85	-0.23	0.05	s	s	OM
71	-1.45	+22.21	-0.82	0.67	–	–	OM
72	-0.29	-5.83	-0.34	0.12	s	s	OM
73	+0.71	-9.05	+0.53	0.28	s	s	OM
74	-0.96	+13.08	-0.77	0.59	s	s	OM
75	-0.66	+8.43	-0.72	0.52	s	s	OM
76	+0.17	-3.75	+0.20	0.04	s	s	OM
77	-2.25	+34.62	-0.84	0.71	–	–	/
78	-0.19	+3.30	-0.30	0.09	s	s	OM
79	-0.13	+4.64	-0.12	0.01	s	s	OM
80	-0.03	+0.14	-0.06	0.00	s	s	OM
81	-0.85	+11.96	-0.84	0.70	s	s	OM
82	-0.68	+9.88	-0.75	0.57	s	s	OM
83	-0.30	+4.80	-0.34	0.12	s	s	OM
84	-0.98	+15.55	-0.78	0.61	s	s	OM
85	-0.20	+2.46	-0.21	0.04	s	s	OM
86	-0.63	+10.37	-0.53	0.28	–	–	/
87	-0.62	+8.87	-0.59	0.35	s	s	OM
88	-0.44	+7.11	-0.61	0.37	s	s	OM
89	-0.05	+0.06	-0.07	0.00	s	s	/
90	-3.24	+51.28	-0.93	0.87	s	s	/
91	-0.49	+8.43	-0.56	0.32	s	s	OM
92	-0.33	+9.37	-0.35	0.12	s	s	/
93	+0.19	-2.92	+0.29	0.08	s	s	OM

continues overleaf

Table 22.5 (cont.)

Stat.	b	a	r	r²	Spear.	M-K	Thom
32	+0.34	−4.97	+0.72	0.52	s	s	OM
33	−0.89	+11.94	−0.44	0.19	s	s	OM
34	−0.66	+9.68	−0.39	0.15	−	−	OM
35	−0.51	+7.71	−0.43	0.19	s	s	OM
36	−0.26	+3.30	−0.80	0.64	s	s	/
37	−0.99	+14.19	−0.78	0.61	−	−	/
38	−0.15	+0.65	−0.37	0.13	s	s	OM
39	−0.45	+6.39	−0.47	0.22	s	s	OM
40	+0.16	−26.95	+0.67	0.44	s	s	/
41	−0.29	+5.04	−0.35	0.12	s	s	/
42	−0.90	+15	−0.51	0.26	−	−	OM
43	−1.27	+22.64	−0.55	0.31	−	−	OM
44	−0.04	+0.21	−0.10	0.01	s	s	OM
45	−0.48	+5.36	−0.47	0.22	s	s	OM
46	−0.08	+0.29	−0.59	0.35	s	s	OM
47	−0.26	+3.09	−0.89	0.80	s	s	/
48	−1.21	+17.28	−0.73	0.53	−	−	OM
49	−0.94	+15.44	−0.64	0.41	s	s	OM
50	−0.25	+3.68	−0.61	0.37	s	s	/
51	−2.06	+29.08	−0.79	0.62	−	−	OM
52	−1.54	+21.88	−0.75	0.56	−	−	OM
53	−0.89	+14.10	−0.47	0.22	s	s	OM
54	−1.43	+20.17	−0.74	0.55	−	−	/
55	+0.55	+9.91	+0.53	0.28	−	−	/
56	−0.09	+2.67	−0.09	0.01	s	s	OM
57	−0.52	+9.04	−0.51	0.26	s	s	/
58	−0.48	+5.60	−0.66	0.44	−	−	/
59	−0.71	+10.47	−0.79	0.62	−	−	OM
60	−0.27	+5.67	−0.32	0.10	s	s	/
61	+0.89	−14.98	+0.72	0.51	+	+	OM
62	−3.00	+45.24	−0.90	0.81	−	−	/

Stat.	b	a	r	r²	Spear.	M-K	Thom
94	−0.44	+7.18	−0.52	0.27	s	s	OM
95	−0.80	+11.77	−0.65	0.42	s	−	/
96	−0.69	+11.46	−0.66	0.43	s	s	OM
97	−0.37	+5.33	−0.47	0.22	s	s	OM
98	−0.31	+3.54	−0.33	0.11	−	−	/
99	−0.23	+3.04	−0.17	0.03	−	−	/
100	−0.48	+7.61	−0.50	0.25	s	s	OM
101	−0.89	+14.85	−0.64	0.41	s	s	OM
102	−0.86	+13.58	−0.74	0.54	s	s	/
103	−0.31	+5.21	−0.43	0.19	s	s	/
104	−1.67	+26.47	−0.91	0.83	−	−	OM
105	−2.17	+33.75	−0.76	0.58	−	−	OM
106	−0.69	+10.79	−0.66	0.44	s	s	OM
107	+0.09	+0.81	+0.11	0.01	s	s	/
108	+0.31	−4.18	+0.45	0.20	s	s	OM
109	−0.90	+13.97	−0.72	0.51	s	s	OM
110	−0.84	+13.51	−0.83	0.68	s	s	/
111	−0.35	+5.08	−0.33	0.11	s	s	OM
112	−0.32	+4.42	−0.39	0.15	s	s	OM
113	−1.58	+24.13	−0.83	0.7	−	−	/
114	−0.19	+3.79	−0.13	0.02	−	−	/
115	−1.43	+21.87	−0.68	0.47	s	s	OM
116	−0.57	+8.79	−0.47	0.22	−	−	/
117	−1.12	+16.63	−0.73	0.54	s	s	OM
118	−1.40	+23.02	−0.80	0.64	s	s	OM
119	−1.54	+20.80	−0.79	0.62	−	−	/
120	−0.32	+6.04	−0.278	0.077	−	−	/
121	−0.39	+5.83	−0.40	0.16	s	s	OM
122	−0.91	+15.48	−0.64	0.41	s	s	OM
123	−0.65	+10.27	−0.62	0.39	−	−	OM

Model for linear regression line: $y = bx + a$.
Spear., Spearman test; M-K, Mann–Kendall test

Table 22.6 Statistical analysis of precipitation data

Stat.	b	a	r	r^2	Spear.	M-K	Thom
6	+4.04	−0.27	−0.22	0.05	s	s	/
8	−8.35	+0.68	+0.43	0.19	+	+	/
11	+3.39	−0.30	−0.36	0.13	s	s	/
14	−13.53	+0.93	+0.52	0.28	+	s	/
16	−11.58	+0.72	+0.57	0.32	+	+	/
24	−16.11	+1.20	+0.57	0.33	+	+	/
34	+2.46	−0.10	−0.05	0.003	s	s	/
48	−6.18	+0.31	+0.20	0.04	s	s	/
63	+4.20	−0.22	−0.22	0.05	s	s	OM
81	+0.91	−0.05	−0.05	0.002	s	s	/
95	−7.87	+0.51	+0.44	0.19	s	s	/
113	−1.77	+0.14	+0.10	0.01	s	s	OM

slopes acted differently, an overall reduction in total rainfall was recorded in the years 1978–1987; it mostly concerns autumn, then winter and summer.

The mean annual precipitation analysis has shown that in the period 1954–1987 1233 mm of rain fell on the Tyrrhenian slope and 1098 mm on the Ionian one; the comparison between data relevant to 1954–1977 and 1978–1987 has shown that during the latter period both slopes recorded a rain average of about 100 mm less than during the former period (1160 mm versus 1264 mm for the Tyrrhenian slope and 1029 mm versus 1126 mm for the Ionian one). Moreover, a strong difference in the rain amount between Tyrrhenian and Ionian stations was evident; the latter stations registered over 130 mm of rain less than the former both over the whole period and in the subsets of time. This different behaviour, which also involves the shoreline dynamics, is tied to the orographic characteristics of Calabria; in fact the Apennine chain, crossing this region longitudinally, acts as an obstacle to the typical western perturbations and makes them release their humidity mostly on the Tyrrhenian slope.

As concerns the statistical processing, analysis of the correlation coefficients (r) has indicated a general decreasing trend (Tables 22.5 and 22.6). On the whole, taking into consideration both slopes, 103 processed series out of 123 have shown negative correlation coefficients and, of these, 72 have supplied a good correlation with r values higher than 0.5.

Analysing the results of the statistical tests on the annual precipitation totals of the 123 stations of both slopes, 85 cases of no change, 37 cases of a negative trend and one case of a positive trend have been shown by the Mann–Kendall method, while 79 cases of no change, 43 cases of a negative trend and one case of a positive trend have been shown by the Spearman method (Tables 22.5 and 22.6; Figure 22.13a). Analysis of rainy days has shown 108 cases of Mann–Kendall negative trend and 109 of Spearman negative trend (Table 22.7, Figure 22.13b).

The decrease of rainy days compared to the reduction in rainfall amount has been noticed in a greater number of stations; it is to be emphasized, however, that a parallel smaller homogeneity in the series of rainy days has been shown by the Thom test (Table 22.7).

By comparing the r columns with the 'Spear' and 'M-K' ones in Tables 22.5 and 22.6, it can be observed that the high correlation coefficients refer to the stations which, according to the two tests, show a negative trend.

The statistical processing performed on the 123 examined series has singled out the existence of a remarkable decrease in rainfall in the 1978–1987 period in comparison with

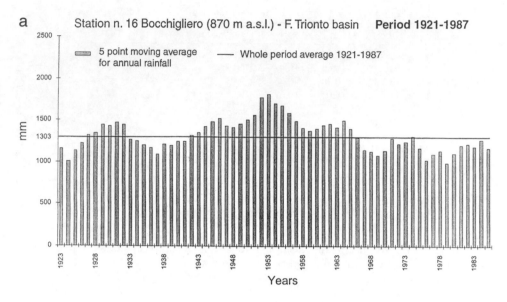

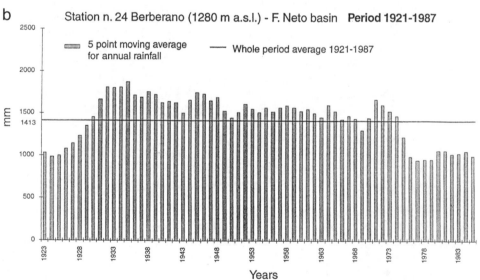

Figure 22.11 Moving average for annual rainfall (period 1921–1987). (a) Station 16 Bocchigliero (870 m a.s.l.), Fiume Trionto basin; (b) Station 24 Berberano (280 m a.s.l.), Fiume Neto basin

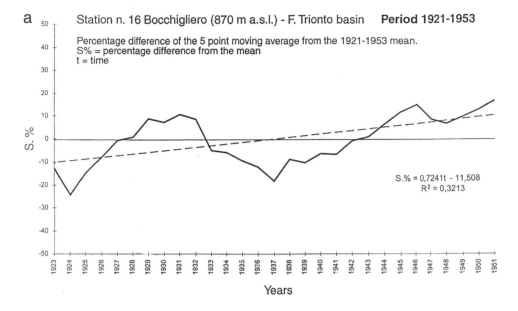

a   Station n. 16 Bocchigliero (870 m a.s.l.) - F. Trionto basin   **Period 1921-1953**

Percentage difference of the 5 point moving average from the 1921-1953 mean.
S% = percentage difference from the mean
t = time

S.% = 0,7241t - 11,508
R² = 0,3213

Years

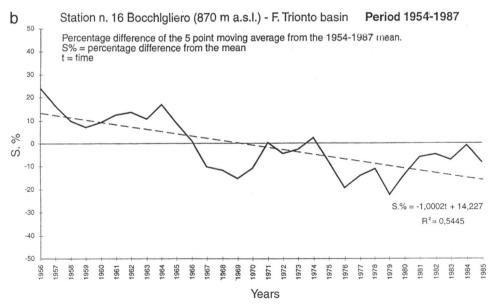

b   Station n. 16 Bocchigliero (870 m a.s.l.) - F. Trionto basin   **Period 1954-1987**

Percentage difference of the 5 point moving average from the 1954-1987 mean.
S% = percentage difference from the mean
t = time

S.% = -1,0002t + 14,227
R² = 0,5445

Years

Figure 22.12   Moving average for annual rainfall; ———, linear regression line; – – –, trend line. Station 16 Bocchigliero (870 m a.s.l.), Fiume Trionto basin: (a) period 1921–1953; (b) period 1954–1987. Station 24 Berberano (1280 m a.s.l.), Fiume Neto basin: (c) period 1921–1953; (d) period 1954–1987

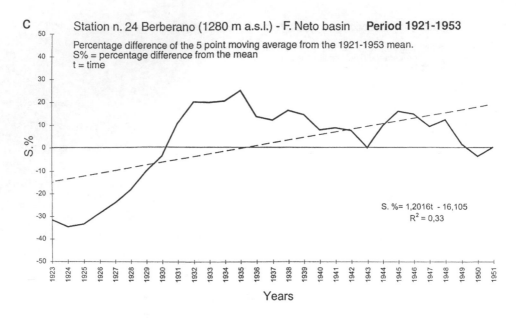

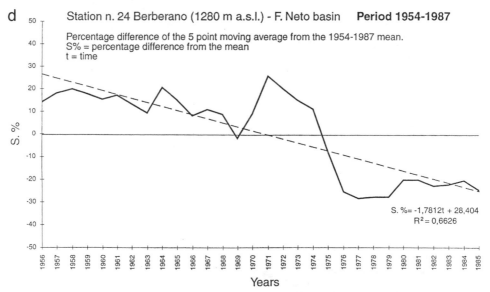

Figure 22.12   *(cont.)*

Table 22.7 Statistical analysis of precipitation data

Stat.	S.%	Spear.	M-K	Thom	Stat.	S.%	Spear.	M-K	Thom
1	−18	−	−	/	63	−2	s	s	OM
2	−8	−	−	OM	64	−16	−	−	/
3	−17	−	−	/	65	−15	−	−	OM
4	−28	−	−	/	66	−9	−	−	/
5	−8	−	−	/	67	−6	−	−	/
6	−14	−	−	/	68	−9	−	−	/
7	−11	−	−	/	69	−16	−	−	/
8	−8	−	−	/	70	−5	−	−	/
9	−47	−	−	/	71	14	s	s	/
10	1	−	−	OM	72	−6	−	−	OM
11	−12	−	−	/	73	−4	s	s	OM
12	−9	−	−	OM	74	−6	−	−	OM
13	−14	−	−	OM	75	−18	−	−	OM
14	−16	−	−	/	76	−1	s	s	/
15	−1	−	−	/	77	−36	−	−	/
16	−11	−	−	/	78	−7	−	−	OM
17	−14	−	−	/	79	−4	−	−	/
18	−14	−	−	/	80	1	−	s	/
19	−3	s	s	OM	81	−25	−	−	/
20	2	s	s	OM	82	−12	−	−	OM
21	−13	−	−	OM	83	−18	−	−	OM
22	−16	−	−	/	84	−7	−	−	OM
23	8	s	s	OM	85	−4	−	−	/
24	−4	s	s	OM	86	−11	−	−	OM
25	−4	−	−	/	87	−14	−	−	/
26	−19	−	−	/	88	−4	−	−	OM
27	−12	−	−	OM	89	12	s	s	OM
28	−10	−	−	OM	90	−20	−	−	/
29	2	s	s	/	91	−7	−	−	/
30	−30	−	−	/	92	−5	−	−	/
31	−11	−	−	/	93	−13	−	−	OM
32	−26	−	−	/	94	−5	s	s	OM
33	−6	−	−	/	95	−17	−	−	/
34	−54	−	−	/	96	−6	−	−	OM
35	−9	−	−	/	97	−12	−	−	OM
36	−2	−	−	OM	98	−14	−	−	/
37	−17	−	−	/	99	−5	−	−	/
38	−5	−	−	/	100	−12	−	−	OM
39	−3	−	−	/	101	−23	−	−	/
40	−2	s	s	/	102	−6	−	−	OM
41	−15	−	−	/	103	0	s	s	OM
42	−13	−	−	/	104	−25	−	−	OM
43	−15	−	−	OM	105	−31	−	−	/
44	−3	−	−	OM	106	−10	−	−	OM
45	−16	−	−	/	107	−8	−	−	OM
46	−4	−	−	OM	108	−33	−	−	/
47	−4	−	−	OM	109	−5	−	−	/
48	−9	−	−	/	110	−3	−	−	OM

continues overleaf

Table 22.7 (*cont.*)

Stat.	S.%	Spear.	M-K	Thom	Stat.	S.%	Spear.	M-K	Thom
49	−13	–	–	/	111	−3	–	–	OM
50	−16	–	–	/	112	−21	–	–	OM
51	−32	–	–	/	113	−33	–	–	/
52	−23	–	–	/	114	−7	–	–	OM
53	−7	–	–	/	115	−33	–	–	/
54	−29	–	–	/	116	−28	–	–	/
55	−10	–	–	/	117	−30	–	–	/
56	−4	–	–	OM	118	−25	–	–	/
57	−36	–	–	/	119	−18	–	–	/
58	−7	–	–	/	120	−9	–	–	/
59	−17	–	–	OM	121	−3	–	–	/
60	−8	–	–	/	122	−10	–	–	/
61	−9	–	–	OM	123	−4	s	s	/
62	−26	–	–	/					

1954–1977 (100 mm a^{-1} less, equal to $1.55 \times 10^9 \, m^3 \, a^{-1}$). Such a decrease is likely to have contributed considerably to the reduction of detrital supply to the beaches, together with the anthropogenic interventions carried out in the drainage basins (landslide-retaining devices, dams, dredgings, etc.).

22.5 FINAL CONSIDERATIONS

In agreement with previous studies carried out on parts of the Calabrian shore, it can be concluded that all the beaches of this region recently underwent a strong erosional crisis which started at different times, involving first the Tyrrhenian beaches and later the Ionian ones.

The investigations carried out have shown the influence on these erosional phenomena of both anthropogenic and natural causes.

Anthropogenic causes that affected the shoreline dynamics include human settlements on the shoreline strip, quarrying activities, landslide-retaining devices, channellization, harbour structures and beach defences.

Concerning the influence of natural causes on the erosional crisis, an important role seems to have been played by the short-term variations, of still uncertain recurrence, in the wind climate and in precipitation amount. These climatic elements had not been previously investigated in enough depth to recognize them as driving forces behind Calabrian beach dynamics.

The results achieved, particularly those on variations in precipitation, suggest that this kind of research should be pursued for the same region and for different Italian areas.

ACKNOWLEDGEMENTS

This work is the result of a series of research projects financially supported by the National Council for Research (CNR) and by the Ministry for University and Scientific and Techno-logical Research (Director of Research: Professor E. Lupia Palmieri).

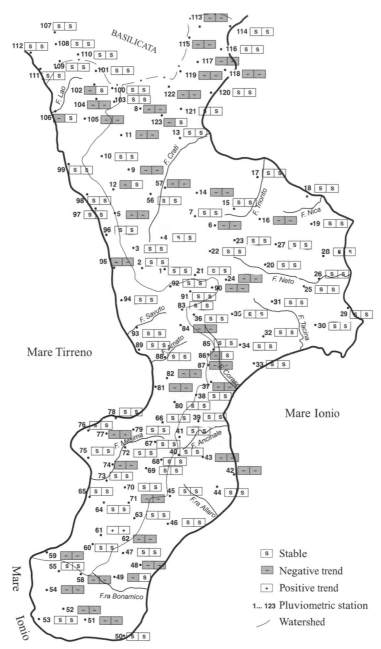

Figure 22.13 (a) Evolutionary trend of mean annual precipitation over the period 1954–1987 estimated by Mann–Kendall and by Spearman tests. (b) Evolutionary trend of number of rainy days in the period 1954–1987 estimated by Mann–Kendall and Spearman tests

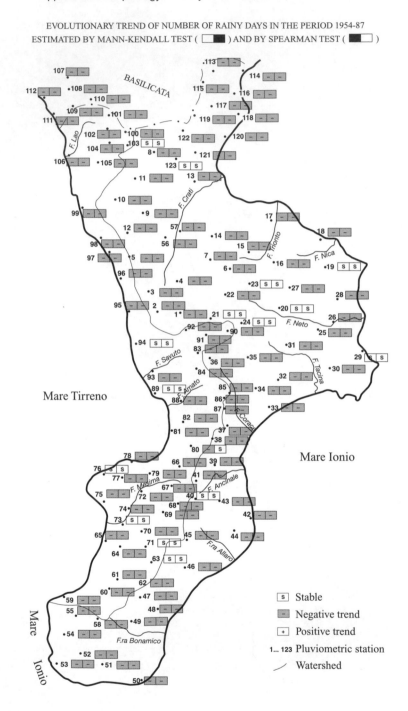

Figure 22.13 (b)

REFERENCES

Caputo, C., D'Alessandro, L., La Monica, G.B., Landini, B. and Lupia Palmieri, E. 1991. Present erosion and dynamics of Italian beaches. *Zietschrift für Geomorphologie*, Suppl. Bd., **81**, 31–39.

Colantoni, P., Genesseaux, M., Vanney, J.R., Ulrege, A., Melegari, G. and Trombetta, A. 1992. Processi dinamici del canyon sottomarino di Gioia Tauro (Mare Tirreno). *Giornale di Geologia*, **54**, 199–213.

Colella, A. and Normark, W.R. 1986. High-resolution side-scanning sonar survey of delta slope and inner fan channels of Crati submarine fan (Ionian Sea). *Mem. Soc. Geol. It.*, **27**, 381–390.

Consiglio Nazionale delle Ricerche, 1985–1996. Atlante delle Spiagge Italiane. Dinamismo – Tendenza evolutiva – Opere Umane. *Prog. Fin. Conservazione del Suolo – Dinamica dei Litorali*, SELCA Firenze.

Cosentino, D. and Gliozzi, E. 1992. Considerazioni sulle velocità di sollevamento di depositi eutirreniani dell'Italia meridionale e della Sicilia. *Mem. Soc. Geol. It.*, **41**, 653–665.

D'Alessandro, L., Davoli, L., Fredi, P. and Lupia Palmieri, E. 1982. Il litorale calabro compreso tra il delta del Fiume Savùto e Capo Bonifati: evoluzione recente della spiaggia e variazione del regime anemometrico. *Prog. Fin. Conservazione del Suolo – Dinamica dei Litorali*, CNR **201**.

D'Alessandro, L., Davoli, L., Fredi, P., Lupia Palmieri, E. and Raffi, R. 1983. Beach erosion on the Tyrrhenian coast of Calabria – Consideration about natural and man induced causes. In Bird, C.F. and Fabbri, P. (eds), *Proceedings of the Symposium 'Coastal problems in the Mediterranean Sea'*. Venice. IGU Commission on the Coastal Environment, 69–81.

D'Alessandro, L., Davoli, L. and Lupia Palmieri, E. 1991. Recent evolution of the beaches belonging to the Tyrrhenian side of Northern Calabria: The harbour of Cetraro. In *Symposium on 'Geomorphology of Active Tectonics Areas'*. Geodata, CNR-IRPI, **39**, 89–92.

D'Alessandro, L., Davoli, L., Fredi, P., Lupia Palmieri, E. and Raffi, R. 1992. Recent dynamics of the Tyrrhenian beaches of Calabria (Southern Italy). *Boll. Oceanol. Teor. Appl.*, **10**, 187–195.

Ghisetti, F. 1980. Evoluzione neotettonica dei principali sistemi di faglie della Calabria Centrale. *Boll. Soc. Geol. It.*, **98**, 387–430.

Giuffrida, A. and Conte, M. 1991. L'evoluzione a lungo termine del clima italiano. *Mem. Soc. Geogr. It.*, **46**, 329–341.

Ibbeken, H. and Scheleyer, R. 1991. *Source and Sediment. A case study of Provenance and Mass Balance at an Active Plate Margin (Calabria, Southern Italy)*. Springer-Verlag, Berlin.

Lupia Palmieri, E., D'Alessandro, L., Raffi, R., Pranzini, E., Amore, C., Giuffrida, E. and Cataldo, P. 1981. Primi risultati delle indagini di Geografia fisica, Sedimentologia e Idraulica marittima sul litorale del golfo di S. Eufemia. *Prog. Fin. Conservazione del Suolo – Dinamica dei Litorali*, CNR **127**.

Pennetta, M. 1992. Morfologia e sedimentazione della piattaforma continentale e scarpata nel tratto di costa compreso tra Punta Alice e Capo Rizzuto (Golfo di Taranto). *Boll. Soc. Geol. It.*, **111**, 149–161.

Rossi, S. and Gabbianelli, G. 1978. Geomorfologia del Golfo di Taranto. *Boll. Soc. Geol. It.*, **97**, 423–437.

23 Effects of Hurricane Roxanne on Coastal Geomorphology in Southeastern Mexico

JOSÉ LUIS PALACIO-PRIETO AND MARIO ARTURO ORTIZ-PÉREZ

Instituto de Geografía, Universidad Nacional Autónoma de México, Coyoacán, México

ABSTRACT

Hurricanes are responsible for important geomorphological changes along coastlines. Along the southeastern Gulf coast of Mexico, the frequent strike of hurricanes is reported to have a major influence on the configuration of the shoreline. This chapter focuses on the geomorphological effects and related damage caused by Hurricane Roxanne (8–19 October 1995) along the shoreline of the State of Campeche, southeastern Gulf coast of Mexico.

The study is based on the interpretation of vertical video images obtained from a helicopter. Results of the stereo video interpretation were integrated on a map showing the location of processes and effects to infrastructure. The surveyed coastline is divided into four sectors according to its litho-geomorphological characteristics and the consequent effects of the hurricane: (a) higher rocky coast (limestone); (b) lower coast consisting of lithified parallel beach ridges; (c) sand-bars and spits; and (d) mainland deltaic depositional coast.

Hurricane Roxanne's effects on coastal geomorphology are strongly influenced by the geomorphological zonation. Effects on infrastructure are related to the topographical position in relation to the coast and sea level. Ancient intertidal channels represent risk zones since their functioning is reactivated during extraordinary events.

23.1 INTRODUCTION

Hurricanes are responsible for major changes in the landscape. Besides the profound effect on human activities, hurricanes are claimed to have a predominant influence on the structure of marine and terrestrial ecosystems, biomass and biodiversity (Lodge and McDowell 1991; Waide 1991; Tanner *et al*. 1991; Brocaw and Walker 1991).

Hurricanes are also responsible for important geomorphological changes along coastlines. For the southeastern Gulf coast of Mexico, the frequent strike of hurricanes is reported to have a major influence on the configuration of the shoreline (Manzano 1989; Ortiz-Pérez 1992) where accumulation and erosion patterns alternate along hundreds of kilometres and marine transgressions have locally been hundreds of metres in the last decades (Ortiz-Pérez 1992). As a consequence of this, damage to infrastructure has been commonly reported in the area (Palacio *et al*. 1996, Matías 1997).

The strongest Caribbean hurricanes originate off the western coast of Africa, gain strength as they pass westward over the Atlantic Ocean and rapidly deteriorate when they reach the mainland somewhere between Panama and New England. 'Weaker' hurricanes and tropical

Applied Geomorphology: Theory and Practice. Edited by R.J. Allison.

Table 23.1 Coverage and spatial resolution of video images. The coverage is referred to one individual image, digitized in a 320 × 240 pixel format

	Coverage (m²)		Pixel resolution (m)
	Horizontal	Vertical	
Camera 1	400	300	1.25
Camera 2	16	12	0.05

storms (wind speed 63–119 km h^{-1}) have their origins throughout the Atlantic Ocean, the Caribbean Sea and the Gulf of Mexico (Walker 1991).

This chapter focuses on the geomorphological effects and related damage caused by Hurricane Roxanne (8–19 October 1995) along the shoreline of the State of Campeche, southeastern Mexico. Special emphasis is given to the interpretation of erosion–accumulation features, palaeogeomorphology, damage to road infrastructure and the use of alternative remote-sensing tools based on a colour video remote-sensing system for rapid assessment.

23.2 METHODS AND MATERIALS

A video system consisting of two colour CCD cameras (8 mm format) installed in a helicopter was used to obtain images from the coast of the State of Campeche. Images were acquired on 6 December 1995, one and a half months after the pass of Roxanne. Both cameras were synchronized using a GPS interface so that the Greenwich Meridian Time was visible on the video tape; this facilitated the visualization of a larger portion of terrain (c. 400 × 300 m²) and a zoom (c. 16 × 12 m²) to identify details such as vegetation types and more local effects. The flight altitude was 500 m above the ground; the general characteristics of the images obtained are described in Table 23.1. The total length of the coastal survey was about 170 km.

Imagery was first interpreted visually on a colour TV monitor (24 inches) to define general processes, landforms and damage. In a second phase, images were selected, digitized and mosaics for stereo visual interpretation were prepared; commercial multimedia software was used for this purpose. Mosaics were georeferenced for measuring purposes using a PC-based GIS; hard copies of key areas and features (scale 1:750, both stereo- and mono-vision) were of great help during interpretation. Individual video images and mosaics were used for estimating distances and areas using the GIS capabilities.

Results of the video interpretation were finally integrated in a map showing the location of processes and affectations and a general interpretation was performed.

23.3 GENERAL CHARACTERISTICS OF HURRICANE ROXANNE

The track of Roxanne was unusually erratic (Table 23.2 and Figure 23.1). The tropical depression no. 19 became a tropical storm and Hurricane Roxanne on 10 October. Then, the speed of winds decreased and increased again. Roxanne stayed about six days leading to an erratic route along the southern Gulf of Mexico, about 200 km offshore of Campeche, causing major flooding and geomorphological changes and damage along the coastal area.

Table 23.2 Activity of Hurricane Roxanne (8–19 October 1995)

Date	Lat. (°N)	Long. (°W)	Atmos. pressure (mb)	Speed (mph)	Category
8 Oct.	16.0	83.2	1003	35	TD19
9 Oct.	17.5	83.0	1003	40	TS Roxanne
10 Oct.	19.6	84.3	987	75	Hurricane
12 Oct.	19.5	92.9	993	60	TS
14 Oct.	22.4	93.4	979	75	Hurricane
17 Oct.	20.7	92.4	990	70	TS
19 Oct.	22.5	96.0	1008	35	TD

Data from Tropical Prediction Center, Miami, 1996
TD, tropical depression; TS, tropical storm.

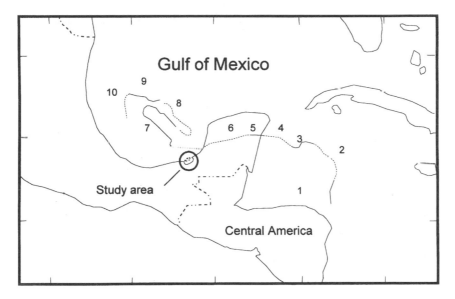

Figure 23.1 Ten-day track of Hurricane Roxanne (10–19 October 1995). After Palacio *et al.* 1999

23.4 THE STUDY AREA AND GEOMORPHOLOGICAL CONDITIONS

The study was carried out along 170 km of the coast of the State of Campeche, southeastern Gulf of Mexico, from the San Pedro river up to the city of Campeche (Figure 23.2).

The northernmost sector geologically belongs to the Cainozoic limestone platform of Yucatan (Figure 23.3); south of this sector, lithified sediments (beach-rock) are arranged in parallel beach ridges (Psuty 1965) along 40 to 50 km, from Ceiba Playa to Sabancuy. To the south, the beach-rock is covered by sediments although locally outcrops exposed by wave action are identified. The western sector is represented by recent deltaic sedimentary deposits of major Mexican rivers draining into the Gulf of Mexico (Yáñez 1971). Sediments in this sector show an arrangement of parallel ridges, locally interpreted as a result of alternate accretion and erosion periods (Yáñez 1971; Coll 1975; Gutiérrez y Castro 1988); Ortiz-Pérez (1992) has reported the occurrence of differential sinking along this sector due to

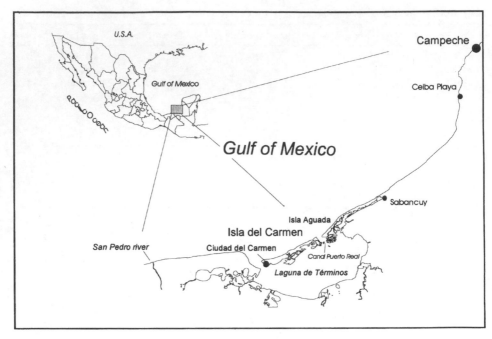

Figure 23.2 Location of the study area; survey was carried out along 170 km of the mainland coast between the San Pedro river and the city of Campeche

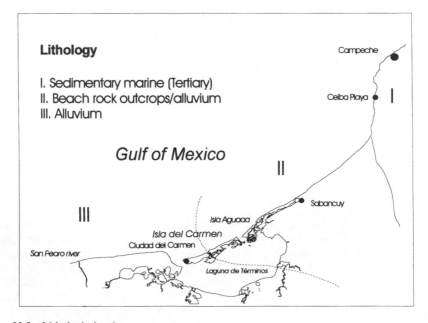

Figure 23.3 Lithological units

sedimentation, resulting in a relative sea-level rise, which in turn may be related to coastal erosion.

This difference in litholgy has a significant effect on the coastal configuration: to the north, sediments are absent, so the coast is rocky, and as a consequence relatively stable. To the south and west, on the contrary, the sedimentary character of the coast explains its high dynamics and morphological changes.

Ancient maps compiled by Antochiw (1994) show important morphological changes when compared to modern cartography (Figure 23.4). Isla del Carmen and Isla Aguada, for example, were in the past (c. 1756) a rosary of islands separated by intertidal channels. Even considering the inaccuracy of the maps of those times (17th and 18th centuries), they are exceptionally descriptive from the hydrological point of view: it was of primary importance to have accurate information about the low bottoms and channels of the Laguna de Términos, especially for trading and shelter reasons. Maps from the 18th century show that the Isla del Carmen consisted of two main islands, not one as nowadays, separated by a channel called Boca Nueva ('new mouth'); to the east, Isla Aguada ('wet island') was then an island, more in accordance with its name, and not a spit as it is today (Figure 23.4, compare A and B).

The location of two main palaeochannels is detected through the presence of remnants of large internal deltas produced by tidal and wave action (Figure 23.4B). These ancient channels were obliterated with sediments transported by littoral drift, so that the deltas stopped functioning under normal conditions, causing the merging of ancient smaller islands into a bigger one in the case of Isla del Carmen, and the build-up of a spit in the case of Isla Aguada. An actual internal intertidal delta is detected in the Canal de Puerto Real, where several low islands are situated showing an anastomosed pattern.

The palaeochannels correspond to a relative depressed topography, in comparison with the surrounding terrain. Through these and other depressions, sediments still flow towards the internal marshes (wash-over) during storm wave action, like that generated by hurricanes.

23.5 RESULTS

The surveyed coast can be divided into four sectors (Figure 23.5) according to its litho-geomorphological characteristics and the consequent effects of the hurricane: (a) higher rocky coast (limestone); (b) lower coast consisting of lithified parallel ridges; (c) sand-bars and spits; and (d) mainland deltaic depositional coast.

23.5.1 Higher Rocky Coast

This sector corresponds to the northernmost 30–35 km of the surveyed coast, from Campeche to Ceiba Playa. Locally, small cliffs (up to 60 m) can be observed; no accumulative features exist. Damage to roads or towns is rare, since infrastructure is commonly built more than 4–5 m above sea level (see Figures 23.6 and 23.7). However, roads were severely damaged when distance from the seashore was less than 80 m and altitude lower than 5 m a.s.l. (see Figure 23.8).

23.5.2 Lower Coast Consisting of Lithified Parallel Beach Ridges

This sector runs from Ceiba Playa to Sabancuy (50 km). Topographically, this sector is considerably lower than the previous one (no cliffs, altitudes commonly lower than 10 m

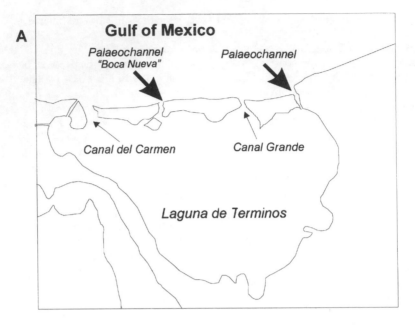

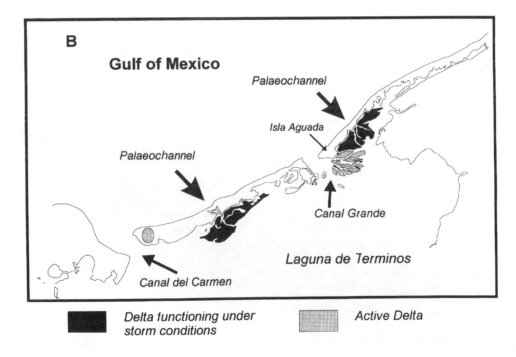

Figure 23.4 Comparison of ancient and modern cartography. (a) Map by Juan Pérez Ramírez, 1756 (Antochiw 1994); (b) actual configuration of the coastline. After Palacio *et al.* 1999

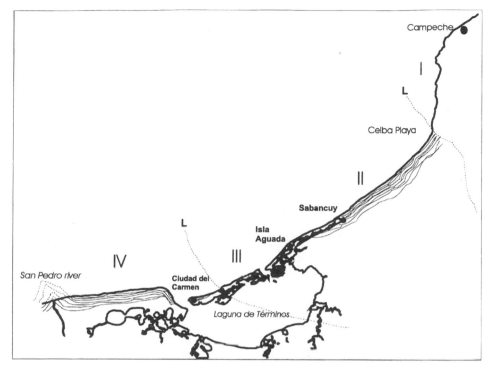

Figure 23.5 Lithogeomorphological sectors: I, sedimentary marine (tertiary) – higher rocky coast; II, beach-rock – lower coast (lithified parallel beach ridges); III, beach-rock outcrops/alluvium – sand-bars and spits; IV, alluvium – mainland deltaic depositional coast

a.s.l.); it is constituted by continuous ridges of organic sediments (shells) heaped up by the action of waves and currents beyond the present limit of storm waves or ordinary tides. These ridges occur as a series of approximately parallel deposits, representing successive positions of an advancing shoreline. Sediments at the coast and accumulative features are rare.

Damage to the road was outstanding along 40 km in this sector (Figure 23.9); locally, the road was completely washed out, so that the authorities considered a new route running farther from the coastline.

23.5.3   Sand-bars and Spits

The third sector (Figures 23.10 to 23.14) is represented by the Isla Aguada spit and the Isla del Carmen bar and is about 60 km long. Along this sector, the beach-rock has been exposed by wave action (Figure 23.13) and at least two areas were significantly affected by wash-over processes (Figures 23.10 and 23.12). Wash-over is associated with ancient channels normally closed but reopened as a result of the storm wave action; the reopening of some sectors of the Isla Aguada spit (Figure 23.11) causes the complete destruction of the road. Some accumulation features clearly show the predominant direction of the littoral drift (NE–SW, Figure 23.14). Damage was concentrated in ancient palaeochannels, mentioned earlier, discernible in ancient cartography.

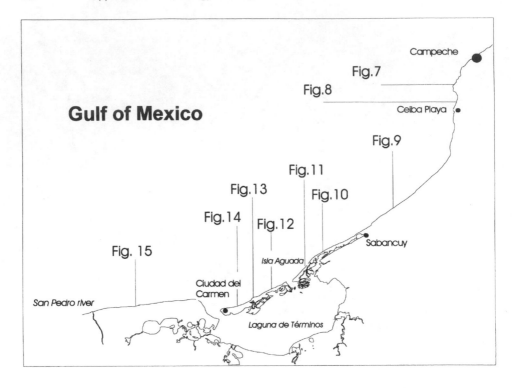

Figure 23.6   Location of video frames (Figures 23.7–23.15)

Figure 23.7   Local damage along the road between Ceiba Playa and Campeche

Figure 23.8 The highway between Ceiba Playa and Campeche was locally destroyed where distance to the seashore is less than 80 m

Figure 23.9 Damage along the highway between Ceiba Playa and Sabancuy

Although included in this sector because of spatial continuity, the western part of the Isla del Carmen could be considered to belong to a separate subunit. The beach-rock is buried by a thicker layer of sediments so that it is not visible in the western portion of the island as it is in the eastern one.

23.5.4 Mainland Deltaic Depositional Coast

This sector runs along 35–40 km, to the San Pedro river, which represents the political boundary between Campeche and the neighbouring state of Tabasco. The sector is formed by a series of parallel ridges locally truncated by erosion. Although there is evidence of alternating deposition and erosion periods, the sector showed accumulative behaviour during

Figure 23.10　Wash-over; notice the new asphalt (top of the figure). Light tones refer to sandy accumulations

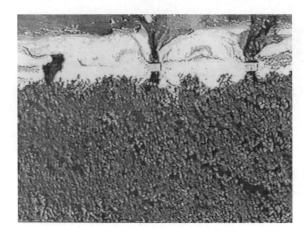

Figure 23.11　Damage to the road in Isla Aguada. Notice the darker water from the marshes flowing into the sea

the event studied here. The video images showed a predominantly accumulative process along the coast of this sector. Locally, the beach prograded more than 300 m; the morphology of the ridges suggests a strong and predominant littoral drift from NE to SW (Figure 23.15).

23.6　DISCUSSION AND CONCLUSIONS

Hurricane Roxanne's effects on coastal geomorphology are strongly influenced by geomorphological zonation (Figure 23.16). Effects on infrastructure and damage location are related to the topographical position in terms of distance to the coast and sea level. Roads located closer than 100 m to the coastline and less than 4 m above sea level were to some

Figure 23.12 Wash-over; notice the new asphalt (top of the figure). Light tones refer to sandy accumulations

Figure 23.13 Beach-rock outcrops in the eastern Isla del Carmen

extent affected and, in some cases, totally destroyed. Ancient intertidal channels represent risk zones since their functioning is reactivated during extraordinary events.

According to this zonation, damage in sectors 2 and 3 (Figure 23.5) is easy to explain and damage is expected to occur in future events. Therefore, it is suggested that alternative routes for road construction should be sought considering the geomorphological dynamics of the different sectors and taking into account the scenario of future relative sea-level rise, associated with the differential sinking of blocks due to sedimentation, reported by some authors (Ortiz-Pérez 1992).

Understanding palaeogeomorphological features and dynamics based on the interpretation of present landforms and ancient cartography represents a valuable tool for territorial planning in relation to natural hazards and eventual future damage mitigation.

Figure 23.14   Spits showing coastal drift from east to west

Figure 23.15   Accumulation took place between Isla del Carmen and the San Pedro River. Notice about 300 m of coastal advance

From the technical point of view, the use of video images provides a low-cost way of providing detailed material for rapid evaluation. Although not mentioned here, the use of an alternative remote sensing tool may be regarded as valuable for rapid assessment and could be considered for coordinating assistance in future events and studies where a rapid response is recommended and needed.

ACKNOWLEDGEMENT

The authors thank Arturo Garrido Pérez for his help in preparing some of the figures included in this work.

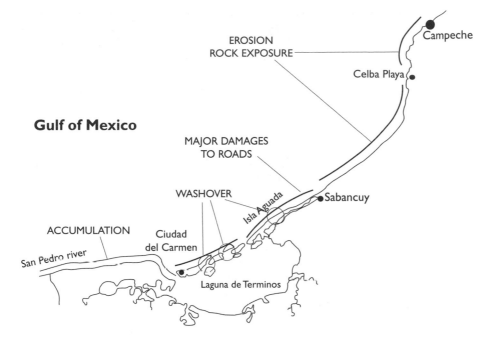

Figure 23.16 Summary of damage and processes as a consequence of Hurricane Roxanne

REFERENCES

Antochiw, M. 1994. Historia cartográfica de la península de Yucatán. Gobierno del Estado de Campeche, Mexico.

Brocaw, N.V. and Walker, L.R. 1991. Summary of the effects of Caribbean hurricanes on vegetation. *Biotropica*, **23**(4a), 442–447.

Coll, A. 1975. El suroeste de Campeche y sus recursos naturales. *Boletín del Instituto de Geografía de la UNAM*, **6**, 1–15.

Gutiérrez, M. and Castro, A. 1988. Origen y desarrollo geológico de la Laguna de Términos. Organización de los Estados Americanos-Instituto de Ciencias del mar y limnología, UNAM, Mexico.

Lodge, D.J. and McDowell, W.H. 1991. Summary of ecosystem level effects of Caribbean hurricanes. *Biotropica*, **23**(4a), 373–378.

Manzano, O. 1989. *Estudio geomorfológico para la zonificación de las áreas de manejo de la reserva de la Biósfera los pantanos de Centla, Tabasco*. BSc thesis, Colegio de Geografía, UNAM, Mexico.

Matías, L.G. 1997. *Climatología del huracán Roxanne y sus efectos en el litoral del Golfo de México*. BSc thesis, Colegio de Geografía, UNAM, Mexico.

Ortiz-Pérez, M.A. 1992. Retroceso reciente de la línea de costa del frente deltáico del río San Pedro, Campeche, Tabasco. *Boletín del Instituto de Geografía de la UNAM*, **25**, 7–24.

Palacio P., Hernández, J., Garrido, A. and Ortiz-Pérez, M.A. 1996. Efectos del huracán Roxanne en la geomorfología costera del estado de Campeche. *IV Reunión Nacional de Geomorfología* (Abstracts), Pátzcuaro, Mich. 23–26 October.

Palacio, P., Ortiz, M. and Garrido, A. 1999. Cambios morfologicos costeros en Isla del Carmen, campeche, por El Paso del huracan Roxanne. *Boletin del Instituto de Geografia, UNAM*, **40**, 48–57.

Psuty, N. 1965. Beach ridge development in Tabasco, Mexico. *Annals of the Association of American Geographers*, **55**, 112–124.

Tanner, E.V.J., Kapos, V. and Realey, J.R. 1991. Hurricane effects on forest ecosystems in the Caribbean. *Biotropica*, **23**(4a), 513.

Waide, R.B. 1991. Summary of the response of animal populations to hurricanes in the Caribbean. *Biotropica*, **23**(4a), 508–512.

Walker, L.R. 1991. Tree damage and recovery from hurricane Hugo in Luquillo experimental forest, Puerto Rico. *Biotropica*, **23**, 379–385.

Yañez, A. 1971. Procesos costeros y sedimentos recientes de la plataforma continental al sur de la Bahía de Campeche. *Boletín de la Sociedad Geológica Mexicana*, **32**(2), 75–115.

24 Dune Generation and Ponds on the Coast of Ceará State (Northeast Brazil)

V. CLAUDINO-SALES
Departamento Geografia, Universidade Federal do Ceará, Fortaleza, Brazil and DEPAM, Université Paris Sorbonne, Paris, France

AND

J.P. PEULVAST
UFR de Géographie, Université Paris Sorbonne, Paris, France and Orsay Terre, Université Paris Sud, France

ABSTRACT

The coast of the Ceará State is characterized by the presence of two systems of coastal dunefields: old fixed dunes, whose accumulation may be associated with regional sea-level and climatic variations from Late Pleistocene to Late Holocene, and younger dunes, which can be fixed or mobile. The new dunes result in part from the erosion of the ancient dunes, which seems to indicate that the coast is going through a period of erosion and sand reactivation. In the middle and behind the dunefields, lakes and duneponds – as distinct from fluvial ponds and lagoons – stretch all over the coastal area and may be considered as a morphological unit. The lakes are temporary or perennial, frequently not larger than $2\,km^2$. They often halt the sand movement, so that mobile dunes become partially or completely fixed by coastal vegetation and peculiar features are created; in other cases the dunes themselves confine superficial water as perennial ponds. These dynamics seem to be relatively autonomous in relation to climatic changes and may repeat themselves frequently in the course of geomorphological evolution of the dunefields, thus producing new arrangements of sand deposits, water and vegetation cover in the area.

24.1 INTRODUCTION

Active coastal dunefields are generally considered as a response to present coastal processes related to climatic parameters (mainly wind and rain), vegetation cover, topography and sediment source. On the Ceará State coast, northeast Brazil, a sequence of different types of coastal ponds and small lakes placed in the middle and behind the coastal dunefields may be considered as one of the elements of local coastal morphological processes.

The presence of perennial and ephemeral ponds and flooding areas at the bases of dunefields is a common feature, caused by the high level of the water table due to poor rainfall run-off and efficient sand water retention (Bagnold 1941), and has already been described and discussed in other coastal areas (e.g. the King Island, Australia (Jennings 1957); west coast of Mexico (Fryberger 1990); west coast of United States (Shenk 1990a)). In Ceará State, as in many other parts of the northeastern Brazil coastal area, this reality becomes peculiar because

Applied Geomorphology: Theory and Practice. Edited by R.J. Allison.

the duneponds are present throughout. These small ponds – generally no more than 2 or 3 km^2 in area – are both intermittent and perennial, creating a specific type of landscape composed of a succession of sand belts and water bodies, where they play a specific role with respect to the morphological dynamics and evolution of dunefields. Unlike fluvial ponds, which have a landwards water source, and lagoons, which are flooded by marine brine, the duneponds have their source of water supplied by the coastal aquifer.

These systems – dunes and ponds – have been poorly studied up to now and the specific dynamics of the whole area still remain unclear. The following analyses aim to contribute to this knowledge, particularly regarding the relation between ponds and dunes in terms of their recent morphological evolution.

24.2 STUDY AREA AND METHODS

The coastal zone of Ceará State, in northeast Brazil, extends for 573 km from SE to NW (Figure 24.1). It presents subhumid climatic conditions with concentrated rainfalls (1000 to 1400 mm during autumn), followed by drought periods of six to nine months. On the hinterland and immediately seaward the climate is semi-arid, strongly influencing the coastal dynamics by fluvial influx of clastics resulting from erosion of the Precambrian rocks that constitute the continental landscape, particularly during periods of torrential rainfall and seasonal floods.

The whole area is sparsely occupied by urban settlements, the demographic density being six to 43 inhabitants per square kilometre, except in the 100 km around its capital, Fortaleza, and in Fortaleza itself (6 in Figure 24.1), where there are respectively 100 and 3800 to 4200 inhabitants per square kilometre. This means that the major features of the coast are relatively preserved in terms of original dynamics, in spite of the present urban changes which have been especially strong since the 1970s, resulting in shoreline erosion, dunefield destruction, degradation of the coastal vegetation and related coastal problems.

The major coastal features are sand beaches, sea cliffs, coastal plains and dunefields.

The sand beaches are flat and large, of a width of 1 km or more. They extend continuously along the shoreline and some of them, especially in eastern Ceará, extend below active sea cliffs 7 to 10 m high, eroded in the sediments of the Barreiras Formation – a continental Tertioquaternary deposit which covers the basement on the coastal area as a tabular layer, locally forming points and capes in the strandline.

The resulting shoreline is a sequence of creeks and headlands limited by active cliffs; the rocky points are also developed in outcrops of Cretaceous sediments, and westward by the basement itself. In places where active cliffs are absent, there may be a line of dead-cliffs strongly masked by Quaternary deposits landward; in other places, the Barreiras Formation seems to have been eroded and shaped down as a flat surface, covered by Quaternary sediments.

Progradation embankments and small coastal plains extend in front of dead-cliffs, where they are often preceded seaward by beach-rock lines. However, the most expressive coastal landforms are the large dunefields. They have heterogeneous morphologies and develop from the beach inwards for tens of metres to several kilometres, covering the Barreiras Formations and cliffs and points. One of the most striking features of the dunefields is the presence of different types of small lakes and duneponds.

Such a geographical diversity suggests a wide variety of dynamics. Thus, the analysis was performed and discussed through the study of representative areas, especially considering the morphological relationship between intermittent ponds and dunefield evolution. The areas chosen to explain these relations are Jericoacoara (18 in Figure 24.1) and Fortaleza (6 in

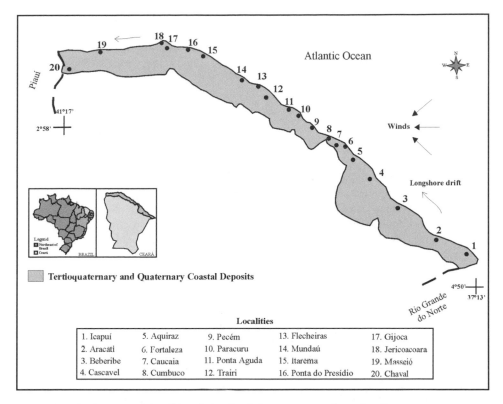

Figure 24.1   Study area and localities of the Ceará State coast, northeast Brazil

Figure 24.1). The discussions involve some considerations about regional sea-level variations and climatic changes in relation to the present system of dunefield dynamics.

The whole study is based on the interpretation of aerial photos at scales of 1:30 000 and 1:25 000, taken in 1975 and 1992 respectively; radar images and maps of radar images at a scale of 1:250 000; thematic maps (geology, geomorphology and soils) and topographic maps at a scale of 1:10 000; and sedimentological analysis of the mobile dune sands. In addition, field trips and flights were made, the last one in May 1996, allowing a survey of most of the 573 km of the coast to be carried out.

24.3   TYPES OF DUNEFIELDS AND DUNEPONDS

The Ceará coastal dunefields are part of a continuous belt of aeolian sands that extend from Rio Grande do Norte to Piaui (over 1000 km) (Figure 24.1). Along the 573 km of the study area, they are composed of two main systems of dunes, vegetated dunes and mobile dune-fields, for which a tentative classification is given in Figure 24.2.

The vegetated dunefields cover the coastal area forming a strip with width varying from a few to 30 km from the coast. Their forms, external characterization and localization in relation to shoreline seem to indicate that they result from at least three periods of sand accumulation.

The first period (D1) would be related to the sandbelt that covers the Barreiras Formation landward for 30 km and more. It seems to represent an older dune and interdune belt now degraded and shaped down as a flat and undulating topographic surface, colonized by shrub vegetation. Sandbelts cover the coastal area in discontinuity and with variable width, but generally they are wider westwards, in places where the Barreiras Formation is not present on the surface.

This development of large old dunes and absence of the Barreiras Formation in the coastline may indicate a structural control of the coastal morphological evolution, concerning the existence of basement heights where the Tertioquaternary deposits were thin enough to have their superficial layers removed by sea erosion of transgressive periods. This possibility is supported by the presence of basement outcrops along the strandline in those sectors and by the existence of a hinterland sequence of Precambrian grabens and horsts whose superficial limits are outlined by the Tertioquaternary deposits (CPRM 1983). In this context the sectors with active Tertioquaternary cliffs would be related to grabens, where old dunefields are less developed, and their absence would be related to horsts, where in contrast the old dune surface is larger; this would probably imply the occurrence of neotectonic activity, which has already been indicated by Maia *et al.* (1993), Bezerra *et al.* (1998) and Saadi and Torquato (1993) on the NE Brazil coastal area.

The second period of sand accumulation (D2) is more recent and corresponds to E–W vegetated dunes that lie over the coastal area for a distance that generally does not exceed 7 km landward. Near to the beach they are exposed to intense deflation where blowouts are present (Figure 24.3), being covered landward for up to 3 or 4 km by mobile dunes, in such a way that their forms are subdued in those sectors. However, forms of parabolic dunes and elongated parabolic dunes are clearly delineated in the sectors free of cover, and 'hairpin dunes' up to several tens of metres long are often found, the large belts being those of Ponta Iguape and Ponta Aguda (5 and 15 in Figure 24.1).

At the back of some of the biggest capes the D2 dunefields assume the character of groups of dunes without defined forms, consisting of undulating deposits of a light colour, up to several metres high and colonized by arboreal vegetation; they extend landward for up to 30 km (e.g. in the area of Icapui: 1 in Figure 24.1), and may correspond to the existence of more than one generation of D2 dune deposition – in this case, there are D2A and D2B periods of dune deposition.

The third period of sand accumulation is related to the youngest dunefields, formed by present-day deposits. Their development is in response to several conditions, such as available sands, supplied by fluvial discharges, sea cliff erosion and erosion of ancient dunefields; important sediment transit landward and in the shoreface, the continuous SE winds, the most important in terms of geomorphological action (the NE winds and the E winds playing the role of secondary winds); strong and permanent SE longshore drift; flat bathymetry and topography of both platform and coastal areas, which exposed to a semi-diurnal tidal range of about 2.4 m (DNH 1997) provides a very wide intertidal zone.

The present-day deposits are composed of vegetated dunes (D3A) and mobile dunefields (D3B). The vegetated D3A dunes do not form defined dunefields but actually correspond to groups of dunes that because of obstacles (such as dead-cliffs and water bodies), more humid weather and even greater distance from shore, have been stabilized and then fixed by vegetation cover, which frequently reaches forest dimensions. They are generally higher than the mobile dunes with which they are sometimes crossed; the forms are not well defined, in spite of the presence of semi-fixed parabolic dunes. They have relatively little spatial expression in comparison with the other types of fixed and mobile dunefields, the best examples being those of Aquiraz and Fortaleza (5 and 6 in Figure 24.1).

Dunes Generations		Present Forms	Dunebelt Width	Position in Relation to Shoreline	Vegetation Cover	Age	Associated Sea Level	Evolution
OLD DUNES D1		Flat and degraded surface	A few to 40 km discontinuous	Inland dunebelt	Shrub	Pleistocene	Last pleistocene transgression	1
	D2A	Indefined	A few to 10 km discontinuous	Inland dunebelt	Arboreal	Pleistocene	Late pleistocene regression	2
D2	D2B	Parabolic, hairpin dune and indefined dunebelt	3 to 7 km continuous	Seaward and landward dunefield evolution	Arboreal	Holocene	Holocene transgression (followed by holocene regression)	3
YOUNG DUNES	D3A	Indefined	A few hundred discontinuous	Seaward and landward dunefield evolution	Arboreal	Present-day dunes	Present-day sea level	4
D3	D3B	Transverse ridge, barchan, parabolic	3 to 6 km continuous	Seaward and landward dunefield evolution	Mobiles dunes	Present-day dunes	Present-day sea level	5

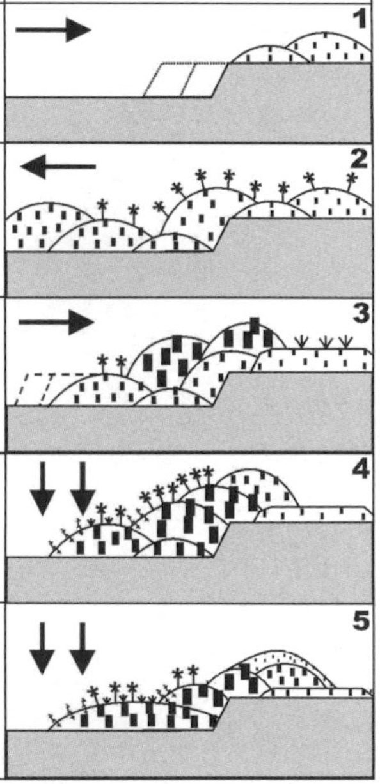

LEGEND

Transgression	Regression	Present-day	Cliff	Tertioquaternary deposits
Sea level				

D1	D2A	D2B	D3A	D3B				
Increasing age					Abrasion	Deflation	Arboreal	Shrub
Dunes					Erosion		Vegetation	

Figure 24.2 Dune generation in the Ceará State coast, northeast Brazil (a preliminary classification)

Figure 24.3   Old dune reactivation and resulting forms, Ponta Aguda coast, Ceará, Brazil. Parabolic and blowout dunes resulted from the present reactivation of old D2 dunes. In the deflation areas the shallow water table forms temporary ponds, which allow the growth of coastal vegetation. The presence of ponds and vegetation associated with the parabolics' low rate of migration partially refixes the recent remobilized sands *in situ* (photo by J.P. Peulvast)

The mobile dunefields (D3B) are composed by assemblages of very well sorted, fine to medium quartz sand deposits that generally lie NE–SW and SE–NW from the shore for 3 or 4 km landward along the whole coastal area. Like the D1 vegetated dunes, they are much more developed westward, i.e. in regions where active cliffs of the Barreiras Formation are not always present. These mobile dunefields have four main types of forms: parabolic, transverse, transverse barchanoid and barchans.

The parabolics, a few metres high, are the first belt of mobile dunes. They clearly result from the reactivation of ancient dunes – in this case, the D2 dunes, which they cover landward – and perhaps from deflation of the Barreiras Formation in areas where the dunes start their development on the top of the cliffs, a frequent situation in eastern Ceará. Westward and especially at the streams ebb, the parabolics give place to wide transverse ridges and transverse barchanoid ridges that extend in strips several kilometres wide along the coastline, with dune heights exceeding 30 m.

In the westward sectors of capes and points (e.g. Paracuru, Jericoacoara, respectively 10 and 18 in Figure 24.1) isolated wide barchans are found. They are large forms with heights of over 50 m that climb the cliff laterally, crossing it as by-pass dunes from the stoss side to be dissipated by the sea waves on the lee shores.

Other dune features of the Ceará coastal area are the foredunes (e.g. Hesp 1984; Paskoff 1989; Psuty 1990), which are found on rectilinear sectors of the coast, colonized by pioneer vegetation. In some sites the foredunes give place to a straight and irregular belt of cemented dunes, subdued forms covered by thin and discontinuous carbonate layers. The origins of these cemented dunes are not clear, but they are probably related to places where frequent marine brines infiltrate, slowing the early diagenesis processes (Shenck 1990). The Ceará continental shelf being almost entirely covered by coarse relict biogenic carbonate sediments (Summerhays *et al.* 1975; Coutinho 1994), these carbonate sands would have lodged more easily in the deep layers of ancient dunes (D1?) and ultimately been subjected to early diagenesis processes, and are now exposed in response to deflation over the dune superficial layers.

The coastal lakes and ponds are associated with all types of dunefields. They belong to three main types: ephemeral, temporary and perennial. A tentative classification is given in Figure 24.4.

The ephemeral ponds result from the flooding of relatively impermeable surfaces during rainfall periods. They are often found in interdune areas and in deflation depressions in the middle of the dunes, but without a spatially defined site of occurrence; in aerial photos, they are especially visible in mobile (D3B) dunefields (Figure 24.5).

The temporary ponds seem to result from the rise of the water table in deflated dune areas, where less permeable surfaces are reached at depth; these conditions are especially found on the D1 surface and deflation depressions of the D2 dunefields, particularly in the blowouts and central areas of the parabolics. There, the water may evaporate during long drought periods and reappear in rainfall seasons on the same site. When placed in the migration path of barchans (D3B) and parabolic dunes (D3A and D2), those ponds are filled by sands in transit, which leave swamp depressions of differentiated ecological character behind them, as exemplified in Beberibe, Aquiraz, Jericoacoara, Massaio and Chaval (respectively 3, 5, 18, 19 and 20 in Figure 24.1) (see Figure 24.6).

The perennial ponds seem to evolve in flooding areas where the water finds favourable conditions to accumulate in large quantities and in a perennial manner, as on the contact between the old D1 dune surface and the Barreiras Formation, and where a barrage (a sand dome) confines superficial water.

The formation of coastal barrages is known from examples where mobile sands moving landward halt the flux of fluvial streams seaward. A similar process has been described by Jennings (1957) to explain the genesis of some perennial dune lakes at King Island, Australia, whose supply of water, like those of the Ceará coast, comes only from groundwater. However, it seems that some other conditions may be related to the origin of these ponds, such as large quantities of sand migration, shallow water table and topographic gradient.

The presence of large perennial ponds near the contact between Quaternary and Tertioquaternary deposits and at the boundary of topographic ruptures (e.g. Beberibe, Aquiraz, Fortaleza and Ponta Aguda–Lagoinha, respectively 3, 5, 6 and 11 in Figure 24.1) reveals that the sites where all these conditions may be best attained are the areas of dead-cliffs; indeed, in the sectors of present active cliffs there are several springs, from which the water flows onto the beach. Furthermore, at the seaward limits of the dead-cliffs all generations of dune deposits have probably been successively accumulated since their erosion and formation (last Pleistocene transgression?), not only allowing them to be masked by constructive relief but also creating different internal permeability in the Quaternary deposits, thus also successively accumulating and confining the water on the surface (see Figures 24.2 and 24.8).

The perennial ponds are often more than 2 m deep (CPRM 1975), the deepest sectors lying seaward, and have mainly elongated forms; this seems to highlight the fact that they are preferentially placed where transverse dunefields are or have been present during their genesis.

From the ponds, outlets flow laterally, frequently running into other perennial lakes that are placed behind the same barrage line before they reach the beach; this situation sometimes leads to a sequence of several ponds connected by small streams, a fact that has always been related to the damming of fluvial streams (in this context triangular forms result, as pointed out by Jennings (1957)). This explanation was accepted particularly for ponds placed on the back berm, whose elongated form has been considered as a result of successive barrages and migration of stream mouths by onshore sands.

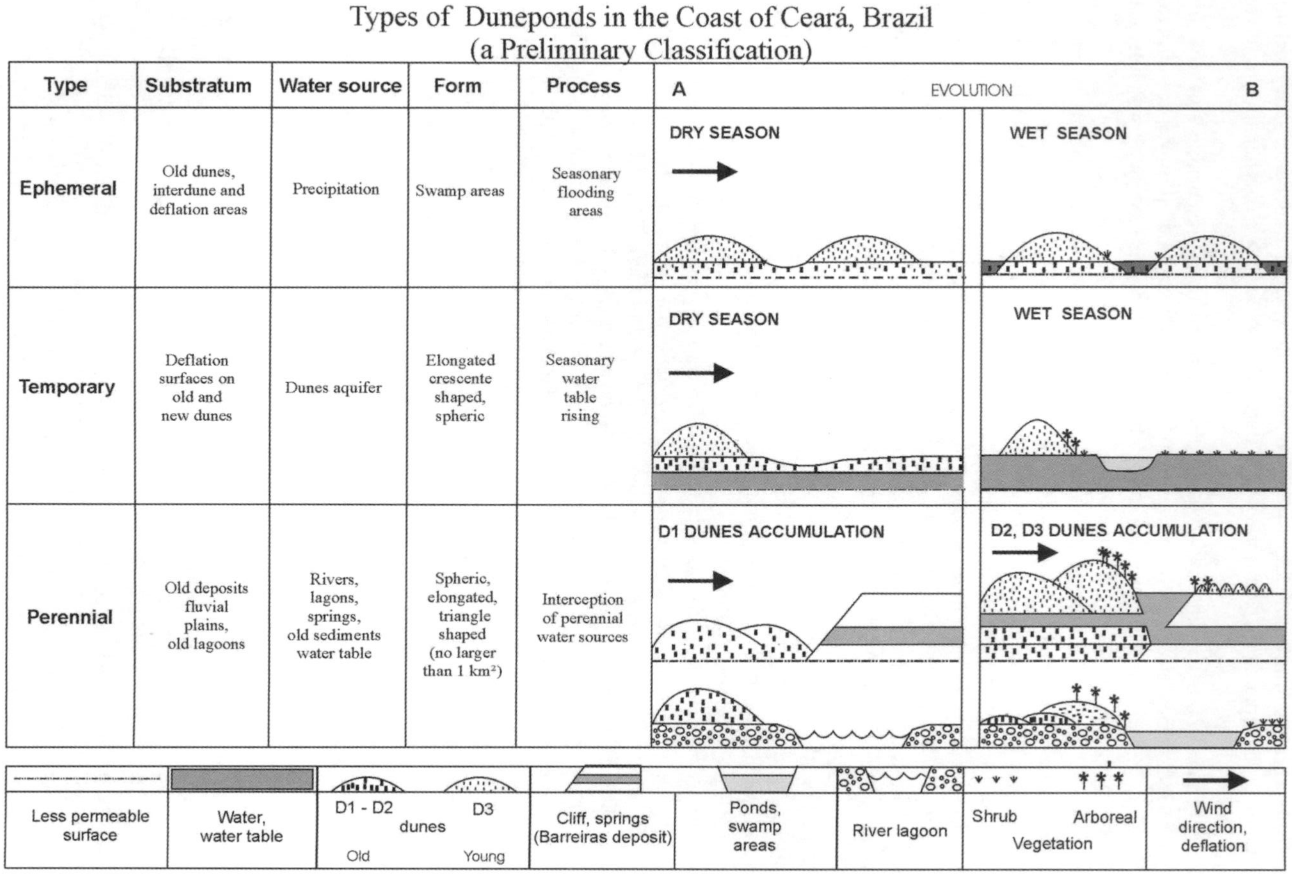

450

Figure 24.4 Types of duneponds on the Ceará coast, northeast Brazil (a preliminary classification)

Figure 24.5 Duneponds at the contact of old and new dunes, Cumbuco coast, Ceará, Brazil. Transverse ridges (D4B dunes) migrating over old D1 and D2 dunes surface where ephemeral, temporary and perennial ponds are present. The new dunes have their migration retarded in places where temporary and perennial ponds are widely present (as seen at the top). The different permeability between new and old dunes allows the presence of ephemeral ponds on the new dune surface (as seen at the bottom), which will be embanked as soon as the wet season is finished (photo by J.P. Peulvast)

On the whole, all the situations explained above indicate that there is a direct relationship between the evolution and morphology of coastal dunes and ponds; this relationship will be analysed in the following section.

24.4 DISCUSSION

24.4.1 Barchans Dunefields and Temporary Ponds

The coastal area of Jericoacoara (18 in Figure 24.1) has been chosen to support the analysis of the relationship between barchan dunes and temporary ponds.

The Jericoacoara area is characterized by a large point formed around a quartzitic hill at the basement to which lines of beach-rocks have accreted. On the east side of the point the coast is marked by an extensive beach line surmounted by typical foredunes; westwards a gentle cove opens up into an extensive sand beach limited by mobile dunes. At the point, an abrupt dome-like sand deposit covered by shrub vegetation, probably an old dune of the D2 dunefields where pedogenesis is now complete, leans on the quartzitic hill (Figure 24.6). The Barreiras Formation is present at the surface in only a few sectors behind the ancient dune, where it has been shaped down, perhaps by a marine terrace now covered by the dune; all these facts indicate that there we are probably facing one of the major coastal structural controls of this coast, related to a basement probably weakly covered by Tertiary deposits, and ultimately exhumed by sea erosion during transgressive periods.

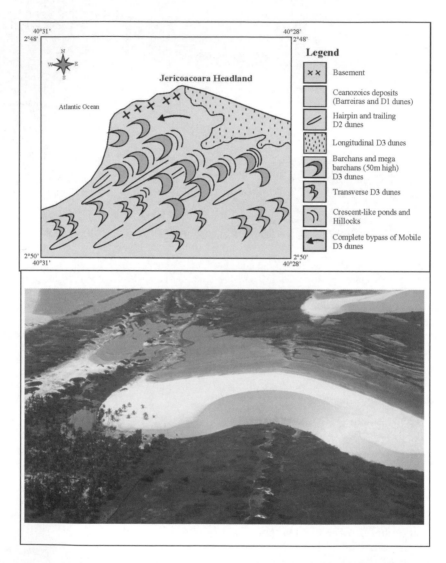

Figure 24.6 The Jericoacoara Point (Ceará, Brazil) coastal features. Top, a schematic map of Jericoacoara coast shows its main features. Bottom, D2 hairpin dunes and trailing dunes lie over an old degraded D1 surface, here flooded during the wet season. The hairpin dunes leave semi-circular humid depressions and residual sand banks behind them in places of flooding water (bottom of photo); this process also occurs in the migration path of D3B barchans (centre photo): these new dunes are partially halted at the base by ephemeral ponds, resulting in the formation of a new generation of crescent-like residual micro-hillocks and humid depressions (see Figure 24.7) (photo by J.P. Peulvast)

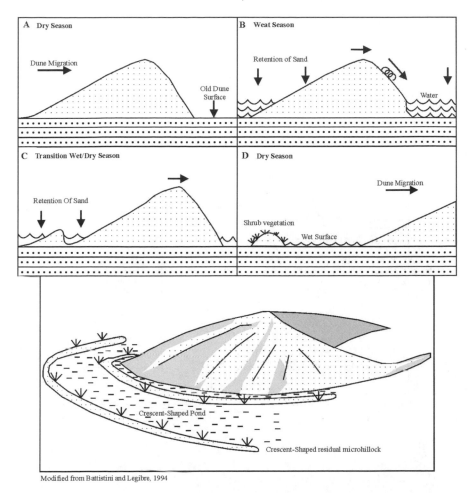

A Dry Season

Dune Migration

Old Dune Surface

B Weat Season

Retention of Sand

Water

C Transition Wet/Dry Season

Retention Of Sand

D Dry Season

Dune Migration

Shrub vegetation

Wet Surface

Crescent-Shaped Pond

Crescent-Shaped residual microhillock

Modified from Battistini and Legibre, 1994

Figure 24.7 Dune migration over flooded areas and residual features. (A) Submersion of the dune base by flooding during wet seasons; (B) a sand crust builds up on the dune base, soon colonized by coastal vegetation; (C) the barchan migrates but deflation does not destroy the vegetated sand crust, which is converted to a residual semi-circular hillock; (D) the continuity of migration leaves a crescent-shaped organic-rich humid depression between the dune and the residual hillock

Landward, the D1 dune surface extends for 30 km as a flat or undulating sand surface fixed by shrub vegetation. It is partially covered by the flat D2 dune surface, where large 'hairpin dunes' are slowly migrating. The area is presently invaded by D3B dunes, represented by isolated barchans of 50 to 54 m high, which migrate from the stoss side of the point crossing it from E to W as by-pass dunes to finally be swept away in the lee-side shores. This barchan field is not continuous and shows a disconnection of several kilometres between inland and seaward dunes, which reveals that at some period of recent coastal evolution the mobilization of a large volume of sand was followed by an interruption of the inward sand transit, whose origin and duration remains to be determined. This situation is also represented in other barchan areas of the Ceará coastal area.

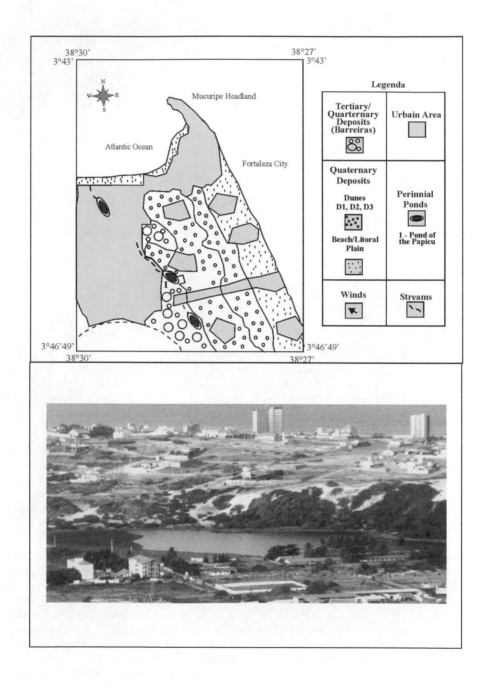

The D1 surface, the area of migration of D2 hairpin dunes and D3B barchan dunes, is strongly flooded during seasonal wet periods, which gives rise to intermittent ponds. In consequence, in the course of the transit of the sands, the bases of the D2 hairpin dunes were submerged by the temporary ponds and the migration of their sediments was halted, while the whole emerged dune body kept moving, leaving behind tiny residual ridges; these micro-ridges kept the original form of the dune, e.g. semi-circular forms, and since they were colonized by grass and pioneer vegetation as soon as the sand was halted, the deflation action over them was inhibited. In the same way, a semi-circular humid depression remained impressed on the field when the dunes were gone, as the presence of organic matter in those areas ensures a high level of moisture even when it has dried out. Presently, the same process is active in respect of the barchans in transit over the area. The situation results in a landscape composed of a succession of half-moon temporary ponds alternating with barrow and low crescent-like sand ridges of different ages (Figures 24.6 and 24.7).

Such a process has been described by Battistini and Legibre (1994) in the region of Toliara, Magadascar, concerning tidal marsh environments. At Jericoacoara, as in many other places of the Ceará coastal area, these patterns seem to be related to surfaces where barchans and parabolic dunes are present and water is not abundant (e.g. where water comes only from the dune aquifer), and hence they are subject to seasonal oscillations or pluriannual periods of drought and humidity. It seems that large quantities of water (e.g. from permanent ponds, a water source that also comes from the continental aquifer) would completely halt the dune migration, as in other places on the coast; on the other hand, large amounts of sand (e.g. transverse ridges) would completely embank the ponds or at least lead the water to drain laterally, which would leave no signs of their former existence on the field.

24.4.2 Dunes and Perennial Ponds

The relationship between coastal perennial ponds and dunes has been closely studied in the Fortaleza area (6 in Figure 24.1).

The coast of Fortaleza is characterized by a point that divides the shoreline into two sectors: the eastern SE–NW coast, formed by a rectilinear shoreface dominated by the Rio Coco stream basin; and the western E–W coastal lines, where a cove formed by sand beaches extends over several kilometres. Landward a topographic rupture seems to indicate the presence of a dead-cliff, already masked by dunes and urban structures.

The present dune features are represented by a vegetated dunebelt several metres high without defined form, placed landward until the undefined limit of this dead-cliff, which probably corresponds to D1 and D2 dunes enlarged by D3 deposits. The mobile (D3B) dunes originally crossed the point from SE to NW as by-pass dunes but are now almost completely absent in response to the intense urbanization of the area.

Figure 24.8 Perennial ponds on the east coast of Fortaleza (Ceará, Brazil). Top, a schematic map of Fortaleza West Coast, where coastal features are widely degraded by urban occupation. Bottom, the perennial pond Lagoa do Papicu resulted from the damming of coastal water by successive D1 and D2 dune accumulations at the base of dead-cliff eroded on the Barreiras Formation. Presently the pond halts the migration of D3 dunes and induces the development of coastal vegetation at their lee side, allowing the increase of dune height; together those processes are responsible for an inversion of relief in the contact between Tertioquaternary and Quaternary sediments, the highest elevations now being found on the dunes' accumulation side (centre) (photo by F. Torres)

The sector analysed here corresponds to Lagoa do Papicu, an elongated SE–NW small perennial lake, 1.4 km long with a depth of 1.8 m (Claudino-Sales 1993) from which an outlet flows toward the E–W coast. It is placed about 1.5 km away from both SE–NW and E–W shorelines and landward immediately above the topographic rupture cited above, at the limit of the contact between Quaternary and Tertioquaternary sediments (DNPM 1988); the dunes are placed limiting the seaward section and the Barreiras Formation, which is a tabular surface no more than 20 m in height (Figure 24.8).

The dune deposits are part of this accumulation of old dunes, very recently enlarged by the D3B dunebelt resulting from urban remobilization of the sand seawards; this dunebelt is 52 m high in the limits of the pond and about 30 m high in the surrounding areas, and is colonized on its southern side by a forest vegetation cover. In the water body at the limits of the dune, sand banks demonstrate that there is some transfer of sand from the dune to the lake.

The analysis of the whole context suggests that this lake probably originated on the limits of the dead-cliff, where spring water flowing from its eroded sectors may have been confined by the successive deposition of D2 dunes, later enlarged by D3 dunes. Recently, the D3B dunebelt has migrated, surpassing the older dunes until they reach the lake, repeating a process now induced by social transformation but which was certainly frequent previously in natural conditions, as found in other sectors of the coast.

The present-day situation shows that the presence of the lake halts the continuity of migration of the D3B dunebelt. In this way, a dense vegetation cover could develop on the lee side of this dunebelt, owing to its protection with regard to wind and high humidity of both air and soil, thus reaching the dimension of a coastal forest (a situation which indicates that, as in the previous example with temporary ponds, the growth of the vegetation was preceded by or is contemporary with the stabilization of the sands). Accordingly, the lake has not been filled by the aeolian sands that arrive from the SE; furthermore, the vegetation cover allowed the dune to increase by trapping the sands in transit through the coastal area. Such a process, as described by Goldsmith (1973) in fixed dunefields of the USA, is responsible for a much higher vertical movement (growth) of the dune in relation to its horizontal movement (migration) and accounts for the increasing height of fixed dunes.

Nevertheless, there is always some quantity of sand that falls from the dune into the lake, as demonstrated by the presence of sandy banks in the water body at the contact with the dune. It is probable that on a short time scale their outlet would be able to transport those sands to the beach, thus preserving the pond from being embanked. However, on a large time scale, the complete obstruction of the lake, followed by vegetation burial, dune destabilization and water drainage seems to be inexorable.

The whole situation (confinement of superficial water by dunes, creating ponds; damming of mobile sands by those water bodies, fixing dunes; finally, obstruction of the lakes followed by reactivation of dunes and water drainage) suggests a context of morphological dynamics in which the entire coastal area is subjected to several changes in dynamics and patterns of water, sand and vegetation cover during present and past times.

Actually this context clearly delineates the presence of a 'two-hand process', expressed by the fact that the ponds halt the transit of sand over the coastal area, but the dunes also dam the drainage of water over the coast; nevertheless the halting and damming is not permanent nor is it responsible for an 'internal' sedimentological and hydrological coastal cycle. This 'internal' cycle would be represented on the one hand by the retention of sand in some sectors of the littoral (followed by its ulterior reactivation and free movement over the whole coastal area). On the other hand, there is a lodging of coastal water on land, which is partially drained into the sea by the action of the outlets: some of the lodged sand is continuously exported back to the shoreline by these outlets, inhibiting an indefinite increase of dunes landward and contributing to the re-establishment of the sedimentologic budget, as has been already suggested by Moraes and Smith (1988).

24.4.3 Sea-level Variations, Climatic Changes and Dunefields Dynamics

Studies of Quaternary regional sea level in the Brazilian shoreline started in the 1970s with observations by Martin and Suguio (e.g. Martin and Suguio 1975, 1978; Martin *et al.* 1979; Suguio and Martin 1984; Suguio *et al.* 1985; Suguio 1993). These studies led the authors to propose a sea-level variation curve for the southern and eastern coast of Brazil since 7000 years BP, based on regional geoid surface variations, radiocarbon dating and coastal feature indicators.

According to this curve, a high sea level of 5 m above the present datum would have been reached around 5100 years BP, followed by two regression–transgression oscillations and a continuous regression regime during the last 2500 years until the present level. Associated with this curve a coastal palaeoenvironmental reconstruction is based on the Bruun (1962) rule, whose interpretation states that this recent marine regression would have led to the progradation of the coast, in response to the transfer of sand from foreshore to backshore. Recently Angulo and Lessa (1997) contested the occurrence of those two oscillations, indicating a maximum high sea level of 3.5 m around 5100 years BP.

Concerning the northeastern Brazilian shoreline, there is no developed research about the subject, the only chronological data being the dates of shells made by Van Andel and Laborel (1964) in the eastern sector (Recife), which indicated an age of 5900 ± 300 years for accumulations about 1 m high in regard to the present level of beach-rock cementation.

In relation to the Ceará shoreline, there are no data concerning sea-level indicators, studies of sea-level variations having begun in the 1980s by Andrade (1986), who pointed out some possible geological and geomorphological evidence of the Martin and Suguio curve to the area, and considered that the present mobile dunefields are an expression of the late Holocene marine regression, which would have allowed the movement of greater quantities of sands over the coastal area. A similar interpretation is found in Moreira and Gatto (1981) and Meireles (1992), who associated the mobile dunefields with dry climatic periods of regression regime.

On the one hand, the vegetated dunefields have been interpreted as resulting from palaeoclimatic conditions related to recent humid periods (Moraes and Smith 1988) that would have allowed the development of denser vegetation cover able to stabilize the mobile dunefields, which they consider were larger in the past than they are today.

However, no accurate study of the palaeoclimatic conditions of the area is available, and the Ceará dunefield morphology and dynamics, as analysed above, indicate a context much more complex than that deriving from such an assessment. The presence of at least three periods of dune accumulation is probably related to sea-level variations and perhaps also to climatic changes, but in a way which is not yet very clear.

It is possible that the oldest D1 dune surface accumulated during the last Pleistocene transgression, which would also have been responsible for the erosion of the dead-cliffs. The dunes would have been fixed during the subsequent regression of the last glacial period, the fixation taking place progressively as the sea receded. This situation would be possible if we consider that the development of vegetated dunes may also be the result of increasing distance from the sand source and/or a lesser intensity of wind action (Pye and Tsoar 1989), both processes being associated with a regression regime. Moreover, in places where there are no signs of cliffs – a fact that seems to respond to a structural control – the D1 dune terrain is more developed landwards. In places where cliffs are present, it forms a straight strip (e.g. Ponta Grossa, Paracuru: 1 and 10 in Figure 24.1) locally covered by younger dunes.

The D2 dunefields are the first dunebelt seaward, presently the theatre of intense wind erosion and deflation. These dunes (especially the D2B dunes) may have been built during the Holocene transgression – and if we consider a scenario based upon the Suguio and Martin curve, it would have been fixed during the subsequent regression, at the same time as new mobile dunes were accumulating. Nevertheless, their foreshore position and present-day erosion reactivation would require another context, involving a slightly transgressive or stable regional sea level during recent periods. The first situation would explain their seaward placement, considering that, as indicated by Goldsmith (1973), an intense erosion of the coast tends to remove the first mobile dunebelt. On the other hand, a stable sea level would restrict the quantity of sand available on the shoreface, because of a reduced supply of the product of the continental and marine erosion, as occurred along many coasts throughout the world after the end of the Holocene transgression (Paskoff 1993).

The lack of data concerning regional sea-level variations in the Brazilian northeast shoreline does not permit verification of the possible divergences in relation to this general sea regression model – such as the Holocene transgression followed by a regressive sea, rapidly stabilized. Moreover, it is known that different sea levels could produce similar coastal features (Psuty 1990; Pirazzoli 1996); besides, this general sea regression model is not sensitive enough to detect little cycles or present-day sea-level variations (R.J. Angulo, personal communication), so such a scenario needs to be evaluated in the light of accurate data.

The D3 dunefields, instead of being the result of dry weather periods or regressive sea-level regime, seem to be completely adapted to the processes that are active at the present time scale, according to a morphological dynamics system controlled by present climatic conditions.

In relation to perennial coastal lakes, it seems that their origins are indirectly associated with the last large transgression and regression, as those events created coastal features such as barriers, dead-cliffs and old dune deposits, where the necessary conditions for their development are found. The occurrence of humid periods would surely influence the presence of all types of ponds in the coastal area, as the water table would be more expressive; nevertheless, their processes of evolution seem to be relatively autonomous in this respect as they are conditioned by the geomorphology of the coast, which they also modify.

24.5 CONCLUSIONS

The analyses of the Ceará dunefields and associated situations as described above lead to the conclusion that the area is undergoing a period of sand reactivation, where the inland sand represented by old fixed dunefields is going through intense erosion and remobilization allowing the formation of new dunes, increased by windborne sands originating from fluvial discharges and cliff erosion. The development of this mobile dunebelt is adapted to the present climatic conditions, while the older vegetated dunefields seem to result from climatic and sea-level variations from Late Pleistocene to Holocene.

Furthermore, the situations analysed above suggest that on this coast the morphological evolution of the dunefields may not be analysed separately from the presence of duneponds. The role of these ponds and the overall effect of their morphological character concern the following situations: the ponds halt the free migration of the dunes over the coast, producing particular coastal forms such as fixed and semi-fixed dune ridges, sand micro-hillocks and other minor features; the ponds and their outlets allow the partial return to the beaches of land-lodged sands, a process that together with the by-pass dunes periodically re-establishes the sedimentary budget along the coastal area. Moreover, the mobile sands confine areas of coastal water on the surface, thus inhibiting its drainage into the ocean and expressing the action of a 'two-hands' coastal process, an element that partially controls the dynamics of this coast.

ACKNOWLEDGEMENTS

This research has been carried out with the agreement of the Brazilian Counsel for Development of Science and Technology – CNPq and DEPAM – Dynamique et Ecologie des Paysages Atlantiques and Meditereneene, Université de Paris-Sorbonne.

REFERENCES

Andrade, E. 1986. Geologia Sedimentar de Aracati – Icapui, Ceará. Dissertaçaõ de UFPE, Recife, Brazil.

Angulo, R.J. and Lessa, G.C. 1997. The relative sea level changes in Brazil in the last 7000 years: criticizing old and adding new data. *Marine Geology*, **140**, 141–166.

Bagnold, R.A. 1941. *The Physics of Blown Sand and Desert Dunes*. Methuen, London.

Battistini, R. and Lebigre, J.M. 1994. Les littoraux tropicaux: enregistreurs de l'évolution de l'environnement. In Maire, R., Pomel, S. and Salomen, J-N. *Enregistreurs et Indicateurs de l'Evolution de l'Environnment en Zone Tropical*. Presses Universitaires de Bordeaux, 255–275.

Bezerra, F.H.R., Lima-Filho, F.P., Amaral, R.F., Caldas, L.H.O., Costa-Neto, L.X. 1998. Holocene coastal tectonics in NE Brazil. In Studart, J.S. and Vita-Finzi, C. (eds). *Coastal Tectonics*. Geological Society, London, Special Publications, **146**, 279–293.

Bruun, P. 1962. Sea level rise as a cause of shore erosion. *American Society of Civil Engineers*, **88**, 177–230.

Claudino-Sales, V, 1993. *Lagoa do Papicu – Cenarios Litoraneos na cidade de Fortaleza, Ce.* Dissertaçaõ de Mestrado, Universidade de São Paulo, Brazil.

Cooper, W.S. 1958. *Coastal Sand Dunes of Oregon and Washington*. Geological Society of America Memoir **71**.

Coutinho, P.N. 1994. Sedimentos Carbonáticos da Plataforma Continental. *Revista de Geologia da UFC*, **6**, 65–24.

CPRM, 1975. *Projeto Diatomito/Argila*, vol. 1. Recife.

CPRM, 1983. *Mapa Geologico do Estado do Ceará*, 1:500.000.

DHN, 1997. *Tabua de Maré*. Porto do Mucuripe, Estado do Ceará.

Fryberger, S.G. 1990. Role of water in eolian deposits. In Fryberger, S.G. *et al. Modern and Ancient Eolian Deposits: Petroleum Exploration and Production*. Rocky Moutain Section, Society of Economic Palaeontologists and Mineralogists (11–1)–17.

Goldsmith, V. 1973. Internal geometry and origin of vegetated coastal sand dunes. *Journal of Sedimentary Petrology*, **43**, 1128–1143.

Hesp, P.A. 1984. Foredune formation in south-east Australia. In Thorn, B.G. (ed.) *Coastal Geomorphology in Australia*. Academic Press, Sydney, 69–97.

Maia, L.P., Morais, J.O. and Torquato, J.R. 1993. Applied geophysics to neotectonics in Aracati, Ceará Region, Northeast Brazil. *Revista de Geologia da UFC*, **6**, 57–65.

Martin, L. and Suguio, K. 1975. The State of São Paulo coastal marine quaternary geology – The ancient strandlines. *Anais da Academia Brasileira de Ciências*, **47**, 249–263.

Martin, L. and Suguio, K. 1978. Excursion route along the coastline between the town of Cananeia (State of Sao Paulo) and Guaratiba outlet (State of Rio de Janeiro). In *International Symposium of Coastal Evolution in the Symposium on Coastal Evolution in the Quaternary*, São Paulo, project **1**, 264–274.

Martin, L., Flexor, J.M., Vilas Boas, G.S., Bittencourt, A.C.P. and Guimaraes, M.M.M. 1979. Courbe du niveau relatif de la mer ou cours des 7.000 dernières années sur un secteur homogene du littoral brésilien (nord de Salvador). In *International Symposium on Coastal Evolution in the Quaternary*, São Paulo, project 1, 264–274.

Meireles, A.J. 1992. *Mapeamento Geologico – Geomorfologico do Quaternario Costeiro de Icapui, Extremo Leste do Estado do Ceará*. Tese de Mestrado, UFPe, Brasil.

Moraes, J.O. and Smith, A.J. 1988. Preliminary studies on coastal zone management in the littoral of Ceará State north-eastern Brasil. *Revista de Geologia da UFC*, 1, 31–40.

Moreira, M.M.M. and Gatto, L.C.S. 1981. Geomorfologia. *Projeto RADAMBRASIL*, Vol. 21, MME, Folha Fortaleza, 213–252.

Paskoff, R. 1989. Les dunes bordières. *La Recherche*, **212**, 888–895.

Pirazzolli, A.P. 1996. *Sea Level Changes – The Last 20 000 Years*. Wiley, Chichester.

Psuty, N. 1990. Foredune mobility and stability. In Nordstrom, K. Psuty, N. and Carter, B. (eds), *Coastal Processes Forms and Processes*. Wiley, Chichester, 159–175.

Saad, A. and Torquato, J.R. Contribuiçao à neotectônica do Estado do Ceará. *Revista de Geologia UFC*, **5**, 5–39.

Shenk, C.J. 1990a. Eolian dune morphology and wind regime. In Fryberger, S.G. *et al. Modern and Ancient Eolian Deposits: Petroleum Exploration and Production*. Society of Economic Palaeontologists and Mineralogists (3–1)–8.

Souza, M.J.N. 1988. Contribuiçao ao Estudo das Unidades Morfo-estruturais do Estado do Ceará. *Revista de Geologia da UFC*, 1, 73–90.

Suguio, K. and Martin, L. 1984. Planicie de cordões litorãneos quaternarios do Brasil: origem e nomenclatura. In Lacerda, L.D. (coord.), *Restingas: Origem, Estruturas e Processos*. CEUFF, Nitéroi, 453–458.

Suguio, K., Martin, L., Bittencourt, A.C.S.P., Dominguez, J.M.L., Flexor, J.M. and Azevedo, A.E.G. 1985. Flutuações do nivel relativo do mar durante o Quaternario Superior ao longo do litoral brasileiro e suas implicaçoes na sedimentação costeira. *Revista Brasileira Geociências*, 15(4), 273–286.

Summerhays, C., Coutinho, P.N., França, A.M. and Ellis, J.P. 1975. Northeastern Brazil. *Contribution to Sedimentology*, Stuttgart, **4**, 77–78.

Van Andel, T.H. and Laborel, J. 1964. Recent high sea-level stand near Recife, Brazil. *Science*, **145**, 580–581.

Index

Note: Page numbers in *italic* refer to illustrations; those in **bold** type refer to tables.